T0358576

Topological Polymer Chemistry
Progress of cyclic polymers
in syntheses, properties and functions

Topological Polymer Chemistry
Progress of cyclic polymers
in syntheses, properties and functions

Edited by

Yasuyuki Tezuka
Tokyo Institute of Technology, Japan

NEW JERSEY · LONDON · SINGAPORE · BEIJING · SHANGHAI · HONG KONG · TAIPEI · CHENNAI

Published by

World Scientific Publishing Co. Pte. Ltd.

5 Toh Tuck Link, Singapore 596224

USA office: 27 Warren Street, Suite 401-402, Hackensack, NJ 07601

UK office: 57 Shelton Street, Covent Garden, London WC2H 9HE

Library of Congress Cataloging-in-Publication Data
Topological polymer chemistry : progress of cyclic polymers in syntheses, properties,
   and functions / edited by Yasuyuki Tezuka, Tokyo Institute of Technology, Japan.
     pages cm
   Includes bibliographical references and index.
   ISBN 978-9814401272 (hardcover : alk. paper)
   1. Cyclopolymerization. 2. Topology. I. Tezuka, Yasuyuki, editor of compilation.
   QD281.R5T67 2013
   547'.28--dc23

                                                            2012046466

British Library Cataloguing-in-Publication Data
A catalogue record for this book is available from the British Library.

Printed in Singapore by World Scientific Printers.

CONTENTS

Part II: Cyclic Polymers - Developments in the New Century - 135

GENERAL INTRODUCTION

Yasuyuki Tezuka

Department of Organic and Polymeric Materials,
Tokyo Institute of Technology, Meguro-ku, Tokyo, Japan
E-mail: ytezuka@o.cc.titech.ac.jp

There are frequent examples in the macroscopic world where the form of objects underlies the basis of their functions and properties. On the other hand, the fabrication of extremely small objects having precisely defined structures has only recently become an attractive challenge, which is now opening the door to nanoscience and nanotechnology.[1,2] In the field of polymer chemistry and polymer materials science, the choice of macromolecular structures has continuously been extended from linear or randomly branched forms toward a variety of precisely controlled topologies by the introduction of intriguing synthetic techniques.[3,4] During the first decade of this century, a number of formidable breakthroughs have been achieved to produce an important class of polymers having a variety of cyclic and multicyclic topologies.[5,6] These developments now offer unique opportunities in polymer materials design to create unprecedented properties and functions simply based on the form, i.e. *topology*, of polymer molecules.

In this book on *topological polymer chemistry*, the recent developments in this growing area will be outlined, with particular emphasis on new conceptual insights for polymer chemistry and polymer materials. The book consists of the two parts. Part-I (**Topological Polymer Chemistry - concepts and practices -**) includes chapters to describe systematically *topological polymer chemistry* from fundamental aspects to the practices. Part-II (**Cyclic Polymers – developments in the new century-**) includes contributing chapters on broad aspects of cyclic polymers covering from the historical overview by Dr. Roovers, a pioneer of ring polymer sciences, to the recent developments on the synthesis, the structure characterization, the basic properties/functions and the eventual applications.

Part-I starts with the chapter (Chapter 1) focusing on the topological geometry (soft geometry) viewpoints toward polymer molecules, in contrast to small chemical compounds conceived upon Euclidian geometry (hard geometry) principles. And key conception of *topological polymer chemistry* is presented by taking into account for the basic geometrical properties of polymer chains uniquely flexible in nature. Topological geometry and graph theory are applied to formulate the systematic classification and notation protocol of the non-linear constructions of polymer molecules, including not only branched, but also single cyclic and multicyclic polymer topologies. The following chapter (Chapter 2) discusses on the geometrical/topological relationship between different polymers having distinctive constructions, to introduce a unique conception of *topological isomerism* in contrast to conventional constitutional and stereoisomerism occurring in small chemical compounds. Followed by these conceptual/theoretical chapters, Part-I will proceed to important practical aspects covering the effective construction of complex polymer topologies including, in particular, cyclic and multicyclic forms. Thus first, Chapter 3 overviews synthetic procedures for nonlinear polymers by telechelic polymer technique, in which linear and branched polymer precursors having a series of cyclic onium salt end groups are employed. The following Chapter 4 details an *electrostatic self-assembly and covalent fixation* (ESA-CF) protocol by employing specifically designed telechelic precursors having selected cyclic ammonium salt groups for designing new polymer topologies including cyclic and multicyclic polymer components. In Chapter 5, an important extension of the ESA-CF process involving dynamic character of electrostatic self-assembly is presented. Unprecedented opportunities to produce tadpole polymers, dicyclic topological isomers, as well as co-polymacromonomers and polymeric catenanes are demonstrated. In Chapter 6, moreover, an additional important feature of the ESA-CF process is discussed with respects to the facile introduction of versatile functional groups at the specific location of single and multicyclic polymer units. These are *kyklo*-telechelics and cyclic macromonomers applicable to construct complex polymer topologies of covalent or mechanical linking with cyclic polymer components. The subsequent Chapter 7 presents an alternative versatile synthetic protocol for ring polymers by employing metathesis condensation reactions. A variety of single to multicyclic polymers and of cyclic block copolymers can be obtained by effective intramolecular metathesis condensation process. Finally, Chapter 8 outlines a number of recent synthetic achievements in *topological polymer chemistry*, including complex multicyclic polymers of either *spiro-*, *bridged-* and *fused-*forms through the ESA-CF process in conjunction with recently developed effective linking chemistries, including alkyne-azide addition (*click*) and olefin metathesis (*clip*) reactions. These are prototypical examples to highlight the current frontier of synthetic organic chemistry and, furthermore, may hint future perspectives and directions of *topological polymer chemistry*.

Part-II starts with a chapter (Chapter 9), *Overview Physical Properties of Cyclic Polymers*, by Dr. Roovers, the pioneer of this research field, to provide comprehensive accounts on basic properties of cyclic polymers with the particular emphasis on developments observed in recent years. The following two chapters by Dr. Grayson and Zhang, *The Ring-Closure Approach for Synthesizing Cyclic Polymers* (Chapter 10), and by Dr. Kubo, *Cyclic Macromonomers: Synthesis and Properties* (Chapter 11), deal with ongoing progress in synthetic chemistry of cyclic polymers. These two chapters discuss on the innovative synthetic protocols developed in a recent decade for a variety of cyclic polymers with a particular emphasis on the end-to-end linking of prepolymer processes (click, metathesis, end-crosslinking etc) or other important alternative processes, and on a novel synthetic opportunity by using a cyclic polymers having a polymerizable group (cyclic macromonomer), respectively. The following chapter by Dr. Deguchi, *Topological Effects on the Statistical and Dynamical Properties of Ring Polymers in Solution* (Chapter 12), reviews ongoing theoretical/simulation studies on the statistical and dynamical properties of ring polymers, in particular, those associated with topological effects. The subsequent Chapter 13 by Dr. Habuchi, *Dynamics of Cyclic Polymers Revealed by Single-Molecule Spectroscopy*, outlines a recently developed single molecule spectroscopy technique, to allow the direct and real time observation of dynamics of cyclic polymers. This breakthrough methodology offers an unprecedented opportunity to reveal the topology effect on the diffusion process with the use of appropriate cyclic and linear model polymers.

The following two chapters by Dr. Vlassopoulos, *Progress in the Rheology of Cyclic Polymers* (Chapter 14), and by Dr. Shiomi, *Crystallization of Cyclic and Branched Polymers* (Chapter 15), cover recent developments on studies of rheology and crystallization of cyclic polymers, which are fundamentally important bulk polymer properties. These two chapters discuss on new insights of rheological properties of cyclic polymers revealed by using rigorously purified samples, and of crystallization kinetics influenced significantly by the topology of polymer molecules, respectively. Finally, Dr. Yamamoto describes in Chapter 16, *Self-Assembly and Functions of Cyclic Polymers*, newly revealed significant topology effects by cyclic polymers against linear counterpart, in particular by their self-assemblies.

We are now experiencing a rapid growth of researches both in the synthesis of diverse cyclic and multicyclic polymers of unprecedented structural control, and in the modeling/simulation as well as experimental verification of fundamental properties uniquely achieved by cyclic and multicyclic polymers. Furthermore, these progresses are now attracting to explore practical applications of cyclic and multicyclic polymer materials exhibiting useful functions simply based on the form, i.e. *topology*, of polymer molecules. We are glad to observe that this book will convey an ongoing progress in the field of an important class of polymers having a variety of cyclic and multicyclic topologies, and will become a standard and comprehensive reference book on all aspects of *topological polymer chemistry*. We hope also this book to serve as a long awaited follow-up of a seminal book on cyclic polymers by Semlyen published a decade before.[7]

Finally, I thank all authors of Chapters in Part 2 for their valuable contributions. In particular, I am most thankful to Dr. Roovers for his continuing encouragements during this Book project, and for his practical assistance by reviewing Chapters in Part 2.

References

1. E. Regis, *Nano! The emerging science of nanotechnology: remaking the world – molecule by molecule*, (Little Brown, Boston, 1995).
2. N. C. Seeman, *Annu. Rev. Biochem.*, **79**, 65 (2010).

3. N. Hadjichristidis, A. Hirao, Y. Tezuka and F. Du Prez, Eds, *Complex Macromolecular Architectures,* (Wiley, Singapore, 2011).
4. A. D. Schlüter, C. J. Hawker and J. Sakamoto, Eds., *Synthesis of Polymers, New Structures and Methods* (Wiley-VCH, Weinheim, 2012).
5. T. Yamamoto and Y. Tezuka, *Polym Chem.*, **2**, 1930 (2011).
6. Y. Tezuka, *Polym. J.*, in press (DOI: 10.1038/pj.2012.92)
7. J. A. Semlyen, Ed., *Cyclic Polymers, 2nd ed.* (Kluwer, Dordorecht, 2000).

PART I

Topological Polymer Chemistry - Concepts and Practices -

CHAPTER 1

SYSTEMATIC CLASSIFICATION OF NONLINEAR POLYMER TOPOLOGIES

Yasuyuki Tezuka

Department of Organic and Polymeric Materials,
Tokyo Institute of Technology, Meguro-ku, Tokyo, Japan
E-mail: ytezuka@o.cc.titech.ac.jp

A systematic classification of nonlinear polymer topologies including branched and cyclic forms is presented by reference to the graph presentation of constitutional isomers in a series of alkanes (C_nH_{2n+2}), monocycloalkanes (C_nH_{2n}) and polycycloalkanes (C_nH_{2n-2}, C_nH_{2n-4}, etc). A systematic notation of nonlinear polymer topologies is also presented based on their principal geometrical parameters of terminus and junction numbers.

1. Introduction

A systematic classification of polymer constructions will provide not only fundamental insights on geometric relationships between different macromolecular structures, but also their rational synthetic pathways. A rational classification might also give a systematic nomenclature, as in the cases of dendrimers, knots, catenanes and rotaxanes.[1-3] However, there have been few attempts on the systematic classification of nonlinear, and in particular cyclic and multicyclic polymer architectures composed of sufficiently long and thus flexible segment components.[4,5]

Herein, a systematic classification process for a series of nonlinear, thus cyclic and branched polymer architectures is proposed by reference to the graph presentation of constitutional isomers in alkanes (C_nH_{2n+2}) and in a series of mono- and polycycloalkanes (C_nH_{2n}, C_nH_{2n-2}, etc).[6,7] The molecular graph of linear, branched, cyclic and multicyclic alkane molecules is taken as a source according to the procedure by Walba,[8] in order to produce a unique topological construction. Thus in this transformation, the total number of termini (chain ends) and of junctions (branch points) are maintained as invariant (constant) geometric parameters. The total number of branches at each junction and the connectivity of each junction are preserved as invariant parameters as well. On the other hand, such Euclidian geometric properties as the distance between two adjacent junctions and that between the junction and terminus are taken as variant parameters, to conform with the flexible nature of the randomly-coiled and constrained polymer segments. Furthermore, topological constructions having five or more branches at one junction can be produced despite the corresponding isomers having the relevant molecular formula are not realized.

In this Chapter, a classification procedure of simple to complex branched and cyclic polymer topologies is presented to group them into different main-classes and sub-classes based on geometrical considerations. A systematic notation method for nonlinear polymer topologies is also presented based on the above classification and on principal geometrical parameters of terminus and junction numbers.

2. Classification of Branched Polymer Topologies

Alkane molecules of generic molecular formula C_nH_{2n+2} with n = 3–7 and selected topological constructions by reference to their corresponding higher alkane molecules of C_nH_{2n+2} with n > 7, together with their relevant topological constructions produced by the procedure described in the preceding section are shown in Table 1. A point from methane (CH_4) and a line construction from ethane (C_2H_6) are not included since the former is not significant with respect to polymer topology and the latter produces an equivalent topological construction from propane (C_3H_8).

As seen in Table 1, two butane isomers of *n*- and *iso*-forms produce a linear and a three-armed star construction, respectively. Likewise, from pentane isomers a new four-armed star construction is produced upon neo-pentane in addition to the two others already produced from butane

Table 1. Topological constructions produced by reference to alkane isomers (C_nH_{2n+2}: n = 3–7) and selected isomers (C_nH_{2n+2}: n = 8–10 and m).

Topology	C_nH_{2n+2} n=3	4	5	6	7	Topology	C_nH_{2n+2} n = 8	Topology	C_nH_{2n+2} n = 10
$A_3(2,0)$						$A_8(5,3)$		$A_{10}(6,4)[3\text{-}3(3)\text{-}3]$	
$A_4(3,1)$						$A_8(6,2)$		$A_{10}(6,4)[3\text{-}3\text{-}3\text{-}3]$	
$A_5(4,1)$						Topology	n = 9	Topology	n = m
$A_6(4,2)$						$A_9(6,3)[4\text{-}3\text{-}3]$		$A_{m+1}[m,1]$ m-Arm Star Polymer	
$A_6(5,1)$			(*)	(*)					
$A_7(5,2)$						$A_9(6,3)[3\text{-}4\text{-}3]$			
$A_7(6,1)$					(*)				

isomers. From five hexane isomers, two new constructions of an H-shaped and of a five-armed star architecture are produced. And further, heptane isomers produce the two new constructions of a super H-shaped and a six-armed star architecture. Thus by this process, a series of branched polymer topologies is relevantly ranked as shown in Table 1.

A systematic notation for a series of branched topologies is also introduced as listed in Table 1. All these constructions are classified as A main-class, since they are produced from alkane isomers. A linear construction is produced from propane (C_3H_8) and this particular topology is ubiquitous in those from all higher alkanes. This sub-class construction is thus termed A_3, or alternatively $A_3(2,0)$ by indicating the total number of termini and of junctions, respectively, in parentheses. Likewise, sub-classes A_4 (or $A_4(3,1)$) and A_5 (or $A_5(4,1)$) are uniquely defined as shown in Table 1. On the other hand, sub-classes A_6, A_7 and A_n with higher n values consist of multiple constructions, and each component can be defined by specifying the total number of termini and junctions, respectively, in parentheses as shown in Table 1. As a typical example, an m-armed star polymer topology is classified $A_{m+1}(m,1)$, as listed in Table 1.

Selected constructions from octane and higher alkane isomers are collected also in Table 1. Thus in sub-class A_8, the two new constructions are distinctively defined again by specifying their total numbers of termini and junctions, respectively, in parentheses, i.e., $A_8(5,3)$ and $A_8(6,2)$. In sub-class A_9, on the other hand, two of the newly produced constructions of $A_9(6,3)$ cannot be distinguished by simply showing the total numbers of termini and junctions. The connectivity of junctions for these two constructions is each distinctive, and can be specified by applying the nomenclature rule for substituted alkanes. That is, first a backbone chain having the most junctions is identified, and the number of branches at each junction is given in brackets in descending order from the most substituted junction. Thus, the above two $A_9(6,3)$'s are designated as $A_9(6,3)[4-3-3]$ and $A_9(6,3)[3-4-3]$, respectively. Another pair of constructions in sub-class $A_{10}(6,4)$ in Table 1, namely one having a dendrimer-like star polymer structure and another having a comb-like branched structure, are defined by specifying their junction connectivity, as $A_{10}(6,4)[3-3(3)-3]$ for the former and $A_{10}(6,4)[3-3-3-3]$ for the latter, respectively.

From a topological viewpoint, it is important to note that a series of branched constructions in Tables 1 is distinct from dendrimers (dendritic polymers possessing well-defined branched structures) and comb-shaped polymers, although they are also referred as model branched macromolecules having well-defined structures. Thus, the topological constructions produced above are based on the assumption that the distance between two adjacent junctions and between the junction and the terminus are variable geometric parameters. This conforms to the flexible nature of sufficiently long polymer chains capable of assuming a random coil conformation. In dendrimers, on the other hand, the distance between two adjacent junctions and between the junction and the terminus are regarded to be invariant. Consequently, they tend to constitute a stiff, shape-persistent molecule possessing a gradient of structural density.[9] For comb-shaped polymers, likewise, the distance between two junctions along the backbone are regarded as invariant, whereas the branch chain is either flexible (polymacromonomers)[10] or stiff (dendron-jacketed polymers).[11]

3. Classification of Cyclic Polymer Topologies

3.1. *Monocyclic polymer topologies*

Monocycloalkane molecules of C_nH_{2n} with up to n = 7 and their relevant topological constructions are listed in Table 2. They are produced according to the procedure applied for the A main-class, branched topologies. Thus, a simple cyclic topology is produced upon the molecular graph of cyclopropane (C_3H_6). Likewise, two constructions, namely one having *a ring with a branch* architecture and another having a simple ring structure, respectively, are produced from the two isomers of C_4H_8, that is methylcyclopropane and cyclobutane. And the latter simple ring topology is already produced upon cyclopropane. Moreover, the two new constructions are produced from C_5H_{10} isomers, in addition to those observed either from C_3H_6 or C_4H_8 isomers. These two constructions are distinguished from each other by their junction and branch structures, i.e., one has two outward branches at one common junction in the ring unit while the other has two outward branches located at two separate junctions in the ring unit. Four new topological constructions are subsequently produced by reference to C_6H_{12} isomers; the one having five branches at one junction is hypothetical and is therefore shown in parentheses in Table 2. Thus by this process, a series of *a ring with branches* constructions has been ranked by reference to the constitutional isomerism in monocycloalkanes.

A systematic notation for a series of *a ring with branches* constructions has been accordingly introduced as given in Table 2. First, these are classified into a I main-class topology, since they are produced from <u>mono</u>cycloalkanes. Then, a simple ring construction produced from cyclopropane is designated as sub-class I_3, or alternatively $I_3(0,0)$ by showing the total number of termini and junctions in parentheses. This topology is ubiquitous among all higher sub-classes in the I main-class. A new construction from C_4H_8 is labeled likewise as I_4 or $I_4(1,1)$. As is

Y. Tezuka

Table 2. Topological constructions produced by reference to cycloalkane isomers (C_nH_{2n}: n = 3–6 and 7).

Topology	C_nH_{2n} n = 3 4 5 6	C_nH_{2n} n = 7
$I_3(0,0)$	▷ ◇ ⬠ ⬡	$I_7(3,2)[1(4)]$
$I_4(1,1)$	▷– ◇– ⬠– ▷⌐ ▷⟋ ◇⟋	$I_7(3,2)[2(3,0)]$
$I_5(2,1)$	⋈ ⋈ ⋈	$I_7(3,3)$
$I_5(2,2)$	⟋△ ◇ ⟋△⟋ ⊓	$I_7(4,1)$ (⋈)
$I_6(2,2)$	▷⋈	$I_7(4,2)[3-1]$ (⋈)
$I_6(3,1)$	(⋈)	$I_7(4,2)[2-2]$ ⋈
$I_6(3,2)$	⋈	$I_7(4,3)$ ⋈
$I_6(3,3)$	⋏	

evident from Table 2, sub-classes I_5, I_6 and other I_n comprise multiple constructions, and each of them can be basically identified by specifying the total number of termini and junctions.

The newly produced constructions from monocycloalkane isomers of C_7H_{14} are shown also in Table 2. Seven new constructions are produced in sub-class I_7, and most of them are uniquely defined again by specifying the total number of termini and junctions, respectively, in parentheses. The two sub-classes $I_7(3,2)$ and $I_7(4,2)$, however, contain two distinctive constructions. Thus, they are identified by specifying their branch modes on a ring unit. First, the number of outward branches at each junction on the ring unit is identified, and is indicated in the brackets placed after the closing parenthesis. The detailed junction architecture of the outward branches is then identified, according to the procedure applied to the A main-class topology, and is indicated in the parentheses enclosed within the brackets. The two constructions of $I_7(3,2)$ are thus defined as $I_7(3,2)[1(4)]$ and $I_7(3,2)[2(3,0)]$, respectively. Likewise, the two constructions of $I_7(4,2)$ are labeled as $I_7(4,2)[3-1]$ and $I_7(4,2)[2-2]$, respectively, where the connectivity of junctions along the ring unit is indicated by connecting the number of the outward branches at each junction with a hyphen. It is also noable that all constructions up to sub-class I_7 are produced initially from the relevant isomer containing a cyclopropane unit. In the higher sub-class I_8, a topological construction containing four or more junctions on the ring unit is first produced by reference to a relevant isomer containing a cyclobutane unit, i.e., $I_8(4,4)$.

3.2. Multicyclic polymer topologies

Bicycloalkanes of C_nH_{2n-2} with up to $n = 6$ and their relevant topological constructions are listed in Table 3, where three basic bicyclic constructions without any outward branches are included; namely, θ-shaped or internally linked (fused) rings, 8-shaped or spiranic rings and manacle-shaped rings, i.e., two rings externally linked (or bridged) by a linear chain.

For dicyclic constructions, a θ-ring construction is first produced from bicyclo[1,1,0]butane, C_4H_6. Thereafter, three new constructions are produced from the five molecular graphs of bicycloalkane isomers of C_5H_8. One of them is a spiranic (8-shaped) ring construction from spiro[2,2]pentane. These three constructions are distinguished from each other by their total number of termini and junctions as compared in Table 3. By reference to bicyclohexane (C_6H_{10}) isomers, moreover, eight new constructions are produced. All constructions but one from bi(cyclopropane) possess outward branches emanated from either the θ-ring or the spiranic (8-shaped) rings shown above. In this manner, a series of internally and externally linked double cyclic topological constructions can be classified into the corresponding categories by reference to the bicycloalkane isomers.

A systematic notation for a series of topological constructions in Table 3 is then introduced, as in the A and I main-classes. These are classified into a II main-class topology, since they are produced from <u>bi</u>cycloalkane isomers. A θ-form construction, which is produced from

Table 3. Topological constructions produced by reference to bicycloalkane isomers (C_nH_{2n-2}: n = 4–6).

Topology	C_nH_{2n-2}			Topology	C_nH_{2n-2}
	n =4	5	6		n =6
$II_4(0,2)$				$II_6(0,2)$	
$II_5(0,1)$				$II_6(1,1)$	
$II_5(1,2)$				$II_6(1,2)$	
$II5(1,3)$				$II_6(2,2)[2^a\text{-}0^a]$	
				$II_6(2,2)[1^a\text{-}1^a]$	
				$II_6(2,3)[2\text{-}0^a\text{-}0^a]$	
				$II_6(2,3)[1^a\text{-}1\text{-}0^a]$	
				$II_6(2,4)$	

bicyclo[1,1,0]butane, and is ubiquitous in those from all higher bicycloalkanes, is defined as sub-class II_4, or $II_4(0,2)$ by showing the total number of termini and junctions, respectively, in parentheses. Sub-classes II_5 and II_6 consist of multiple constructions as shown in Table 3, namely three for II_5 and eight for II_6, respectively. While they are basically distinguished by specifying their total number of termini and junctions in parentheses, sub-classes $II_6(2,2)$ and $II_6(2,3)$ comprise distinctive constructions that require further structural specifications. Thus in the brackets after the closing parenthesis, the number of not only outward branches but also internally linked branches on the ring unit is indicated, and these numbers (thus 0 for the latter, and 1, 2, $\cdots$ for the former) are linked by hyphens. Moreover, the positions of the two specific junctions internally linked to each other are indicated by giving superscripts (a, b, etc) at the relevant junction numbers. For examples, the two $II_6(2,2)$'s are designated as $II_6(2,2)[2^a\text{-}0^a]$ and $II_6(2,2)[1^a\text{-}1^a]$, and the two $II_6(2,3)$'s are designated as $II_6(2,3)[2\text{-}0^a\text{-}0^a]$ and $II_6(2,3)[1^a\text{-}1\text{-}0^a]$, respectively. Accordingly, any branching modes in an internally linked ring unit can be uniquely defined.

All constructions in the II main-class contain one of either three types of internally or externally linked ring units, i.e., the θ-ring like a construction $II_4(0,2)$, the spiranic (8-shaped) ring like $II_5(0,1)$ or the manacle-shaped ring like $II_6(0,2)$. A variety of *rings with branches* constructions is also obtainable by emanating from these ring units. They are produced primarily from the corresponding isomers containing cyclopropane units, as was noted in the I main-class topologies. Indeed, all constructions up to sub-class II_6 are produced first from the relevant isomers containing exclusively cyclopropane units, as seen in Table 3.

Furthermore, 15 basic topological constructions produced from the relevant tricycloalkanes of C_nH_{2n-4} with n = 4–10 are shown in Table 4. A large number of *rings with branches* constructions are also produced in this group, but are omitted for brevity. Thus, eight topological constructions are produced as the fused- or spiro-forms as shown in Table 4. The six of them are produced directly from the corresponding tricycloalkanes, and the additional two from hypothetical molecular formula possessing five or six branches at a junction as given also in Table 4. Moreover, the seven topological constructions of externally

Table 4. Fused, spiro and bridged ring constructions produced by reference to tricycloalkane isomers (C_nH_{2n-4}: n = 4–10).

Topology	C_nH_{2n-4}				Topology	C_7H_{10}
	n = 4	5	6	7		
$III_4(0,4)[0^a\text{-}0^b\text{-}0^a\text{-}0^b]$					$III_7(0,3)$	
					$III_7(0,4)$	
$III_5(0,2)[0^{a,b}\text{-}0^{a,b}]$					Topology	C_8H_{12}
$III_5(0,3)[0^{a,b}\text{-}0^a\text{-}0^b]$					$III_8(0,2)$	
					$III_8(0,3)$	
$III_6(0,4)[0^a\text{-}0^a\text{-}0^b\text{-}0^b]$					Topology	C_9H_{14}
$III_6(0,2)$					$III_9(0,3)$	
$III_6(0,3)$					$III_9(0,4)$	
$III_7(0,1)$					Topology	$C_{10}H_{16}$
$III_7(0,2)$					$III_{10}(0,4)$	

linked (or bridged) rings with a simple ring and the fused- or spiro-dicyclic forms are obtainable in reference to the relevant tricycloalkane isomers as listed in Table 4.

A systematic notation for a series of topological constructions in Tables 4 is then introduced as in the previous A, I and II main-classes. These are classified into III main-class, since they are produced from tricycloalkane isomers, respectively. A doubly-fused (internally double-linked) ring construction is produced, in the first place, from tetrahedrane and is defined as sub-class III_4 or $III_4(0,4)$ by showing the total number of termini and junctions, respectively, in parentheses. The two constructions in sub-class III_5 and the three in sub-class III_6 shown in Table 4 are a few of the large number of possible constructions, in particular those

Table 5. Selected topological constructions produced by reference to tetracycloalkane isomers (C_nH_{2n-6}: n = 6).

Topology	C_6H_6
$IV_6(0,3)[0^{a,c}-0^{a,b}-0^{b,c}]$	
$IV_6(0,6)[0^a-0^b-0^a-0^c-0^b-0^c]$	
$IV_6(0,6)[0^a-0^b-0^c-0^a-0^b-0^c]$	$K_{3,3}$

possessing additional branches. Constructions in sub-classes $III_4(0,4)$, $III_5(0,2)$, $III_5(0,3)$ and $III_6(0,4)$ commonly possess a doubly-fused ring unit, while their linking modes on the ring unit are distinctive. In mathematical graph theory, they are named as α, β, γ and δ ring, respectively.[4] Their notations are accordingly given in Table 4. For instance, the architectures of $III_4(0,4)$, $III_5(0,2)$ and $III_5(0,3)$ are further specified by indicating their linking modes as $[0^a- 0^b- 0^a- 0^b]$, $[0^{a,b}- 0^{a,b}]$ and $[0^{a,b}- 0^a- 0^b]$, respectively. On the other hand, a group of bridged-ring constructions listed in Table 4 are defined distinctively by showing merely the total number of termini and junctions respectively in parentheses.

Finally, three particularly notable constructions produced by reference to the relevant tetracycloalkane isomers of C_nH_{2n-6} with n=6 (main class IV) are listed in Table 5. These include prisman, $IV_6(0,6)[0^a- 0^b- 0^a- 0^c- 0^b- 0^c]$, and $K_{3,3}$ type, $IV_6(0,6)[0^a- 0^b- 0^c- 0^a- 0^b- 0^c]$, constructions, which commonly possess an internally triply fused ring, while their linking modes are distinctive.

4. Ongoing Challenges and Future Perspectives

This Chapter presented a systematic classification procedure of branched and cyclic polymer topologies into different main-classes and sub-classes based on geometrical consideration of the graph presentation of alkane and cycloalkane isomers. Besides, a systematic notation of nonlinear polymer topologies is also presented based on principal geometrical parameters of terminus and junction numbers. Further elucidation of the geometrical and topological basis over the proposed classification and notation procedures will provide deeper insights to understand the intrinsic properties of any of nonlinear polymer constructions, and mutual geometrical relationship. These will be also intriguing to the modeling and simulation studies of polymers of complex structures, which are now becoming obtainable as will be shown in the following Chapters.

References

1. G. R. Newkome, C. N. Moorefield and F. Vögtle, in *Dendritic Molecules-Concepts, Syntheses, Perspectives* (Wiley-VCH, Weinheim, 1996), p.37.
2. T. Schucker, *New J. Chem.*, **17**, 655 (1993).
3. J.-C. Chambron, C. Dietrich-Buchecker and J.-P. Sauvage, in *Cyclic Polymers, 2nd ed.*, Ed. J. A. Semlyen, (Kluwer, Dordorecht, 2000), p.155.
4. A. T. Balaban, in *Chemical Application of Graph Theory*, Ed. A. T. Balaban, (Academic Press, London, 1976), p.63.
5. R. B. King and D. H. Rouvray, Eds., *Graph Theory and Topology in Chemistry* (Elsevier, Amsterdam, 1987).
6. Y. Tezuka and H. Oike, *J. Am. Chem. Soc.*, **123**, 11570 (2001).
7. Y. Tezuka and H. Oike, *Prog. Polym. Sci.*, **27**, 1069 (2002).
8. D. M. Walba, *Tetrahedron*, **41**, 3161 (1985).
9. M. Fischer and F. Vögtle, *Angew. Chem. Int. Ed.*, **38**, 884 (1999).
10. K. Ito and S. Kawauchi, *Adv. Polym. Sci.*, **142**, 129 (1999).
11. A. D. Schlüter and J. P. Rabe, *Angew. Chem. Int. Ed.*, **39**, 864 (2000).

CHAPTER 2

TOPOLOGICAL ISOMERS IN POLYMER MOLECULES

Yasuyuki Tezuka

Department of Organic and Polymeric Materials,
Tokyo Institute of Technology, Meguro-ku, Tokyo, Japan
E-mail: ytezuka@o.cc.titech.ac.jp

The unique isomerism occurring in flexible nonlinear macromolecules, in particular those including cyclic constructions, is presented by comparing with the classical constitutional isomerism and stereoisomerism, in which the Euclidian geometric properties are hypothesized. The isomerization process of these macromolecules is subsequently discussed based on the geometrical or topological transformation of the graph constructions of polymeric topological isomers.

1. Introduction

Isomerism, termed from Greek, *isos* (equal) and *meros* (part), has been a fundamental concept in chemistry since Berzelius.[1] Isomers are a set of compounds having the same chemical constitution (thus molar mass) but different properties. The three-dimensional structure of the isomers is incapable of transforming each other to bring the distinctive properties. The recognition of diverse conception in the isomerism since Kekulé[2] has continuously brought about deeper insights in both static and dynamic structures of chemical substances. In organic chemistry, isomers are basically classified into constitutional (structural) isomers and stereoisomers. The former are the isomers of distinctive *connectivity* with respect to atoms or atomic groups, while the latter are the isomers of indistinguishable *connectivity*, but distinctive each other based on the Euclidian geometric rigidity of molecules, such as the restriction of bond angle bending and of bond rotation. Recently, an increasing number of topologically unique molecules such as cyclics,[3] knots[4] and catenanes[4] have been synthesized not merely from scientific attraction by their forms but also from technological quest for molecular devices and molecular machines.[5] Topological properties uniquely attainable by these molecules have been conjectured by mathematic theories of knots, links and graphs,[6] and remarkably topological chirality of a trefoil knot molecule has been experimentally verified.[4]

In this Chapter, the unique isomerism occurring in flexible macromolecular cyclics, knots and catenanes is presented in reference with the classical constitutional isomerism and stereoisomerism,[7,8] in which the Euclidian geometric properties are hypothesized. The isomerization process of these macromolecules is subsequently discussed based on the geometrical or topological transformation of the graph constructions of polymeric topological isomers.

2. Constitutional and Stereoisomers in Polymer Molecules

First, the *constitutional* isomerism in macromolecules of flexible segment components is considered in reference with small molecules, for which the Euclidian geometric properties are hypothesized. In constitutional isomers of polymer molecules, by definition, the *connectivity* of compositional atoms or atomic groups is distinctive to each other. For a typical example, poly(vinyl alcohol) and poly(ethylene oxide) are distinctive in the connectivity of atoms within their repeating units (Scheme 1), relevant to the case of organic compounds, like dimethyl ether, CH_3OCH_3, and

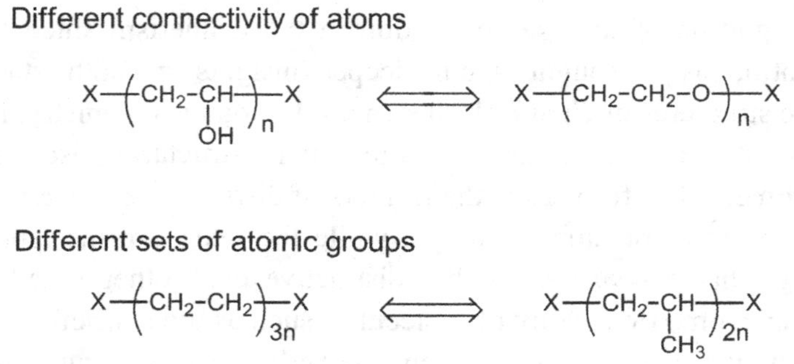

Different connectivity of atoms

$$X\text{---}(CH_2\text{-}CH)_n\text{---}X \quad \Longleftrightarrow \quad X\text{---}(CH_2\text{-}CH_2\text{-}O)_n\text{---}X$$
$$\qquad\qquad |$$
$$\qquad\qquad OH$$

Different sets of atomic groups

$$X\text{---}(CH_2\text{-}CH_2)_{3n}\text{---}X \quad \Longleftrightarrow \quad X\text{---}(CH_2\text{-}CH)_{2n}\text{---}X$$
$$\qquad\qquad\qquad\qquad\qquad\qquad\qquad |$$
$$\qquad\qquad\qquad\qquad\qquad\qquad\qquad CH_3$$

Scheme 1. Constitutional isomers in polymers by different connectivity of atoms and by different sets of atomic groups.

ethanol, CH_3CH_2OH. Likewise, poly(ethylene) and poly(propylene) could be regarded as a constitutional isomer pair, as they are comprised of a different set of atomic groups (Scheme 1), relevant to the case of *n*-butane and *iso*-butane consisting of distinctive compositional atomic groups, namely two CH_3 and two CH_2 groups for the former, and three CH_3 and one CH groups for the latter. Moreover, and to be significant in topological viewpoint, a different type of constitutional isomerism is observed in star polymers having the same arm numbers and the same total arm length, but having different sets of arm length composition (Scheme 2), where the total chain length of the molecular graphs, corresponding to the molar mass, should be maintained as well as the number of junctions and terminus during this topological transformation process. This isomerism is relevant to the specific constitutional isomer pair of hexanes, in which 2-methyl and 3-methyl pentanes are comprised of common atomic groups, i.e. three CH_3, two CH_2 and one CH groups. And the isomerism arises from the distinctive linking mode of these groups. These reflect diverse classes of the *connectivity* conception in the classical constitutional isomerism. It is notable, moreover, that the molecular graphs of 2-methyl and 3-methyl pentane are topologically equivalent, i.e. convertible each other through the topological transformation by conceptual continuous deformation of lines (chains) in the molecular graph. This is, notably, not exactly conforming with the theorem of topological geometry. The molecular graphs of other hexane isomers are, on the other hand, topologically distinctive each other, as they are distinctive in the compositional atomic groups (CH_3, CH_2, CH and C groups).

In addition, another class of constitutional isomers is introduced with the common compositional atomic groups but having topologically distinctive molecular graphs. A prototypical example observed in alkanes is a pair of a dendritic type and a comb type decane molecule shown in Chapter 1 (Scheme 2). These are isomers both comprised of six CH_3 and four CH groups, but possess topologically distinctive molecular graphs. Thus, we designate this particular type of constitutional isomers as *topologically distinctive constitutional isomers* in contrast to another set of constitutional isomers of 2-methyl and 3-methyl pentanes as *topologically equivalent constitutional isomers*. As we will show later, *topologically distinctive constitutional isomers* are often observed in macromolecular cyclics, knots and catenanes (Scheme 2).

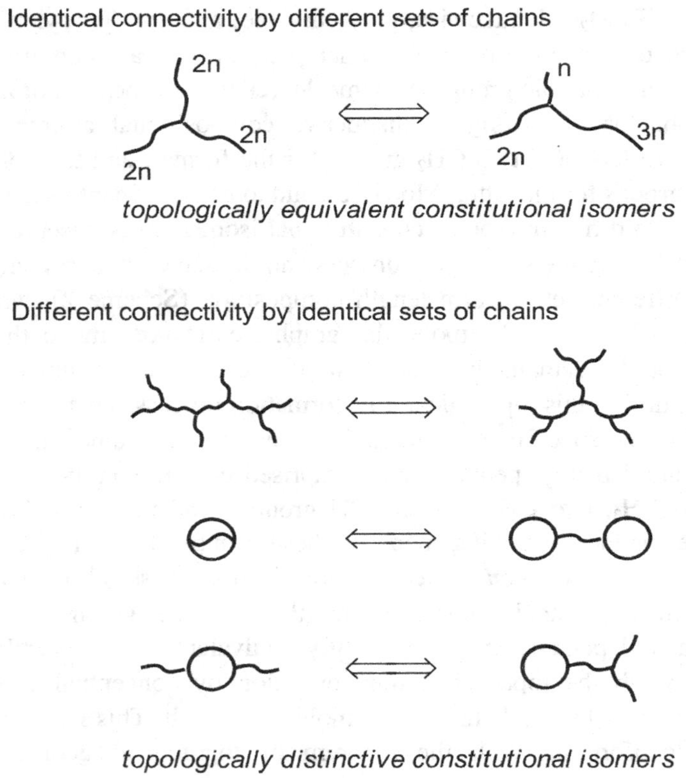

Identical connectivity by different sets of chains

topologically equivalent constitutional isomers

Different connectivity by identical sets of chains

topologically distinctive constitutional isomers

Scheme 2. Constitutional isomers in polymers with identical connectivity by different sets of chains and different connectivity by identical sets of chains.

It should be stressed that the mutual conversion (or isomerization) process in classical constitutional isomers and in classical stereoisomers intrinsically differs each other in the expression of the molecular graphs. Since the former is distinctive each other in their *connectivity*, the chain-breaking with at least *two* appropriate positions is required, followed by the chain-rearrangement and the chain-recombination to complete the process. On the other hand, for the mutual conversion in the molecular graphs of a pair of classical stereoisomers, the chain-breaking process is not required but the conceptual unrestricted deformation of bond angles and the freedom of bond rotation are sufficient to complete the process.

3. Topological Isomers in Polymer Molecules

A pair of compounds possessing a simple ring (or a trivial knot by the topology term) molecular graph and a knot (typically a trefoil knot) counterpart has been referred as a typical example of *topological stereoisomers* (Scheme 3).[4] It should be noted that an open ring compound is also a stereoisomer of a triangle, a square or a randomly constrained ring compound (Scheme 3). These two isomer conception are recognized

topologically distinctive diastereomers

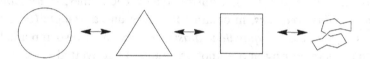

topologically equivalent diastereomers

topologically distinctive enantiomers

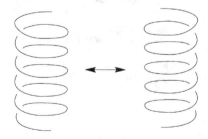

topologically equivalent enantiomers

Scheme 3. Topologically distinctive and equivalent diastereomers and enantiomers.

distinctive each other by comparing the topological property with their molecular graphs. It is intuitively understood that a ring and a triangle (and a square etc.) are topologically equivalent, as they are capable of converting each other through the conceptual continuous chain deformation with retaining the total chain length, but without chain-breaking process. In contrast, a ring and a knot can only be converted each other by applying the chain-breaking process. Also notably, the chain-breaking with a SINGLE position and the subsequent chain-rearrangement/chain-recombination should be completed in order to mutually convert this isomer pair. This geometrical manipulation is fundamentally inconsistent with that applied to the constitutional isomers, where the chain-breaking with at least TWO positions (appropriately chosen) is required to complete the process. By this distinction, we call a ring and knot as *topologically distinctive stereoisomers* or more specifically, since they are not mirror images each other, as *topologically distinctive diastereomers*, in contrast to a ring and a triangle (or a square etc.) as *topologically equivalent diastereomers*. It is also remarkable to note that a simple ring and a knot are mutually converted when they are handled in 4-dimensional space rather than 3-dimensional counterpart, by a topological geometry theorem.[6]

Moreover, a trefoil knot compound is known to be resolved into a right-handed and a left-handed enantiomeric isomer (Scheme 3). It is

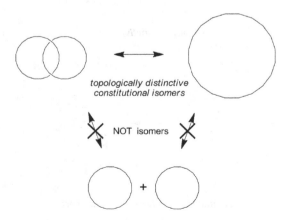

topologically distinctive constitutional isomers

NOT isomers

Scheme 4. Topological relationships between a simple ring and their links.

remarkable again that these isomers are mutually converted only by applying the chain-breaking process with a SINGLE position, in contrast to the classical enantiomeric isomers, where the molecular graphs are converted each other simply by the conceptual continuous chain-deformation, but without applying the chain-breaking process. Thus, we designate a right and a left-handed knot as *topologically distinctive enantiomers* in contrast to another relevant isomer pair, such as a right and a left handed open-helix chain compounds as *topologically equivalent enantiomers* (Scheme 3).

The isomerism in a catenane (more precisely [2]-catenane, or Hopf link by topology term) and two separated rings has often been quoted since the pioneering study by Wasserman (Scheme 4).[9] This particular isomerism is analogous to the isomerization between a ring and a knot, as a [2]-catenane is capable of converting into two rings by applying the chain-breaking with a SINGLE position and the subsequent chain-rearrangement/chain-recombination process. This definition is again reflecting a topological geometry theorem that these two constructions are mutually converted when they are placed in 4-dimensional space rather than 3-dimensional counterpart. It must be emphasized, however, in chemistry that the isomerism refers, in itself, to a set of compounds possessing the same chemical formula. Hence for example, a [2]-catenane molecule consisting of two rings of 30 methylene units is an isomer of a simple ring molecule of 60 methylene units, and more importantly, either is not an isomer but a product of two molecules of a ring of 30 methylene units. This is conceivable in chemistry just as because a cyclohexane is not an isomer of TWO cyclopropanes! Therefore, this process-based analogy is chemically inappropriate for a basis of the isomerism involving a

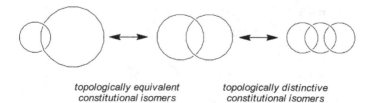

topologically equivalent
constitutional isomers

topologically distinctive
constitutional isomers

Scheme 5. Topological isomers by catenated ring polymers.

catenane. Likewise, a rotaxane molecule is a product but not an isomer of a separated axle and a ring component.

Moreover, a large single ring and a catenane are mutually converted each other by applying the chain-breaking with at least appropriate TWO positions followed by the chain-rearrangement/chain-recombination. This exactly conforms to the criterion of the constitutional isomers. Therefore, a set of a ring and a catenane is classified as *topologically distinctive constitutional isomers* as in the previous dendritic and comb-type branched decane isomers (Scheme 4). Likewise, a [2]-catenane is a *topologically distinctive constitutional isomer* of a [3]-catenane. On the other hand, a set of [2]-catenanes possessing different ring sizes is classified into *topologically equivalent constitutional isomers* (Scheme 5).

4. Polymeric Topological Isomers and Homologues

The geometrical relationship between different polymer architectures could be elucidated on the basis of the classification of nonlinear polymer topologies into A, I, II and III main-classes and further into sub-classes, as detailed in Chapter 1.[7,8]

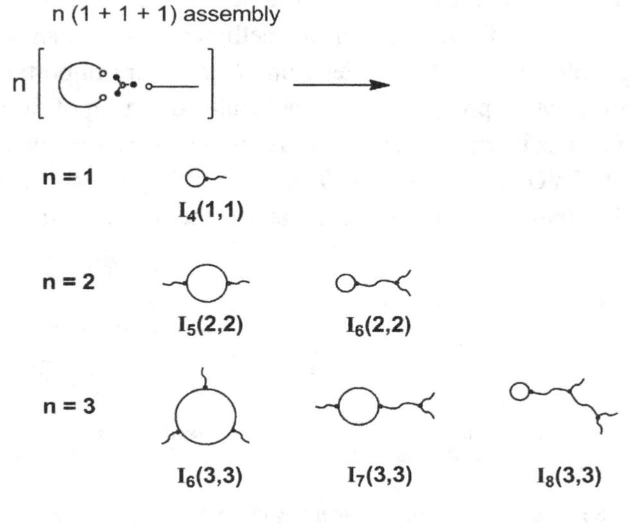

Scheme 6. Topological isomers by *a ring with branches* polymers.

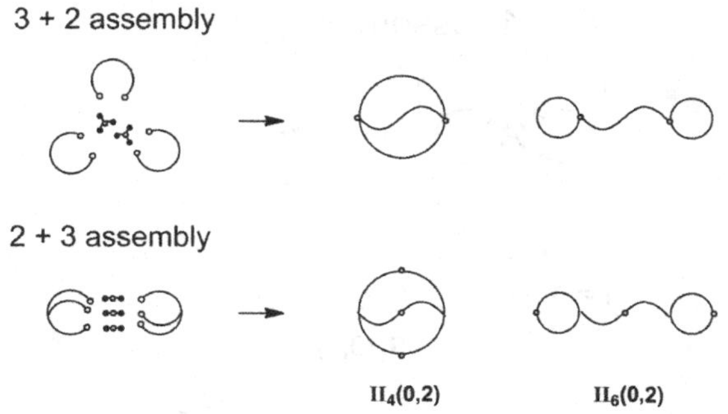

3 + 2 assembly

2 + 3 assembly

$II_4(0,2)$ $II_6(0,2)$

Scheme 7. Topological isomers by *dicyclic* theta- and manacle-shaped polymers.

First, a set of different topological constructions belonging to different sub-classes, but possessing identical terminus and junction numbers, typically a pair of $I_5(2,2)$ and $I_6(2,2)$, is considered (Scheme 6). It is intuitively recognized that these two constructions are produced from an identical precursor set of telechelic (end-reactive) polymers and end-linking reagents, i.e., two bifunctional and two monofunctional polymer precursors and two trifunctional end-linking reagents (Scheme 6). Since they are produced from the identical precursors by common chemical reactions, their chemical compositions are identical with each other, and they possess an identical molecular weight. They are, on the other hand, distinctive from each other with respect to their topologies. They are thus regarded as a pair of constitutional isomers, and more specifically as *topologically distinctive constitutional isomers*. Such topological isomerism frequently occurs among constructions in the I, II or III main-classes; though also observed in the A main-class topologies, as in $A_9(6,3)$ and $A_{10}(6,4)$ shown in Table 1 in Chapter 1. A number of topological isomer pairs are observed in the I main-class listed in Tables 2 in Chapter 1, including the three in sub-classes $I_6(3,3)$, $I_7(3,3)$ and $I_8(3,3)$ (Scheme 6).

Another notable pair of topological isomers is observed in the II main-class, namely the two constructions of $II_4(0,2)$ and $II_6(0,2)$ (Scheme 7). These are again the product of an identical set of telechelic (end-reactive) polymer precursors and end-linking reagents, i.e., three bifunctional polymer precursors and two trifunctional end-linking

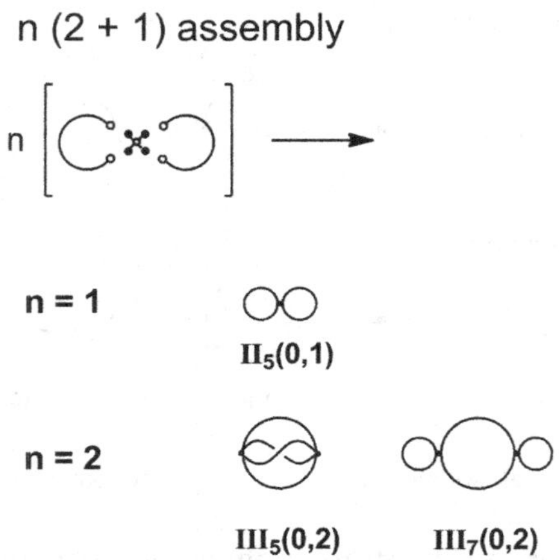

n (2 + 1) assembly

$$n \left[\text{C} \bullet \text{O} \right] \longrightarrow$$

n = 1   ∞

$II_5(0,1)$

n = 2

$III_5(0,2)$ $III_7(0,2)$

Scheme 8. Topological isomers by self-assembly of linear polymer precursors.

reagents, or two trifunctional star polymer precursors and three bifunctional end-linking reagents as detailed in Chapter 6. Other pairs of topological isomers in the II and III main-classes listed in Tables 3 and 4 in Chapter 1 include: the two constructions of $II_5(1,2)$ and $II_6(1,2)$, the two of $III_4(0,4)$ and $III_6(0,4)$, and the two of $III_5(0,2)$ and $III_7(0,2)$.

Secondly, a series of constructions belonging to different sub-classes, while their terminus and junction numbers are in a regular order from one to another, are of an interest from topological point of view. Thus, a group of sub-classes $I_4(1,1)$, $I_5(2,2)$, $I_6(2,2)$, $I_6(3,3)$, and $I_7(3,3)$ is an typical example to be considered (Scheme 6). It is recognized that the two topological isomers of $I_5(2,2)$ and $I_6(2,2)$, the three of $I_6(3,3)$ and $I_7(3,3)$ are the products of twice and three times, respectively, of a precursor set for the production of the $I_4(1,1)$, namely one monofunctional and one bifunctional polymer precursors and one trifunctional end-linking reagent (Scheme 6). This reveals a hierarchy of a *homologous* series of polymer topologies deduced from the classification of nonlinear topological constructions.

Thirdly, a group of topological constructions possessing the relevant terminus and junction numbers, while belonging to different main-classes in either II or III are to be considered. Thus, the two constructions of $III_5(0,2)$ and $III_7(0,2)$, are produced from twice of a precursor set for

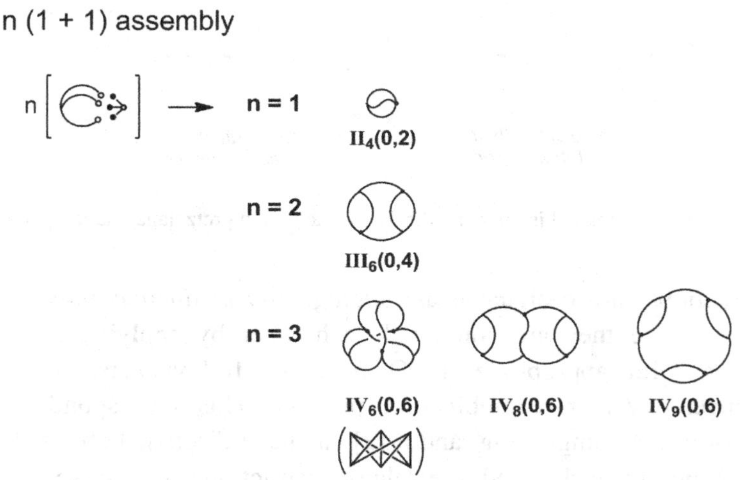

n (1 + 1) assembly

Scheme 9. Topological isomers by self-assembly of star polymer precursors.

the production of a construction of $II_5(0,1)$, namely two bifunctional polymer precursors and one tetrafunctional end-linking reagent (Scheme 8). Likewise, the construction of $III_6(0,4)$ and another $K_{3,3}$ type of $IV_6(0,6)$ are produced twice and three times, respectively, of a precursor set for the production of a construction of $II_4(0,2)$, namely one trifunctional star polymer precursor and one trifunctional end-linking reagent (Scheme 9).

5. Topological Isomers in Dicyclic Polymer Molecules

Finally, the isomerism of three bicyclic polymer molecules, namely a θ-shaped, a manacle-shaped and a pretzelan-shaped, catenated ones is of interest to note (Scheme 10). As shown in Chapter 1, θ-shaped and a manacle-shaped constructions are designated as $II_4(0,2)$ and $II_6(0,2)$, respectively. It is also evident that they are isomeric forms each other, as they are products of an identical set of the chain precursors and the linking reagents, namely a set of three linear chains and two trifunctional linking reagent or of two three-armed chains and three bifunctional linking reagents. The θ- and manacle-constructions (or θ- and pretzelan-

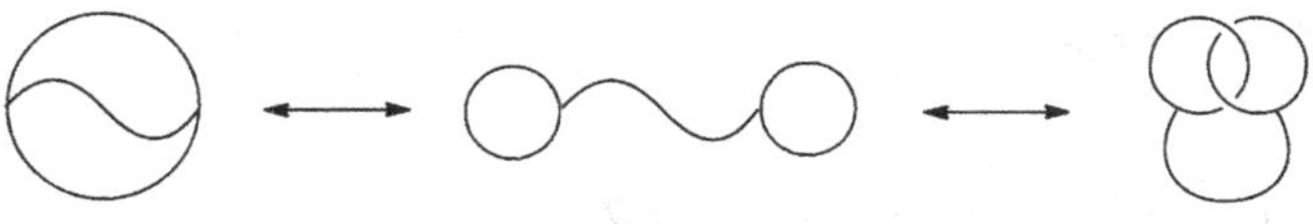

topologically distinctive
constitutional isomers

topologically distinctive
diastereomers

Scheme 10. Topological isomers by theta-, manacle- and pretzelane-shaped polymers.

constructions) are designated as *topologically distinctive constitutional isomers*, since they are converted each other by applying the chain-breaking with appropriate TWO positions followed by the chain-rearrangement/chain-recombination process. This corresponds to the cases of a large single ring, and a [2]-catenane discussed above. On the other hand, manacle- and pretzelan-constructions are converted each other by applying the chain-breaking with an appropriate SINGLE position. Thus, they are classified into, as in the case of a ring and a knot, *topologically distinctive diastereomers*. It is also notable that, in topological geometry, a *pretzel transformation* refers a process to this interconversion, in which not only a pair of manacle- and pretzelan-constructions but also an 8-shaped counterpart are mutually transformed through unrestricted deformation of chains with the number of junction point as variable geometrical parameter.

6. Ongoing Challenges and Future Perspectives

This Chapter presented the geometrical and topological approaches toward the unique isomerism occurring in flexible nonlinear macromolecules, in particular those including cyclic constructions in comparison with the classical constitutional isomerism and stereoisomerism, in which the Euclidian geometric properties are hypothesized. The isomerization process of these macromolecules is subsequently discussed based on the geometrical or topological transformation of the graph constructions of polymeric topological isomers. Further elucidation of the geometrical and topological basis over the proposed topological isomerism conception will provide deeper insights to understand the intrinsic structural relationship of nonlinear polymer constructions, and their transformations. It is noteworthy that a modified theorem of topological geometry is awaited to cope with real polymer molecules, which are considered to be restricted in some essential geometrical parameters in 3D space, like the total chain length, the number of junctions, as well as spatial free crossing of chains, which are presumed unrestricted in a genuine topological geometry. These will provide useful basis on designing and synthesizing novel polymers of complex structures. Indeed in the following Chapters, a synthetic strategy for complex polymer topologies is discussed upon topological insights obtained above on polymer architectures. A protocol of an *electrostatic self-assembly and covalent fixation* process provides an efficient means for the end-linking reaction of various telechelic (end reactive) polymer precursors of either mono-, bi-, and trifunctionalities having moderately strained cyclic ammonium salt groups.

References

1. J. J. Berzelius, *Pogg. Ann.*, **19**, 305 (1830).
2. F. A. Kekulé, *Ann. Chem.*, **106**, 129 (1858).
3. J. A. Semlyen, Ed., *Cyclic Polymers, 2nd ed.* (Kluwer, Dordrecht, 2000).
4. J.-P. Sauvage and C. Dietrich-Buchecker Eds., *Molecular Catenanes, Rotaxanes and Knots* (Wiley-VCH, Weinheim, 1999).
5. V. Balzani, A. Credi, F. M. Raymo and J. F. Stoddart, *Angew. Chem. Int. Ed.,* **39**, 3348 (2000).
6. E. Flapan, *When Topology Meets Chemistry: A Topological Look at Molecular Chirality* (Cambridge University Press, Cambridge, 2000).
7. Y. Tezuka and H. Oike, *J. Am. Chem. Soc.*, **123**, 11570 (2001).
8. Y. Tezuka and H. Oike, *Prog. Polym. Sci.*, **27**, 1069 (2002).
9. H. L. Frisch and W. Wasserman, *J. Am. Chem. Soc.*, **83**, 3789 (1961).

CHAPTER 3

TELECHELICS HAVING CYCLIC ONIUM SALT GROUPS

Yasuyuki Tezuka

Department of Organic and Polymeric Materials,
Tokyo Institute of Technology, Meguro-ku, Tokyo, Japan
E-mail: ytezuka@o.cc.titech.ac.jp

A new class of telechelic polymers having specifically cyclic onium salt groups is presented. A series of telechelic polymer precursors having different cyclic onium salt groups with tunable reactivity were introduced to produce effectively model branched polymers, such as star polymers, polymacromononers and graft copolymers, as well as model networks of single and multiple polymer components.

1. Introduction

The term *telechelic* was first introduced in 1960 in order to describe polybutadienes having acetyl and hydroxyl end groups, which were designed for use in curing reactions to achieve improved elastic properties.[1] Polymers having reactive end groups had already been known as polyesters and polyamides produced under non-stoichiometric polycondensation. However, with the telechelic polybutadienes, the importance of prepolymers having purposely introduced reactive end groups was demonstrated for the synthesis of multidimensional and multicomponent polymer materials with controlled and predictable properties. This new term of *telechelic polymers* or *telechelics* has since widely been used to define a group of polymers with common chemical properties.[2,3]

In addition to *telechelics*, the relevant concepts of *oligomers* and *macromonomers* were introduced in the 1950s and in the 1970s, respectively.[3] Also, the term *prepolymer* has frequently been used. A brief description of the historical background introducing these terms will be instructive to overview this field of polymer science.

Linear oligomers and cyclic oligomers are two major groups of oligomers based on their structural features. The latter have become particularly important since the discovery of crown ethers,[4] which have opened up an entirely new discipline of chemistry based on the molecular recognition behavior of cyclic oligomers from synthetic as well as natural sources.[5] The term *reactive oligomers* is, on the other hand, based on their functional feature of deliberately introduced reactive groups either along or at the end of the oligomer chain, which are utilized for further reactions. Hence, telechelics are obviously within the class of reactive oligomers.

Major progress in the preparation of telechelics was made along with the understanding of the fundamental reaction processes in a radical polymerizations to introduce the specific end groups by the appropriate choice of either an initiator or a chain transfer reagent.[6] Further progress was achieved through the discovery and subsequent developments of a living polymerization technique,[7] which provided an opportunity to control the chain length of the polymer product in addition to the introduction of various functional end groups.[2]

The term *telechelics* initially referred to oligomers possessing reactive groups at both chain ends, applicable as a cross-linker or a chain extender. Thus, the structure of telechelics is defined by indicating the nature of the end groups, such as α-hydroxy-ω-carboxy polybutadiene. Telechelics possessing only one reactive group at a single end of the polymer chain, named *semi-telechelics*, have also gained increasing attention. In particular, those having a polymerizable end group have become more and more important for the preparation of well-defined graft copolymers. This led to the introduction of the term of macromers (or *macromonomers*).[8] Macromonomers are subjected to a copolymerization reaction with conventional monomers, in which polymerization forms the main chain segment. The advantage of the macromonomer technique is that the graft segment component can be fully characterized prior to the preparation of the graft copolymers. The macromonomers initially referred primarily to oligomers having a group copolymerizable by a chain polymerization mechanism, namely a radical, ionic or coordination procedure. Macromonomers possessing an end group applicable for a step polymerization mechanism, namely polycondensation and polyaddition, to produce polyesters, polyamides and polyurethane-based graft copolymers, were subsequently introduced.[8]

It must be recalled here that common telechelics having reactive groups at both chain ends have been also used in step polymerization systems for the synthesis of polyester, polyamide and polyurethane-based block copolymers. These graft and block type copolymers are thus closely relevant with respect to the synthetic procedure in practice and complementary each other in their structural features. Nevertheless, the branched graft copolymer is capable of having more structural versatility and requires geometrical parameters to specify its structural characteristics than the linear block copolymer.

In this Chapter, telechelic polymers having specifically cyclic onium salt groups are presented. A series of telechelic polymer precursors having different cyclic onium salt groups with tunable reactivity were introduced to produce effectively model branched polymers, such as star polymers, polymacromononers and graft copolymers, as well as model networks of single and multiple polymer components.

2. Telechelics Having Various Cyclic Onium Salt Groups

The choice of an appropriate reactive end group is critical in the preparation of precisely designed multidimensional and multicomponent polymer structures through a telechelic polymer technique.[2] As the reaction proceeds under very low initial concentration of the reactive end groups in telechelic polymer reactions, the apparent reaction rate suffers further decrease along with the progress of the reaction. This kinetic suppression is particularly eminent in the synthesis of star polymers and model networks, in which the reaction between a mono- and bifunctional telechelic prepolymer and a multifunctional coupling reagent is performed in a strictly equimolar condition to completion in order to avoid structural defects in the final product. Although the use of a highly reactive living end group may overcome this problem and considerable

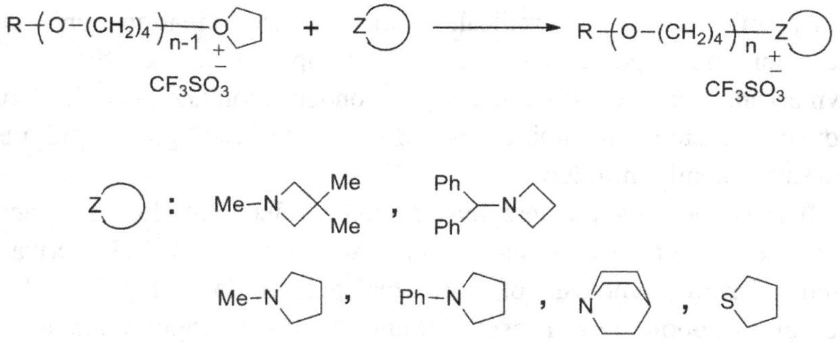

Scheme 1. Telechelic poly(THF)s having cyclic onium salt groups.

progress has indeed been achieved, the reaction at the living end group is frequently beyond control and handling the living polymer requires a cumbersome experimental procedure, further limiting its practical applications. Also in the synthesis of block and graft copolymers by the telechelic polymer system, the incompatible nature of different polymer chains causes a significant decrease of the apparent reactivity of the end groups due to the repulsive interaction and the eventual phase separation of the two incompatible polymer components.

A new type of telechelic polymers were thus introduced to overcome above difficulties in polymer reactions, typically a series of uniform-size poly(tetrahydrofuran)s, poly(THF)s, having such moderately strained cyclic onium salt groups as 4-membered (azetidinium), 5-membered cyclic (pyrrolidinium), and 6-membered bicyclic ammonium (quinuclidinium) salts as well as 5-membered cyclic sulfonium (tetrahydrothiopheniums) salt groups.[3] These cyclic onium salt groups are readily introduced either at a single chain end or at both chain ends through the reaction of the corresponding cyclic amines or a sulfide with the oxonium salt end group of a living poly(THF), which is produced with a triflic acid ester or a triflic anhydride as initiator (Scheme 1). The following unique features are noted for these poly(THF)s with a series of cyclic onium salt end groups;

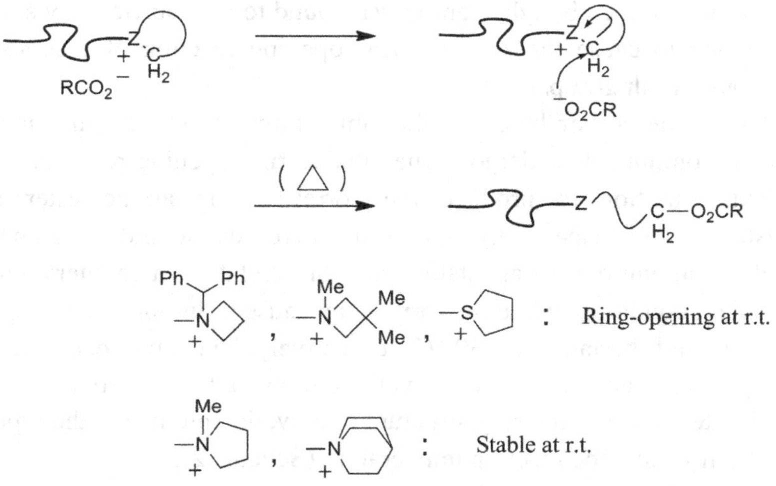

Scheme 2. Reactivities of cyclic onium salt groups on telechelic poly(THF)s.

1. They are sufficiently stable during common isolation and characterization procedures, and no specific precautions are required for handling and storage.

2. When the molecular weight exceeds a few thousand, they become readily soluble in organic solvents but insoluble in water, since the lipophilic nature of the polymer chain dominates the solubility property despite the presence of ionic end groups.

3. On the other hand, the aggregation of the ionic end groups takes place in nonpolar organic solvents to increase the local concentration of the ionic end group despite the low overall concentration.

4. The cyclic onium salt end group initially accompanies a counteranion with a weak nucleophilic reactivity, such as triflate or tosylate anion, but which can be replaced by another through an anion-exchange reaction. The anion-exchange reaction indeed takes place simply by the precipitation of a THF solution of the telechelic poly(THF) into an aqueous solution containing an excess amount of another anion as a salt form.

5. In particular, a carboxylate anion was found to be a sufficiently strong nucleophile to cause the selective ring-opening reaction of a series of cyclic onium salt groups.

Thus, the 4-membered cyclic ammonium and the 5-membered cyclic sulfonium salt undergo a quantitative ring-opening reaction at an ambient condition to produce the corresponding amino ester and thioester groups, respectively. The 5-membered cyclic and 6-membered bicyclic ammonium salts are stable with carboxylate as a counteranion at an ambient condition, while the former can cause a ring-opening reaction at 100 °C and the latter at 130 °C, respectively. Thus, one can tune the ring-opening reaction of these cyclic onium salt end groups by an appropriate choice of the ring structure of a cyclic onium salt, the type of counteranion, and the reaction temperature (Scheme 2).

By making use of these unique characteristics of telechelic polymers with cyclic onium salt groups,

1. The modification of the polymer end group was performed by the simple precipitation of telechelic poly(THF)s into an aqueous solution containing a carboxylate compound with another functional group, such as sodium methacrylate, to produce a poly(THF) macromonomer.[9]

2. The synthesis of various model polymers such as star polymers and model networks was performed by the simple precipitation of mono- and bifunctional telechelic poly(THF) into an aqueous solution containing an excess amount of a plurifunctional carboxylate salt.[10]

3. The grafting reaction onto water-soluble polymers containing carboxylate salt groups, such as sodium carboxymethylcellulose, was carried out by the precipitation of monofunctional telechelic poly(THF) into an aqueous solution containing a polymeric carboxylate.[11]

4. The macromolecular ion-coupling reaction of telechelic poly(THF)s with a hydrophobic polymer containing a carboxylate salt group, namely, poly(styrene-co-sodium acrylate) of less than 5 mol% carboxylate content was conducted. The strong ionic interaction between the end group of the telechelic poly(THF) and carboxylate groups on the polystyrene chain could overcome the repulsive interaction between these two incompatible polymer segments to cause an efficient coupling

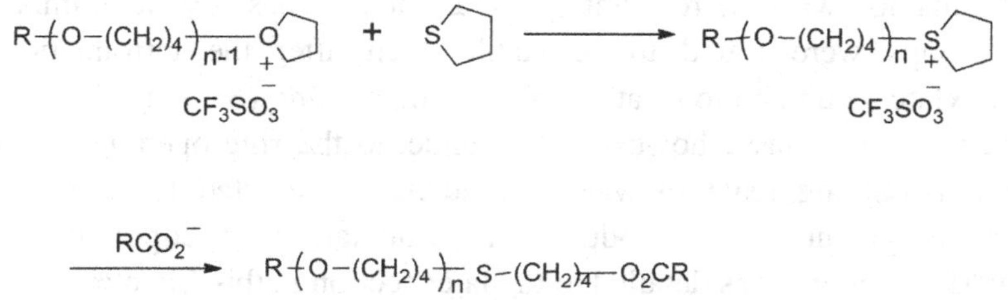

$$R\text{-}(O\text{-}(CH_2)_4\text{-})_{n-1}\text{-}O\overset{+}{\bigcirc} \quad + \quad S\bigcirc \quad \longrightarrow \quad R\text{-}(O\text{-}(CH_2)_4\text{-})_n\overset{+}{S}\bigcirc$$

$$\underset{CF_3SO_3^-}{\qquad\qquad\qquad\qquad} \qquad\qquad\qquad\qquad\qquad \underset{CF_3SO_3^-}{}$$

$$\xrightarrow{RCO_2^-} \quad R\text{-}(O\text{-}(CH_2)_4\text{-})_n\,S\text{-}(CH_2)_4\text{-}O_2CR$$

Scheme 4. Synthesis and ring-opening reaction of poly(THF) having tetrahydrothiopjhenium salt groups.

homopolymerization but was found to cause a selective ring-opening reaction by carboxylate anion at ambient condition (Scheme 4). This is just relevant to the case of azetidinium salts described above.[15]

Thus, poly(THF) having tetrahyrothiophenium salt groups was prepared to be stable at ambient condition to allow full characterization and subsequent storage until use without any precautions.[9] The ion-coupling reaction was also conducted with these telechelic precursors to produce, in particular, polymacromonomers as detailed in the later section in this Chapter.

2.3. Telechelics having 5-membered cyclic ammonium (pyrrolidinium) and 6-membered bicyclic ammonium (quinuclidinium) salt groups

A class of 5-membered cyclic amines, pyrrolidines, are known to be incapable of undergoing homopolymerization though they have ring strains. This is also the case of quinuclidinium, 6-membered bicyclic ammonium salts, having ring strains due to their boat-form conformations. These modestly strained cyclic ammonium salts could, therefore, undergo the ring-opening reaction with appropriately reactive nucleophiles at suitable conditions. Thus, telechelic polymers of poly(THF), polystyrene, polyethylene oxide and poly(dimethylsiloxane) having 5-membered ammonium (pyrrolidinium) and 6-membered bicyclic ammonium (quinuclidinium) salt groups were prepared either by direct end-capping of oxonium propagating species for Poly(THF), or by the endgroup transformation from hydroxyl group to tosylate, followed by the quaternization with the relevant cyclic amines.[16] These cyclic ammonium salt groups were found to be stable even after the introduction of carboxylate counteranions at ambient temperature (Scheme 2). At the elevated temperatures, however, they undergo the ring-opening reaction. The ion-coupling reaction was subsequently conducted by using these telechelic precursors to produce, in particular, graft copolymers and network copolymers as detailed in the later section in this Chapter.

3. Ion-Coupling Reactions with Telechelics Having Cyclic Onium Salt Groups

3.1. *Star polymers and polymacromonomers*

A star polymer, i.e. a macromolecule possessing a defined number of segments at the branch points and controlled branch segment length, is a model for branched polymers. The preparation of star polymers can be performed by one of the following methods:

1. A coupling reaction of a monofunctional living polymer with a multifunctional coupling reagent.

2. The reaction of a living polymer with a multifunctional monomer to form a microgel at a single end of a polymer chain.

3. Living polymerization initiated by a multifunctional initiator, including a reactive microgel.

4. Homo- and copolymerization of a macromonomer.

Method **1** can provide a star polymer with a precisely controlled number of branches. Star polymers having even dendritic branches were synthesized by this method.[17] In practice, however, an excess amount of a living prepolymer is required against a multifunctional coupling reagent in order to avoid an incomplete coupling reaction. Therefore, subsequent fractionation to remove excess prepolymer is usually unavoidable. Method **2** can produce a star polymer having a large number of branches, but rigorous control over the number of branches is difficult to achieve. Method **3** provides an opportunity to produce a block copolymer segment in star polymers. However, the initiation reactions by a multifunctional initiator hardly take place simultaneously, resulting in poor control over the branch length. Method **4** is an alternative to provide new architectures in pseudo-star polymers.

By making use of the unique features of telechelic polymers having cyclic onium salt groups, star polymers can be synthesized in high yield by a facile experimental procedure (Scheme 5).[18] The reaction procedure is simply the precipitation of the telechelic polymer solution into an aqueous solution containing an excess of a multifunctional carboxylate salt. Thus, when a monofunctional telechelic poly(THF) having a 4-membered cyclic ammonium salt group was precipitated into an aqueous solution containing an excess of trisodium 1,3,5-benzenetricarboxylate, the product recovered by filtration was a mixture of the original prepolymer and a fraction of three times the molecular weight as evidenced by GPC analysis. The latter corresponds to a star polymer with a branch number of three, equal to the functionality of the carboxylate used in the reaction. The precipitation procedure with other carboxylate salts with different functionality accordingly produced star polymers with a variety of branch numbers. The yield of the star polymer fraction increases with repeated precipitation of the mixture of the uncoupled prepolymer and the star polymer into an aqueous solution containing an excess amount of a carboxylate salt. Thus, by a similar manner, a star polymer of four branches was obtained in an almost pure form after three precipitations.

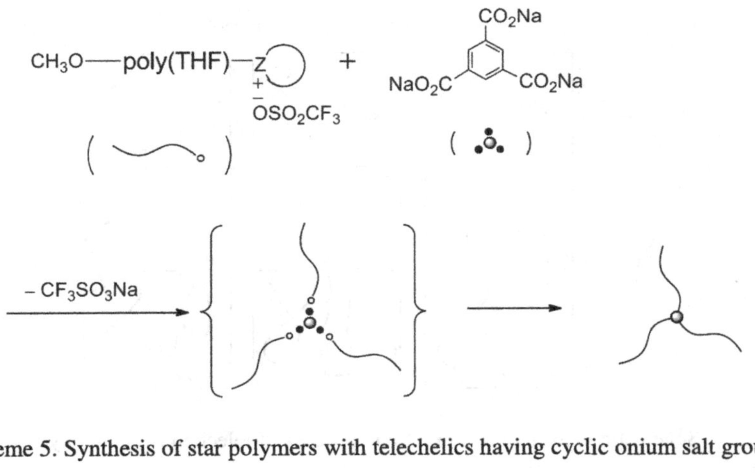

Scheme 5. Synthesis of star polymers with telechelics having cyclic onium salt groups.

This process is unique since after one functional group in the multifunctional carboxylate has reacted with a cyclic onium salt group of the telechelic polymer, the remaining carboxylate functions are rapidly consumed one after another in subsequent reactions. As schematically shown in Scheme 6, a plausible reaction path is the following. First, the hydrophilic cyclic onium salt end groups are organized to face the aqueous phase during the precipitation treatment. Once one of the carboxylate functions of the multifunctional carboxylate in the aqueous phase undergoes the ion exchange reaction, the multifunctional carboxylate anion is immobilized at the interface of the precipitated telechelic polymer. Thus, the remaining carboxylate functions can easily react with the surrounding cyclic onium salt groups by eliminating the water soluble sodium triflate into the aqueous phase.

It is remarkable that the present system employs an excess of a multifunctional carboxylate in the reaction system, and this is in contrast to the conventional process of producing a star polymer by coupling reactions, in which an excess of prepolymer reagent is usually required to avoid structural defects in the final product. It must be noted, moreover, that the direct precipitation of a living poly(THF) itself into an aqueous solution containing a multifunctional carboxylate failed to produce a star polymer, but gave a poly(THF) having a hydroxyl end group.

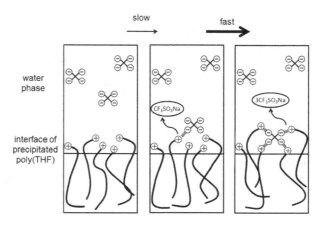

Scheme 6. Schematic picture of ion-exchange process of telechelics having cyclic onium salt groups.

Polymacromonomers (or comb-shaped polymers) are a class of highly branched macromolecules and have attracted growing interests due to their anisotropic cylindrical molecular shapes directly observable by AFM technique.[19] The synthesis of polymacromonomers has so far been performed either by a polymerization of a macromonomer under rigorously controlled conditions, or by a coupling reaction of uniform size polymer precursors for both polyfunctional backbone and monofunctional branch segment components.[8] It is notable, however, the homopolymerization of a macromonomer is frequently circumvented by the unusually low concentration of polymerizable groups, such as vinyl groups, which decreases the apparent rate of polymerization, resulting in a product of only a limited degree of polymerization with a limited conversion of macromonomer. Besides, the control over the degree of polymerization, i.e., the number of the branch segments, of the polymacromonomer is difficult to achieve. A special reaction condition is thus required to compensate these disadvantages through the decrease of the apparent rate of termination due to the viscosity effect. In addition, the copolymerization of different macromonomers has scarcely been reported presumably due to the incompatibility of antagonistic polymer segments, resulting in a macroscopic phase separation in the reaction system.

An alternative means to produce polymacromonomer was developed by making use of monofunctional poly(THF) having either an azetidinium or a tetrahydrothiophenium salt group (Scheme 7).[9,14] Thus, the macromolecular ion-coupling reaction of the telechelic precursor having a 5-membered cyclic thiolanium salt group with poly(sodium acrylate)

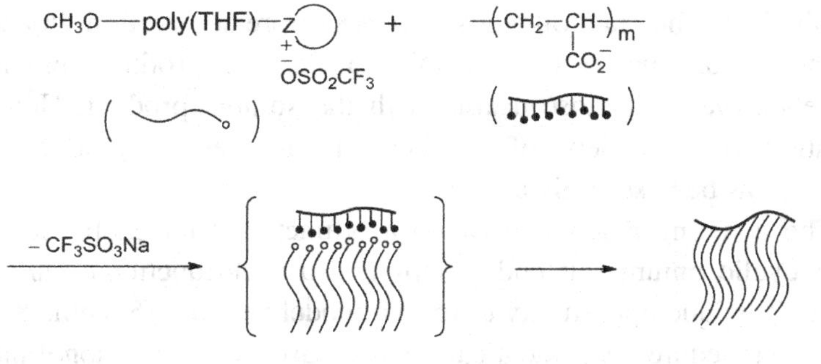

Scheme 7. Synthesis of polymacromonomers with telechelic poly(THF) having cyclic onium salt groups.

having degrees of polymerization of ca. 20 and 60. A simple precipitation of a THF solution of the telechelic precursor into an aqueous solution containing poly(sodium acrylate) promotes the ion-exchange reaction between the two prepolymers, and the subsequent ring-opening reaction of a cyclic sulfonium salt group produces poly(THF) polymacromonomer of predefined branch segment numbers. The ion-coupling process using telechelic polymers having cyclic onium salt groups was also applied to reaction with carboxymethylcellulose sodium salt.[11] The precipitation of a monofunctional poly(THF) having a series of cyclic onium salt groups into an aqueous solution containing such polymeric carboxylates caused the grafting reaction onto the polymeric carboxylates through subsequent heating treatment at 100 °C for the poly(THF) with a 5-membered cyclic ammonium salt end group and at 130 °C for that with a 6-membered bicyclic ammonium salt end group, respectively.

3.2. *Model networks*

A model network, i.e. a crosslinked macromolecule possessing defined branch numbers at the crosslinking points and controlled segment length between the crosslinking points, can be prepared, in principle, by using a bifunctional telechelic polymer in a procedure similar to that used for the production of star polymers. In practice, however, the reaction between a bifunctional prepolymer and a multifunctional coupling reagent is strongly suppressed once the reaction reaches the gel point. Thereafter, the diffusion of the intermediate polymer products becomes extremely difficult. Thus, the reaction tends to cease before completion. In addition, characterization methods applicable to the gel product remain less comprehensive than those usable with the soluble product. Hence, the evaluation of a variety of synthetic techniques to produce model networks has been scarcely achieved.

The coupling reaction between a bifunctional telechelic poly(THF) having cyclic onium salt end groups and a multifunctional carboxylate provides a unique opportunity to form a model network (Scheme 8).[10] The reaction procedure was identical to the case with the monofunctional poly(THF) to produce the corresponding star polymers of defined arm numbers. Thus with a bifunctional carboxylate, *segmented* polymer products were obtained through the chain coupling reaction. No trace of the unreacted prepolymer was detected in contrast to the case of the star polymer synthesis with a monofunctional telechelic poly(THF). This ion-coupling reaction of telechelic poly(THF) with difunctional carboxylates was successfully applied for the synthesis of macromolecular azo initiators.[20] Thus, a dicarboxylate, disodium, 4,4'-dicyano-4,4'-azodivalerate, reacted with a bifunctional poly(THF) having 4-membered cyclic ammonium salt groups to produce a *segmented* poly(THF) having azo functions in the main chain. The azo groups thus introduced could generate a radical species upon heating to 60 °C in the presence of styrene, methyl methacrylate or mixtures thereof to produce the corresponding block copolymers consisting of ionically obtained poly(THF) and the radically obtained polystyrene or polymethacrylate segments.

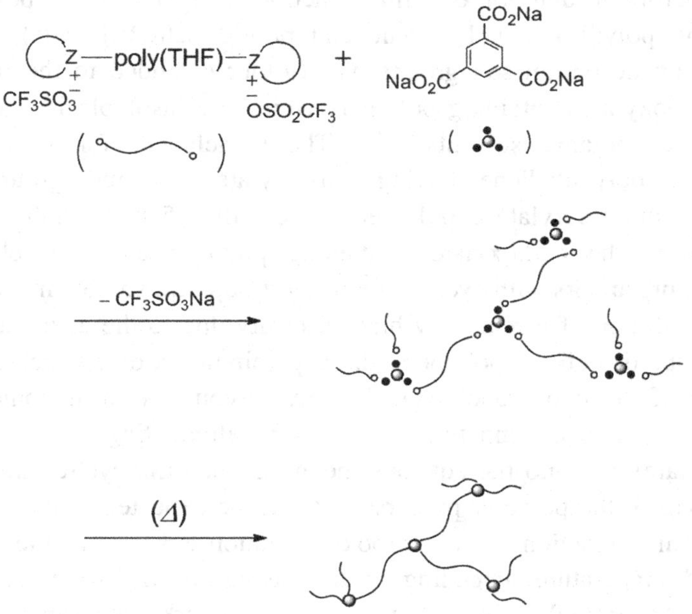

Scheme 8. Synthesis of model networks with telechelic poly(THF) having cyclic onium salt groups.

The reaction of tri- and tetrafunctional carboxylates with a bifunctional telechelic poly(THF) produced a gel product in an almost quantitative yield (Scheme 8). The obtained gel product was reasonably assumed to be a model network with well-defined crosslinking points and segment length, which correspond to the functionality of the carboxylate and the molecular weight of the telechelic poly(THF), respectively. The swelling behavior of the gel products obtained with a trifunctional carboxylate was remarkably different from that of those obtained with a tetrafunctional one, indicating the structural difference at the crosslinking point between the two products. Similarly, poly(dimethylsiloxane) having 5-membered cyclic ammonium salt groups were applied to produce polydimethylsiloxane model networks.[21]

3.3. Graft copolymers and network copolymers

The macromolecular ion-coupling reaction using telechelic polymers, including poly(THF), polystyrene and poly(dimethylsiloxane), having various cyclic onium salt groups was further extended to the reaction with carboxylate-containing polymers which are insoluble in water but soluble in organic solvents.[12,22,23] These included, for instance, a telechelic polybutadiene having carboxylate salt end groups and styrene/sodium acrylate copolymers of less than 5 mol% salt content. Since both the carboxylate containing polymer and the telechelic polymer precursors with cyclic onium salt groups are soluble in THF and form a solution of a relatively high viscosity due to the aggregation of salt groups of both prepolymers, the coprecipitation of a stoichiometric mixture of the two prepolymers into water could form an ionic bond between the two incompatible polymers by eliminating sodium triflate (or tosylate) salt into the aqueous medium. Then the cyclic onium salt end group of the polymer precursors would be expected to undergo the ring opening reaction by the carboxylate anion either at ambient or an elevated temperature depending on the structure of the cyclic onium salt group. Accordingly, block, graft and network copolymers were conveniently obtained by this very simple procedure.

Indeed, the simple copreeipitation of poly(THF) having 4-membered cyclic ammonium salt end groups with a telechelic polybutadiene having sodium carboxylate groups produced a poly(THF)-polybutadiene block copolymer as confirmed by GPC analysis of the products. In addition, the ion coupling reaction of styrene/sodium acrylate copolymer with a monofunctional and a bifunctional poly(THF) having various cyclic onium salt groups produced polystyrene-poly(THF) graft and network copolymers, respectively (Scheme 9).[3] This reaction was monitored by IR to show that the ester carbonyl absorption appeared immediately after the isolation of the products in the reactions using both poly(THF) having 4-membered cyclic ammonium and 5-membered cyclic sulfonium salt groups. On the other hand, the IR ester absorption appeared only after the heat treatment at 100°C or 130°C for

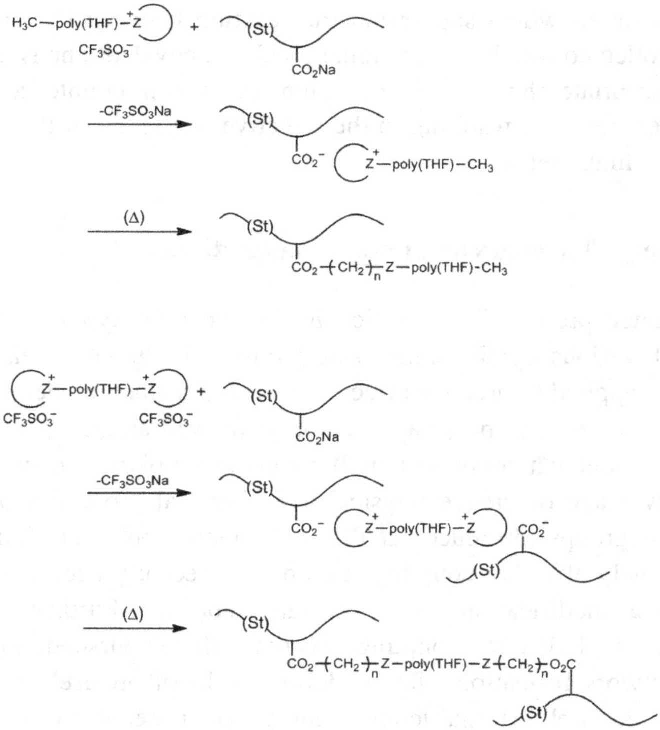

Scheme 9. Synthesis of graft copolymers and two-component network copolymers with telechelics having cyclic onium salt groups.

the products from poly(THF), polystyrene or poly(dimethylsiloxane) having 5-membered cyclic or 6-membered bicyclic ammonium salt end groups, respectively.[12,22,23] Thus, the nucleophilic reaction of the cyclic onium salt group by the carboxylate anion located in the polymer segments took place as in the case with a low molecular weight carboxylate of benzoate anion.

Hence, the present macromolecular ion-coupling process, making use of the unique reactivities of telechelic polymers having a series of cyclic onium salt groups, is regarded as a simple and efficient method to produce various types of block, graft and network copolymers consisting of inherently incompatible segments. The ionic interaction between the salt groups located in the two different prepolymers plays an important role in overcoming the repulsive interaction of the incompatible segments. This repulsion is otherwise likely to cause a macroscopic phase separation which suppresses the reaction between the polymers. The controlled conversion of the ionic bond to a covalent one is achieved by an appropriate choice of cyclic onium salt group, counter anion and reaction temperature, resulting in the selective formation of the coupling products in high yields.

4. Ongoing Challenges and Future Perspectives

This Chapter presented telechelics having specific types of reactive groups of various cyclic onium salt groups with tunable reactivities. They were applied to produce effectively such branched model polymers like star polymers and polymacromononers as well as graft copolymers, and also model networks and multicomponent polymer networks, by taking advantage of unique physical and chemical properties of cyclic onium salt groups introduces at the hydrophobic polymer chain ends. The extremely effective coupling reaction has recently been applied for the surface modification of solids and fabrics.[24] Further synthetic opportunities of this ion-coupling process will be pursued, especially toward network formations like model networks of precisely controlled structures, as well as multicomponent polymer networks, which are difficult to be obtainable in conventional step-by-step polymer reactions.

References

1. C. A. Uraneck, H. L. Hsieh and O. G. Buck, *J. Polym. Sci.*, **46**, 535 (1960).
2. E. J. Goethals, *Telechelic Polymers* (CRC Press, Boca Raton, 1989).
3. Y. Tezuka, *Prog. Polym. Sci.*, **17**, 471 (1992).
4. C. J. Pedersen, *Science*, **241**, 536 (1988).
5. J.-M. Lehn, *Supramolecular Chemistry, concepts and perspectives* (VCH, Weinheim, 1995).
6. K. Matyjaszewski and T. P. Davis, Eds., *Handbook of radical polymerization* (John Wiley and Sons, Hoboken, 2002).
7. K. Matyjaszewski and A. H. E. Mueller Eds., *Controlled and Living Polymerization, from mechanism to applications* (Wiley-VCH, Weinheim, 2009).
8. Y. Yamashita, *Chemistry and Industry of Macromonomers* (Hüthig & Wepf, Basel, 1993).
9. Y. Tezuka and S. Hayashi, *Macromolecules*, **28**, 3038 (1995).
10. Y. Tezuka and E. J. Goethals, *Makromol. Chem.*, **188**, 791 (1987).
11. Y. Tezuka, K. Yaegashi, M. Yoshino and K. Imai, *Kobunshi Ronbunshu*, **49**, 809 (1992).
12. Y. Tezuka, T. Shida, T. Shiomi, K. Imai and E. J. Goethals, *Macromolecules*, **26**, 575 (1993).
13. J. E. McGrath Ed., *Ring-opening Polymerization, kinetics, mechanism and synthesis*, ACS Symp. Ser. 286, (ACS, Washington, D. C. 1985).
14. H. Oike, T. Yaguchi and Y. Tezuka, *Macromol. Chem. Phys.*, **200**, 768 (1999).
15. F. D'haese and E. J. Goethals, *Br. Polym. J.*, **20**, 103 (1988).
16. Y. Tezuka and E. J. Goethals, *Makromol. Chem.*, **188**, 783 (1987).
17. H.-S. Yoo and A. Hirao, in *Complex Macromolecular Architectures*, Eds., N. Hadjichristidis, A. Hirao, Y. Tezuka and F. Du Prez (Wiley, Singapore, 2011), p.133.
18. H. Oike, H. Imamura, H. Imaizumi and Y. Tezuka, *Macromolecules*, **32**, 4819 (1999).
19. M. Schappacher and A. Deffieux, in *Complex Macromolecular Architectures*, Eds., N. Hadjichristidis, A. Hirao, Y. Tezuka and F. Du Prez (Wiley, Singapore, 2011), p.647
20. H. Kazama, Y. Tezuka and K. Imai, *Macromolecules*, **24**, 122 (1991).
21. Y. Tezuka, T. Iwase and T. Shiomi, *Macromolecules*, **30**, 5220 (1997).
22. T. Shiomi, K. Okada, Y. Tezuka, H. Kazama and K. Imai, *Makromol. Chem.*, **194**, 3405 (1993).
23. Y. Tezuka, Y. Murakami, T. Shiomi and K. Imai, *Polymer*, **39**, 2973 (1997).
24. M. Foston, C. Hubbell, D.-Y. Park, F. Cook, Y. Tezuka and H. W. Beckham, *Angew. Chem. Int. Ed.*, **51**, 1849 (2012)

CHAPTER 4

ELECTROSTATIC SELF-ASSEMBLY AND COVALENT FIXATION (ESA-CF) PROCESS

Yasuyuki Tezuka

Department of Organic and Polymeric Materials,
Tokyo Institute of Technology, Meguro-ku, Tokyo, Japan
E-mail: ytezuka@o.cc.titech.ac.jp

An effective polymer reaction process has been introduced by using telechelic polymer precursors having selected cyclic ammonium salt groups of unique reactivity. Thus, an electrostatic self-assembly and covalent fixation (ESA-CF) protocol was developed through either the ring-opening of 5-membered or the ring-emitting of 6-membered cyclic ammonium end groups of telechelic precursors by pluricarboxylate counteranions at prescribed temperatures to produce a variety of cyclic and multicyclic polymers having unprecedented topologies.

1. Introduction

Electrostatic interaction is ubiquitous in nature, and playing crucial roles in diverse biological events.[1] Also, hydrophobic polymers containing a small amount of ionic groups are known to exhibit unique properties both in bulk and in solution. This is a basis for their application to mechanically tough coating materials known as *ionomers*.[2] In an organic solution, these polymers tend to aggregate to form clusters (self-assembly) through Coulombic interaction between ionic groups located along hydrophobic polymer backbone. The content and location of ionic groups dictate their aggregation (self-assembly) behaviors. In this regard, telechelics having ionic groups exclusively at the chain ends have drawn an attention as a model of *ionomers*. The concentration of ionomers in solution and the solution temperature were shown to influence

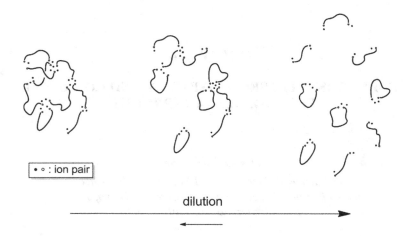

• ◦ : ion pair

dilution

Scheme 1. Polymer self-assembly by telechelics having ionic groups in solution.

significantly on their aggregation dynamics. In a diluted solution, ionic aggregates tend to be transformed into a small assembly comprised of a smallest number of polymer precursor units, while an ion-pair of cations and anions always balance the charges.

By this remarkable feature of polymer self-assembly by telechelics having ionic end groups, the structure formed under dilution by the self-assembly should be directed by the two physicochemical factors, namely the osmotic dispersion of polymer molecules and the balance of the charges between cations and anions. Thus, a linear polymer having two ionic end groups carrying two monofunctional counterions or one bifunctional counterion tends to dissociate into a dispersed state of a single polymer component (Scheme 1). On the other hand, unique self-assemblies consisting of the two or three linear telechelic polymers could be formed with a tetra- or a hexafunctional counterion, respectively, in a dispersed state, where the cations and anions balance the charges to maintain the smallest number of the components (Scheme 2).

Based on these characteristics of electrostatic polymer self-assemblies, a novel polymer reaction process has been developed, in which electrostatically self-assembled telechelic precursors having cyclic onium salt groups, described in Chapter 3, are utilized for the

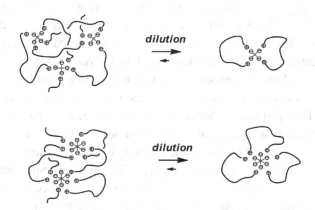

Scheme 2. Electrostatic polymer self-assemblies by telechelics having cationic end groups accompanying tetra- and hexafunctional counteranions.

construction of unusual and complex polymer structures by the subsequent covalent fixation treatment. This Chapter first presents key principles of the organic chemistry for the *electrostatic self-assembly and covalent fixation* (ESA-CF) process, followed by the synthetic application for polymers having various complex topologies.

2.   Control of the Reactivity by the Ring Size of Cyclic Onium Salts

2.1. *Ring-opening reactions*

Telechelic polymers having ionic groups tend to form a self-assembly both in bulk and in solution under dynamic equilibrium. Therefore, the electrostatic self-assembly is subjected to transform into thermodynamically favored forms depending on the conditions of medium. Thus, a controlled conversion of electrostatic (ionic) interactions into non-equilibrium covalent linkage is a key process in the ESA-CF protocol to become useful synthetic means. As described briefly in Chapter 3, a variety of cyclic onium salt groups could be utilized for such controlled covalent conversion of ionic species introduced in hydrophobic polymer segments.

Accordingly, a systematic study was conducted on the reactivity of a series of 3-, 4-, 5-, 6-membered cyclic as well as 6-membered bicyclic ammonium salts at the polymer chain ends toward nucleophiles, in particular carboxylates of different reactivity.[3] As shown in Chapter 3, the 3-membered cyclic ammonium salt end groups are highly reactive to undergo a spontaneous ring-opening reaction with 3-membered cyclic amine itself to produce a block copolymer *in situ*.[4] The 4-membered cyclic ammonium salt end groups, on the other hand, are incapable of the ring-opening reaction with 4-membered cyclic amines at ambient temperature, and the telechelic poly(THF) with 4-membered cyclic ammonium salt groups was isolated for the spectroscopic and chromatographic characterization, and was subsequently utilized as a reactive polymer precursor to produce block copolymers by the reaction with 3- and 4- membered cyclic amines at an elevated temperature.[5] Moreover, the 4-membered cyclic ammonium salt end groups underwent the ion-exchange reaction from an originally accompanied weak nucleophile of a triflate to others like sulfonates or carboxylates through a simple precipitation of the telechelics solution into an aqueous solution containing desired anions as a salt form. And when a strong nucleophile such as a carboxylate counteranion was introduced, a spontaneous ring-opening reaction of the 4-membered cyclic ammonium salt groups took place at an ambient temperature as discussed in Chapter 3. It is notable that polymer substrates, possessing the ionic end groups, are readily soluble in appropriate organic medium, and the nucleophilic substitution by various carboxylates in organic medium proceeds spontaneously, as the anion is dehydrated to form a *naked* nucleophile to enhance the nucleophilic reactivity.

Telechelic poly(THF)s having modestly strained 5-membered cyclic (pyrrolidinium) as well as 6-membered bicyclic (quinuclidinium) ammonium salt groups were isolatable even with carboxylates as a counteranion, and underwent the ring-opening reaction at appropriately elevated temperatures. On the other hand, the strain-free 6-membered cyclic ammonium salt group failed to cause the selective ring-opening reaction. Thus, telechelic poly(THF)s having 5-membered cyclic and 6-membered bicyclic ammonium salt groups have been chosen as polymer

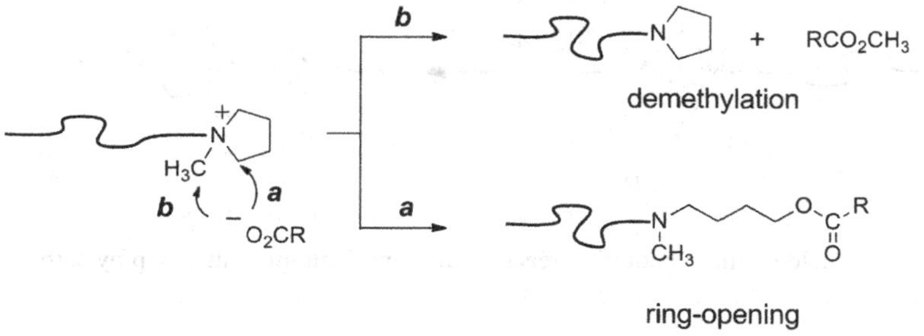

Scheme 3. Nucleophilic substitution reaction of *N*-methylpyrrolidinium salt group by carboxylate anion.

precursors for the *electrostatic self-assembly* with appropriate carboxylate counteranions, and for the subsequent *covalent fixation* by the ring-opening reaction to produce covalently linked products.[6]

The ring-opening reaction of a series of moderately strained cyclic ammonium salt groups, i.e., *N*-alkyl and *N*-arylpyrrolidinium salt groups as well as quinuclidinium salt groups was compared in more detail (Scheme 3). Indeed, the reaction of the *N-methyl*pyrrolidinium salt end groups with a variety of carboxylate counteranions resulted in not only the ring-opening reaction but also a noticeable concurrent demethylation reaction.[7] This is caused by the nucleophilic attack of the carboxylate counteranion at the *N-methyl* carbon. Remarkably, on the other hand, no Hofmann-type elimination products were detectable. And, the reaction on the *N-methyl*pyrrolidinium groups was notably dependent on the type of the counteranion having different pKa values.[7] Thus, an electron-donating *p*-methoxybenzoate (pKa = 4.47) produced the ring-opening product in a high yield and with high ring-opening selectivity, while an electron-withdrawing *p*-nitrobenzoate (pKa = 3.42) failed to cause any reaction at the relevant reaction condition. It was also observed that the higher the reaction temperature, the lower the selectivity for the ring-opening reaction over the demethylation.

In the case of the 6-membered bicyclic, quinuclidinium salt groups, the nucleophilic attack of the carboxylate counteranion took place exclusively at the *endo*-methylene position to cause the ring-opening reaction of a strained azabicyclo unit, and a reaction at the alternative *exo*-methylene position was excluded.[8] And indeed, the quinuclidinium

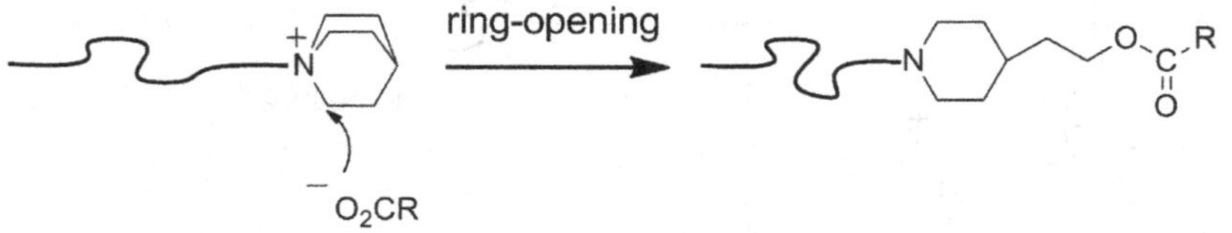

Scheme 4. Nucleophilic substitution reaction of quinuclidinium salt group by carboxylate anion.

salt group of a telechelic poly(THF) underwent a selective ring-opening reaction by a benzoate counteranion, though a higher temperature of 130 °C was required to complete the process (Scheme 4).

An alternative poly(THF) precursor having *N-phenyl*pyrrolidinium salt groups was then introduced in place of *N-methyl*pyrrolidinium salt groups.[9] Since the nucleophilic substitution reaction on the phenyl group is apparently suppressed, the ring-opening reaction on aliphatic *endo*-methylene groups was found to occur preferentially at 60-70 °C. Thus the ring-opening reaction was promoted by introducing an aniline derivative group, which is a better leaving group than the alkylamino group in the nucleophilic substitution reaction. It was shown, indeed, that the *N-phenyl*pyrrolidinium salt end group of poly(THF) underwent a complete ring-opening reaction even with such a weak nucleophile of *p*-nitrobenzoate counteranion, exclusively at the *endo*-position of the 5-membered pyrrolidinium ring, in contrast with the *N-methyl*pyrrolidinium counterpart (Scheme 5).

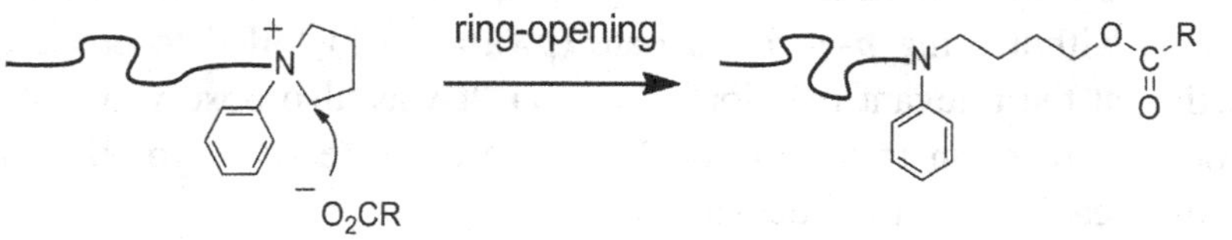

Scheme 5. Ring-opening esterification of N-phenylpyrrolidinium salt group by carboxylate anion.

2.2. *Ring-emitting reactions*

An alternative covalent conversion process to form a robust ester linking structure was disclosed by using a telechelic precursor having *unstrained* cyclic ammonium salt end groups.[10] Inspired by the regioselective demethylation reactions of six-membered, *N,N*-dimethylpiperidinium and *N*-methy1-*N*-phenylpiperidinium salts by an alkoxide as a nucleophile,[11,12] a telechelic poly(THF) having 6-membered, *N*-phenylpiperidinium salt groups carrying either monofunctional benzoate or bifunctional biphenyldicarboxylate counteranions was prepared. Upon the subsequent heating treatment, the carboxylate counteranions were found to attack predominantly (80–90%) at the exo position of a cyclic ammonium salt group to form a simple ester group by eliminating *N*-phenylpiperidine from the polymer chain ends (Scheme 6).

Through the *ring-emitting* covalent conversion using the telechelics carrying a bifunctional carboxylate under dilution, cyclic polymers containing predominantly a simple ester linking group were formed by the ESA-CF process. The relevant cyclic polymers might be obtained but by an equimolar reaction between a polymeric diol and a dicarboxylic acid under dilution. In practice, however, such a bimolecular process is hardly applicable as a routine synthetic means because the reaction becomes suppressed under high dilution. Hence, this process appears useful for the practical synthesis of cyclic polymers containing a simple ester linking group.

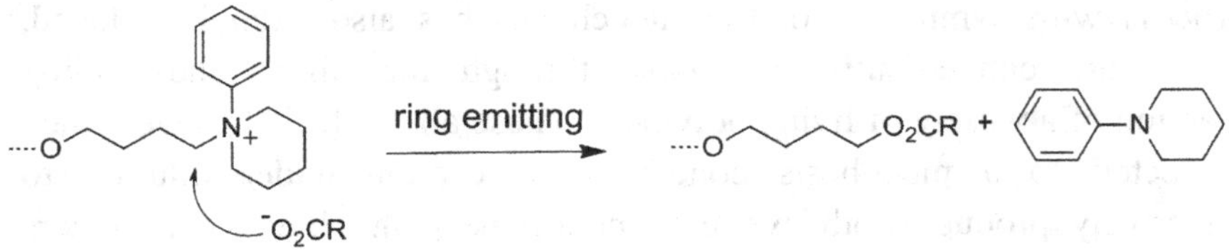

Scheme 6. Ring-emitting esterification of *N*-phenylpiperidinium salt group by carboxylate anion.

3. Electrostatic Polymer Self-Assembly and Covalent Fixation for Complex Polymer Topologies

3.1. *Ring (simple cyclic) polymers*

A direct end-to-end linking reaction of an α-ω-bifunctional linear polymer precursor with a bifunctional coupling reagent is a straightforward method to prepare ring polymers.[13] In practice, however, this *bimolecular* process has rarely been applied to produce high purity ring polymers in high yields. This is because the asymmetric telechelics formed as the product of the first step is subjected to react again with the coupling reagent to form a symmetric telechelics, that is any more unable to undergo the cyclization. Secondly, a high dilution condition is required to complete the intramolecular cyclization process and to avoid concurrent intermolecular chain-extension by employing the strictly stoichiometric balance of the large polymer and the small coupling compound. However, this dilution inevitably causes serious suppression of the reaction rate obeyed by the second-order kinetics. Accordingly, a tedious fractionation procedure is usually required to remove the linear side products that have the same chain length.[14]

An alternative *unimolecular* polymer cyclization process has been introduced that uses α-ω-heterobifunctional polymer precursors, which are obtainable through living polymerization techniques. In particular, a highly efficient alkyne-azide addition reaction (a *click* process) has been demonstrated as a remarkably improved polymer cyclization process in which the telechelic precursor is obtainable through the ATRP protocol. (Chapter 10 in Part II) An effective *unimolecular* polymer cyclization process with symmetric olefinic-telechelics has also been introduced, which are conveniently obtainable through the direct end-capping reaction of a variety of living polymers. These allyl-telechelics were then subjected to a metathesis condensation reaction under dilution to effectively produce both cyclic homopolymers and block copolymers (Chapter 7).

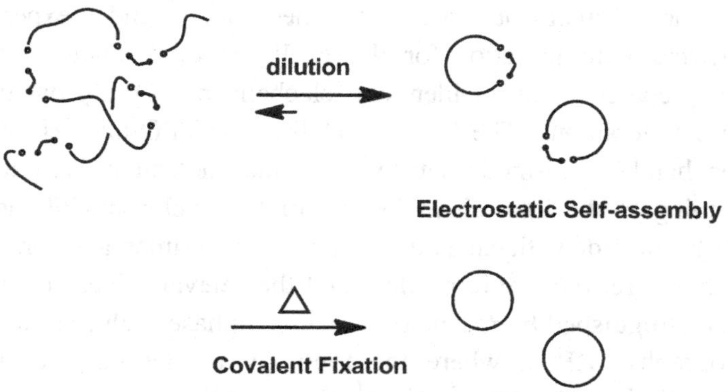

Scheme 7. Ring polymer formation by the electrostatic self-assembly and covalent fixation (ESA-CF) process.

The ESA-CF process was successfully applied to prepare a variety of ring polymers.[15] The polymer cyclization reaction was extensively studied by using the telechelic poly(THF) having *N-phenyl*pyrrolidinium salt groups.[16] The telechelics initially accompanied triflate counteranions, which were subsequently replaced by a desired carboxylate counteranion, through a simple precipitation of a poly(THF) precursor into an aqueous solution containing excess amount of a carboxylate as a sodium salt form. The ^{1}H NMR spectra of the ion-exchange products with dicarboxylate showed signals due to the corresponding carboxylate counteranions. The charges between the quaternary ammonium cations at the ends of the polymer precursor and the carboxylate anions balance each other, corresponding to the molar ratio of exact stoichiometry.

This ionic polymer precursor tends to self-assemble electrostatically in an organic solution. The dissociation/association of ionic species depends on the solution concentration, while the cations and anions balance the charges throughout the process (Scheme 7, see also Scheme 1). A cyclic poly(THF) with a narrow size distribution (PDI < 1.10), corresponding to the polymer precursor, was obtained after the heat treatment in an appropriately diluted THF solution toward 0.2 g/L. The cyclic structure of the product was unequivocally confirmed by means of SEC, and viscosity measurements, as well as ^{1}H NMR, VPO and later MALDI-TOF mass analyses.[14] Thus, the SEC and viscosity measurements showed the agreement between theoretical and experimental hydrodynamic volume ratios for the cyclic product against its linear analogue, prepared from an identical telechelic poly(THF) but carrying benzoate counteranions. The MALDI-TOF mass, VPO and ^{1}H NMR, on the other hand, confirmed that the absolute molecular weights (thus corresponding to the total chain length) of the cyclic and of the linear poly(THF)s coincide with each other within an experimental error.

Furthermore, the cyclic product and the relevant linear counterpart could be distinguished by means of a reversed-phase high pressure liquid chromatography (RPC), where the topology of the polymer product directs the elution property.[15,16] The RPC technique was also found to separate a mixture of cyclic products produced by multiple polymer precursor units from a cyclic product by a single polymer precursor unit, and was shown to provide more accurate yield values than the conventional SEC. Thus, it was confirmed that the content of the latter was found to increase along with the dilution of the reaction solution. Typically, the polymer cyclization proceeded with as high as 93% yield at the concentration of 0.2 [g/L], and the cyclic poly(THF) of more than 98% purity was obtained in 74% yield by the subsequent purification with preparative thin-layer-chromatography (TLC) technique.[16]

Based on the statistics for an intramolecular cyclization and intermolecular chain extension processes by the end-reactive polymer precursor,[17] the ratio of the respective products formed at different concentrations can be estimated from the corresponding probability ratios, P_c and P_e, respectively, and is given by the following equation,

$$P_c/P_e = [3/2\pi<r_0^2>]^{3/2} \times MW/[C]N_A$$

where $<r_0^2>$ (cm^2), [C] (g/mL), and N_A are a mean square end-to-end distance of the polymer precursor, a polymer concentration in the cyclization reaction, and the Avogadro's number, respectively. For instance, the theoretical yields of cyclic poly(THF) at different concentrations from the linear precursor (MW = 4700 g/mol) were thus estimated by assuming the end-to-end distance per unit mass for poly(THF), $r_0/MW^{1/2}$, to be 0.09 nm (mol/g)$^{1/2}$,[16] at a theta condition. The experimental yields were indeed very close to the theoretical ones, confirming with the cation/anion stoichiometry in the electrostatically self-assembled polymer precursor system, and the quantitative and selective chemical conversion process at the polymer precursor chain ends.

The effective polymer cyclization was found to proceed under appropriate dilution (0.2 [g/L]) in THF for a series of telechelic poly(THF)s having the molecular weight ranging from 4000 to 12,000, corresponding to the 260–800-membered ring.[16]

The polarity of the reaction solvent was found to affect the cyclization process.[16] Thus, the covalent conversion reaction proceeded quantitatively in toluene, chloroform, THF and in acetone. In contrast, no reaction was observed at all in ethanol. The solvation of ethanol to carboxylate anions through the hydrogen bond is presumed to suppress their nucleophilic reaction. The SEC showed that the products obtained in chloroform and in acetone possess narrow size distributions as observed in THF. On the other hand, the SEC peak profile was comparatively broader (PDI > 1.5) for the product obtained in toluene. And the cyclic polymer yield of a single polymer precursor unit estimated by RPC was slightly lower (80% against 89–93% in THF, chloroform and acetone). These imply that the ionically aggregated polymer precursor is reluctant to dissociate in a less-polar toluene solution.

Furthermore, the type of dicarboxylate counteranion was found to affect the extent of the ion-exchange reaction of telechelic polymer precursors.[16] Thus, a series of aromatic and aliphatic dicarboxylate counteranions, i.e., terephthalate, *o*-phthalate, succinate, adipate and *trans*-2-butene-1,4-dicarboxylate, in addition to 4,4'-biphenyldicarboxylate, were examined. The ion-exchange reaction of a telechelic poly(THF) having *N*-phenylpyrrolidinium salt groups proceeded easily with 4,4'-biphenyldicarboxylate and terephthalate, while ineffective with *o*-substituted phthalate. And for aliphatic dicarboxylates, the ion-exchange yield reached off at most 70–80% after the five-times precipitation treatment. From these results, aromatic dicarboxylates are preferred to aliphatic ones in the ion-exchange process, and 1,2-dicarboxylates are less efficient presumably due to the steric hindrance. On the other hand, the ring-opening reaction of pyrrolidinium salt groups proceeded quantitatively by each of the carboxylate counteranion.

Finally, notable feature of the present covalent conversion process is stressed. That is, the cyclic poly(THF) products having narrow size distributions were obtained in proportional yields even in the reaction employing a partially ion-exchanged polymer precursor, i.e., carrying mixed counteranions of a dicarboxylate anion and a triflate anion. This is uniquely ascribed to the exclusive formation of an electrostatically self-assembled single polymer precursor unit in dilution, in which cations and anions balance the charges (Scheme 1). In consequence, the remarkably efficient polymer cyclization has been achieved by the present the ESA-CF process.

The ESA-CF process was subsequently applied for the synthesis of ring polymers by using polystyrene and poly(ethylene oxide) telechelic polymers having the relevant cyclic ammonium salt groups described in Chapter 3 for the synthesis of graft and network copolymers.[18,19] In addition, the electrostatic self-assemblies comprised of not only telechelic polymer having cyclic onium salt groups and small carboxylate counteranions, but also polymeric (telechelic) precursor having carboxylate groups are applicable to the ESA-CF process to produce simple ring to complex polymer topologies.[20]

3.2. *Multicyclic and cyclic-linear hybrid polymers*

Dicyclic and tricyclic polymer topologies were also constructed by using the telechelic poly(THF) having *N*-phenylpyrrolidinium salt groups.[15] Initially accompanied triflate counteranions of this telechelics were replaced by a tetra and hexacarboxylate counteranions, through a simple precipitation of a poly(THF) precursor into an aqueous solution containing excess amount of a carboxylate as a sodium salt form. The ¹H NMR analysis of the ion-exchange products with tetra- and hexacarboxylate confirmed that the balance of the charges between the ammonium salt groups in the polymer precursor and the carboxylate anions was maintained, corresponding to the molar ratio of 2:1 and 3:1, respectively as shown previously in Scheme 2.

2 + 1 assembly   **3 + 1 assembly**

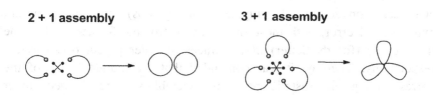

Scheme 8. Construction of spiro-dicyclic and spiro-tricyclic polymer topologies by the ESA-CF process.

These polymer precursors were then subjected to the heat treatment in solution at different concentrations (Scheme 8). When the reaction was conducted either in bulk or in concentrated solution, insoluble gel products were produced. Along with dilution, on the other hand, the product became totally soluble. And remarkably, the three-dimensional size in solution (hydrodynamic volumes observed by SEC) of the quantitatively recovered crude products, approached to be as uniform as in the linear polymer analogue, which was independently prepared from the bifunctional polymer precursor carrying benzoate counteranions. These results implicate, as was the case of the simple polymer cyclization described in the preceding section, that a unique form of an electrostatic self-assembly was produced upon appropriate dilution from the aggregated polymer precursors, while cations and anions balance the charges.

The obtained di- and tricyclic poly(THF)s were readily purified by the preparative thin layer chromatography, and were fully characterized by spectroscopic and chromatographic techniques. Thus, the ^{1}H NMR confirmed that pyrrolidinium salt groups were totally converted into amino-ester groups through the ring-opening reaction. The IR spectra showed the absorption assignable to ester carbonyl groups. The SEC analysis of the dicyclic and tricyclic poly(THF)s showed that they possess significantly larger hydrodynamic volumes than that of the linear analogue from a single polymer precursor unit, but smaller than those of the twice (for the former) and of the three times (for the latter), respectively. They retained narrow size distributions. On the other hand, the actual molecular weights of the dicyclic and tricyclic poly(THF)s determined either by the VPO or by the ^{1}H NMR (assuming quantitative chemical conversion of polymer end groups) coincided, within experimental error, with those of the twice (for the former) and of the three times (for the latter) of the linear polymer precursor analogue, respectively. Thereby, it was concluded that the unique forms of the corresponding electrostatic polymer assemblies were formed under dilution, and they are comprised of two units of the polymer precursor and one unit of tetracarboxylate, and of three units of the polymer precursor and one unit of hexacarboxylate, respectively. Subsequent heat treatment of these precursors could lead to dicyclic and tricyclic polymer products, through the covalent fixation by the ring-opening reaction of pyrrolidinium salt groups.

The ESA-CF process was subsequently applied for the synthesis of a dicyclic 8-shaped polymer by using a telechelic polystyrene having the relevant cyclic ammonium salt groups described in Chapter 3 for the synthesis of graft and network copolymers.[18]

As an extension of the ESA-CF process, a trifunctional star-shaped telechelic precursor having cyclic ammonium salt end groups was employed to produce a fused-dicyclic θ-shaped polymer (Scheme 9).[15,21] Thus, a trifunctional star-shaped poly(THF) precursor was first prepared with a trifunctional initiator for the living polymerization of THF, formed in situ by the reaction of 1,3,5-tris(hydroxymethyl) benzene with trifluoromethanesulfonic anhydride in the presence of a proton trap. The subsequent polymerization of THF followed by the termination with N-phenylpyrrolidine affords a star-shaped poly(THF) precursor having N-phenylpyrrolidinium salt groups. The anion-exchange reaction from a triflate toward a trifunctional trimesate proceeded by the repeated precipitation of the poly(THF) precursor solution into an aqueous solution containing sodium trimesate. An ion-exchanged product was thus isolated in high recovery yield.

1 + 1 assembly

Scheme 9. Construction of a fused-dicyclic polymer topology by the ESA-CF process.

The subsequent heating treatment was conducted in various solvents under dilution at 0.1–0.2 g/L. While the gelation was observed in bulk or at higher concentration, the reaction at dilution proceeded homogeneously. Thus, the deassembly of an aggregate tended to proceed, and a unique assembly was formed, comprising a single star polymer precursor unit carrying a single trimesate counteranion to balance the charges between cations and anions. The subsequent heat treatment under dilution produced a θ-shaped polymer product.

A remarkable solvent effect was observed on the covalent conversion process, i.e., the ring-opening reaction of *N*-phenylpyrrolidinium salt group by a trimesate counteranion in comparison with the previous polymer cyclization by the relevant bifunctional linear poly(THF) carrying a biphenyldicarboxylate counteranion discussed in the previous section. First, the ring-opening reaction failed to occur at all in ethanol as previously shown. In chloroform, moreover, the reaction by trimesate did not proceed at all, in contrast with a quantitative reaction by biphenyldicarboxylate. The choice of chloroform as solvent is thus decisive either to suppress or to cause the ring-opening reaction, by recognizing lower nucleophilicity of a trimesate (pKa = 2.98) than a biphenyldicarboxylate (pKa = 3.77).

The ring-opening reaction by a trimesate took place both in THF and in acetone, as in the previous biphenyldicarboxylate system. It is remarkable that the reaction by the trimesate proceeded notably faster in THF than in acetone. Thus, the quantitative ring-opening reaction took place in THF after 3 h, while the conversion remained at 48% in acetone. On the contrary, such kinetic discrimination was not observed in the previous system by a biphenyldicarboxylate, where the reaction completed within 3 h both in THF and in acetone. The lower nucleophilic reactivity of a trimesate than a biphenyldicarboxylate will be again attributable to this solvent effect.

SEC also showed another notable solvent effect by comparing the products obtained in THF, in acetone, and in their mixtures. Thus, a SEC profile of the product obtained in THF showed a noticeable shoulder fraction at higher molecular weight region. On the other hand, the products obtained in THF/acetone mixtures showed a narrower peak profile by eliminating higher molecular weight fractions. This accords with that the de-assembly process to a single polymer unit tends to proceed more effectively in a polar solvent of acetone ($\varepsilon = 21$) than in THF ($\varepsilon = 7.5$), where oligomeric self-assemblies remain and subsequently converted to covalently fixed products.

A series of θ-shaped poly(THF)s of different molecular weights were isolated in 40–49% yields after purification by preparative TLC to remove any residual ionic species. The products were then characterized by means of IR, ^{1}H NMR, SEC as well as RPC and MALDI-TOF mass techniques. First, IR and ^{1}H NMR showed a quantitative ring-opening reaction to produce aminoester groups at each chain ends of star-shaped polymer precursors.

SEC showed that the θ-shaped product was significantly smaller in its size than the relevant star-shaped polymer possessing nearly equal molecular weights (thus total chain lengths). The relative size ratio, corresponding to their hydrodynamic volumes, between the θ-shaped and the star-shaped polymers was within the range of 0.71–0.75, as estimated from their SEC peak molecular weights. The three-dimensional size of a θ-shaped polymer is constricted by linking three chain ends of a star polymer precursor, and the extent of the constriction is more significant than in the cyclic polymer formation from the linear polymer precursor. Since the star-shaped polymer is constricted in its size in comparison with its linear analogue, θ-shaped polymers are significantly compact in their three-dimensional structures. Thus, indeed, the relative size ratio between the θ-shaped and the linear polymers was estimated in the range 0.60–0.63.

RPC was also conducted on the obtained θ-shaped poly(THF)s, in comparison with their star-shaped polymer analogues, to check the purity of the isolated products. The isolated θ-shaped polymer showed a nearly unimodal RPC profile with only a trace of side products, whose elution volume corresponds to that of the star-shaped poly(THF) analogue. This is indicative of the high uniformity of the product both in the chemical composition and in the topological structure. Moreover, the RPC elution behavior of a pair of θ-shaped and star-shaped poly(THF)s of different molecular weights was compared with that of the relevant pair of simple cyclic and linear counterparts.

It was shown that θ-shaped and single cyclic poly(THF)s eluted after the corresponding star-shaped and linear analogues, respectively, and the difference in the elution volumes in the pair of θ-shaped and star-shaped polymers was larger than that in the pair of cyclic and linear counterparts. The topological change in the former pair was more significant than in the latter pair by decreasing the number of chain ends from 3 to 0 in the former against 2 to 0 in the latter. In addition, the constriction of the three-dimensional size from the star to θ-shaped polymers is more significant than from linear to cyclic polymers.

Finally, the θ-shaped poly(THF) and its star-shaped precursor analogue were examined by a MALDI-TOF mass spectroscopy. The θ-shaped polymer product showed a uniform series of peaks corresponding to poly(THF) (peak interval of 72 mass units); each peak corresponds exactly to the molar mass summing up the linking structure produced by the ring-opening reaction of three N-phenylpyrrolidinium groups by a trimesate anion. As an example, the peak (assumed to be the adduct with Na^+) at 5836.7 corresponds to the product with the DP_n of 70, $(C_4H_8O) \times 70 + C_{48}H_{51}N_3O_6$, plus Na^+ as 5836.451. The star-shaped product, obtained from the identical poly(THF) precursor but carrying benzoate counteranions, also showed a series of the peaks corresponding to the Na^+ adduct. Thus, the peak (assumed to be the adduct with Na^+) at 5992.2 corresponds to the product with the DP_n of 70, $(C_4H_8O) \times 70 + C_{60}H_{63}N_3O_6$, plus Na^+ as 5992.679. Since both θ-shaped and star-shaped poly(THF)s are produced from the identical precursor after the ion exchange by either a trimesate or three benzoate counteranions, their molecular weights differ by 156 mass units. This was confirmed by the two MALDI-TOF mass spectra.[21]

The ESA-CF process was also applied to prepare polymers having *a ring with branches* topologies (Scheme 10).[22] Thus, an interiorly functionalized, *eso*-telechelic, polymer precursor was prepared by the end-cappling of a bifunctional living poly(THF) with a pyrrolidine-terminated poly(THF). Subsequently, a dicarboxylate, i.e., a terephthalate counteranion was introduced at the interiorly located pyrrolidinium salt groups in the *eso*-telechelic poly(THF), through a simple precipitation of a prepolymer solution into water (< 5 °C) containing an excess amount of sodium terephthalate. The ion-exchange reaction from the initial triflate

1 + 1 assembly

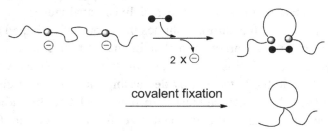

2 x ⊖

covalent fixation

Scheme 10. Construction of *a ring with branches* polymer topology by the ESA-CF process.

anions toward the dicarboxylate anion was readily confirmed by IR and ^{1}H NMR spectroscopic analyses, as in the previous cases.

The subsequent heat treatment of the polymer precursor was performed in a toluene solution, to cause a ring-opening reaction of the interiorly located pyrrolidinium salt group by a nucleophilic attack of a carboxylate counteranion. The ^{1}H NMR analysis of the quantitatively recovered product showed that the nucleophilic attack of a carboxylate counteranion occurred exclusively at the *endo*-methylene position of the pyrrolidinium salt group. Thus, the *exo*-methylene position on the interiorly located pyrrolidinium salt group was intact during the present nucleophilic reaction by carboxylate counteranions. The IR spectrum of the product also showed absorption assignable to the ester carbonyl group of the ring-opening product.

The SEC of the isolated poly(THF)s of *a ring with two branches* structure showed narrow size distributions (PDI = 1.08–1.13), and the smaller hydrodynamic volume than those of their linear analogues, prepared from the identical polymer precursors but carrying benzoate counteranions. These results indicate that a unique form of the electrostatic self-assembly was selectively produced under dilution at 1.0 g/L, with balancing the charges between cations and anions (Scheme 2), and was effectively converted into a permanent polymer structure of *a ring with two branches* topology.

In comparison with the case of the *end*-functionalized poly(THF), the selective cyclization with a single polymer precursor unit was found to proceed even at the higher concentration. This implies that the complete deassembly took place for polymer precursors having *interiorly located* pyrrolidinium salt groups at the higher concentration than the *end*-functionalized telechelics. The location of ionic groups either at the interior or at the end positions of the polymer chain appears to direct the equilibrium state of the ionically aggregated polymer precursors, in which the steric (exclusion-volume) effect by polymer chain segments is considered to be crucial.

Furthermore, the ESA-CF process *in dilution* by using a series of cationic polymer precursors and anionic counterparts (thus *polymeric* carboxylate counteranions) was applied for the effective construction of a variety of loop and branch polymer topologies, including a two-tailed tadpole construction, which is not directly obtainable by the dilution process using the electrostatically self-assembled polymer precursor carrying *small* counteranions (Scheme 11).[20] Thus, the corresponding electrostatically assembled polymer precursors were produced by the coprecipitation of a cationic and an anionic telechelic precursors of the relevant architectures by removing small anionic and cationic counterions initially accompanied with the telechelic precursors. The subsequent dilution will produce a unique self-assembly, in which cationic and anionic polymer precursors are combined to balance the charges. This is contrary to the mixture of the cationic and anionic telechelic precursors

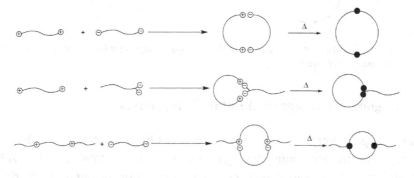

Scheme 11. Construction of *a ring with branches* polymer topologies by the ESA-CF process with polymeric precursors.

keeping small anionic and cationic counterions, where the dilution causes the complete de-assembly to give a smallest number of polymer species (Scheme 12).

Thus, the two types of cationic polymer precursors having two pyrrolidinium salt groups either at both chain ends or at the designated interior positions were prepared. Also, the two types of anionic counterparts having two carboxylate groups either at both chain ends or at a single chain end were prepared. A series of electrostatic polymer-polymer self-assemblies have been prepared by the combination of these cationic and anionic precursors, respectively, through simple coprecipitation of the precursor polymer mixture into an aqueous medium. The subsequent covalent conversion of the electrostatically self-assembled products by heating under dilution could produce either a ring polymer, a tadpole polymer, and a two-tail tadpole, having a loop and two branches, polymer, respectively by the combination of appropriate telechelic precursors.[20]

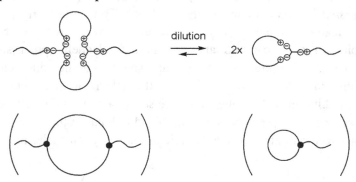

Scheme 12. Construction of *a two-tail tadpole* polymer topology by the ESA-CF process with polymeric precursors.

4. Ongoing Challenges and Future Perspectives

This Chapter presented the ESA-CF process by using telechelics having specific cyclic ammonium salt groups. The electrostatic polymer self-assembly and the subsequent covalent conversion through either ring-opening or ring-emitting of cyclic ammonium end groups by pluricarboxylate counteranions at prescribed temperatures could afford a variety of cyclic and mylticyclic polymers having unprecedented topologies. A further extension of this extremely effective polymer reaction process includes; the wider choice of polymer types other than prototypical example of poly(THF) and preliminary examples of polystyrene, polyethylene oxide and poly(dimethylsiloxane). Thus, a versatile method to allow the introduction of cyclic ammonium salt groups at the designated position of wider variety of polymer precursors is awaited. In addition, the ESA-CF process could become more useful when the applicable precursor molecular weight becomes higher than 10,000. As the covalent conversion process currently requires dilution and heating, still restricting synthetic scope of the ESA-CF process. Further improvements of the process to allow non-dilution process will extend the application potentials. Moreover, the ring-emitting covalent conversion process is still its infancy for a simple and effective esterification to make a robust covalent linkage in any complex polymers and organic materials.

References

1. J. M. Thornton, *Nature*, **295**, 13 (1982).
2. M. Pineri and A. Eisenberg, Eds., Structure and Properties of Ionomers, (Kluwer, Dordrecht, 1987).
3. Y. Tezuka and E. J. Goethals, *Makromol. Chem.*, **188**, 783 (1987).
4. H. Kazama, Y. Tezuka, K. Imai and E. J. Goethals, *Makromol Chem.*, **189**, 985 (1988).
5. Y. Tezuka, *Prog. Polym. Sci.*, **17**, 471 (1992).
6. Y. Tezuka, *Prog. Polym. Sci.*, **27**, 1069 (2002).
7. H. Oike, H. Hatano and Y. Tezuka, *React. Funct. Polym.*, **37**, 57 (1998).
8. Y. Tezuka, T. Shida, T. Shiomi, K. Imai and E. J. Goethals, *Macromolecules*, **26**, 575 (1993).
9. H. Oike, H. Imamura, H. Imaizumi and Y. Tezuka, *Macromolecules*, **32**, 4819 (1999).
10. K. Adachi, H. Takasugi and Y. Tezuka, *Macromolecules*, **39**, 5585 (2006).
11. G. Cerichelli, G. Illuminati and C. Lillocci, *J. Org. Chem.*, **45**, 3952 (1980).
12. G. Cerichelli and L. Luchetti, *Tetrahedron*, **49**, 10733 (1993).
13. J. A. Semlyen, Ed., *Cyclic Polymers, 2nd ed.* (Kluwer, Dordorecht, 2000).
14. Y. Tezuka, *Polym. J.*, in press (DOI: 10.1038/pj.2012.92)

15. H. Oike, H. Imaizumi, T. Mouri, Y. Yshioka, A. Uchibori and Y. Tezuka, *J. Am. Chem. Soc.*, **122**, 9592 (2000).
16. H. Oike, T. Mouri and Y. Tezuka, *Macromolecules*, **34**, 6592 (2001).
17. H. Jacobson and W. H. Stockmayer, *J. Chem. Phys.*, **18**, 1600 (1950).
18. H. Oike, M. Hamada, S. Eguchi, Y. Danda and Y. Tezuka, *Macromolecules*, **34**, 2776 (2001).
19. Y. Tezuka, K. Mori and H. Oike, *Macromolecules*, **35**, 5707 (2002).
20. K. Adachi, H. Irie, T. Sato, A. Uchibori, M. Shiozawa and Y. Tezuka, *Macromolecules*, **38**, 10210 (2005).
21. Y. Tezuka, A. Tsuchitani, Y. Yoshioka and H. Oike, *Macromolecules*, **36**, 65 (2003).
22. H. Oike, M. Washizuka and Y. Tezuka, *Macromol. Rapid Commun.*, **22**, 1128 (2001).

CHAPTER 5

DYNAMIC CONTROL OF POLYMER TOPOLOGIES BY THE ESA-CF PROCESS

Yasuyuki Tezuka

Department of Organic and Polymeric Materials,
Tokyo Institute of Technology, Meguro-ku, Tokyo, Japan
E-mail: ytezuka@o.cc.titech.ac.jp

A dynamic control in electrostatic polymer self-assemblies was implemented in the ESA-CF process to produce a variety of complex polymer topologies, including *ring with branch* (tadpole) polymers, θ- and manacle-shaped dicyclic topological isomers, as well as comb-type multicomponent co-polymacromonomers. The ESA-CF process was also applied to produce polymer catenanes by taking advantage of orthogonal electrostatic and hydrogen-bonding interactions in the polymer self-assembly.

1. Introduction

A remarkable feature of the electrostatic self-assembly and covalent fixation (ESA-CF) process is reversible dynamic equilibrium nature of the non-covalent polymer assembly, which is subsequently subjected to an irreversible covalent conversion (chemical) reaction. Therefore, this process is considered relevant to the dynamic combinatorial library (DCL) system to extract specific or targeted species from an equilibrium product mixture in reversible non-covalent interaction.[1,2] A supramolecular polymer system is considered another closely related process, in which a mixture of non-covalent polymeric products is formed under dynamic equilibrium to be applied for adaptive polymer materials.[3]

In this Chapter, a unique potential of the ESA-CF process is presented by taking advantage of a dynamic nature of electrostatic polymer self-assemblies to produce a variety of complex polymer topologies, including *ring with branch* (tadpole) polymers, θ- and manacle-shaped dicyclic topological isomers, as well as comb-type multicomponent co-polymacromonomers. The ESA-CF process is also applied to produce polymer catenanes by taking advantage of orthogonal electrostatic and hydrogen-bonding interactions in the polymer self-assembly.

2. Dynamic Equilibrium in Electrostatic Polymer Self-Assembly

The ESA-CF process could produce a ring polymer product in proportional yields even by the reaction employing a partially ion-exchanged polymer precursor, i.e., carrying mixed counteranions of a dicarboxylate anion and a triflate anion, as described in Chapter 4.[4] This is uniquely ascribed to the exclusive formation of an electrostatically self-assembled single polymer precursor unit in dilution, in which cations and anions balance the charges. The dynamic selection and the subsequent control of the polymer topologies by the ESA-CF process was further demonstrated by employing a mixture containing two types of telechelic poly(THF) precursors having pyrrolidinium salt groups, carrying benzoate and 4,4-biphenyldicarboxylate.[5] In THF solution, an electrostatic polymer self-assembly carrying randomly distributed monofunctional and bifunctional counteranions (1/1 in mol/mol for the counteranions) was resulted through the cation–anion reshuffling under thermodynamic equilibrium (Scheme 1). The THF solution of the polymer self-assembly was subsequently heated at different polymer concentrations to cause the quantitative ring-opening reaction of the pyrrolidinium salt end groups. SEC and RPC confirmed the exclusive formation of a mixture of the linear and the cyclic poly(THF), both consisting of a single polymer precursor unit.

These results indicate a remarkable dynamic selection in the mono- and bifunctional counteranions of the relevant structures from electrostatic polymer self-assembly involving a polymeric precursor having pyrrolidinium salt groups. The dynamic selection proceeds in such a way that the polymer chain ends combine with either two benzoates or with one 4,4'-biphenyldicarboxylate counteranion.

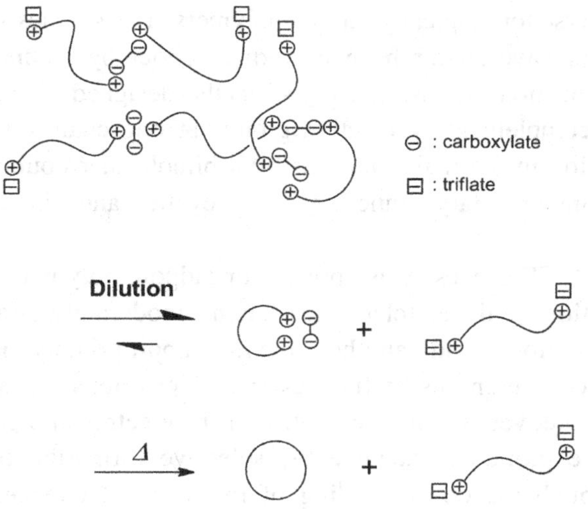

Scheme 1. Dynamic equilibrium by polymeric precursors having triflate and dicarboxylate counteranions in the ESA-CF process.

Therefore, this process allows to discriminate different molecules by the plurality of functional groups but not by the reactivity, with the use of polymeric self-assembly even in the absence of an additional template component. This chemical selection process is in contrast to the formation of a mixture of linear and cyclic products of various chain lengths from kinetic or random selection in bifunctional polymer precursors (telechelics) with coexisting monofunctional and bifunctional reagents. This dynamic selection from a mixture of relevant mono- and bifunctional counteranions was directed by the balance of the charges between the cations and anions during the de-assembly of the ionic random aggregates consisting of the multiple polymer precursors carrying a mixture of carboxylate counterions.

3. Tadpole Polymers by Dynamic Selection in Electrostatic Polymer Self-Assembly

A tadpole topology is regarded as a basic form of a series of *a ring with branches* or ring-linear hybridized constructions and as a useful building block for diverse topologically unique polymers. The synthesis of tadpole-type polymers have so far been performed either by an intramolecular polymer cyclization by using a specifically designed linear precursor having two complementary reactive groups at one chain end and at an interior position in the main chain or by a bimolecular coupling reaction between complementary functionalized cyclic and linear polymer precursors.[5]

The ESA-CF process was applied for tadpole polymer synthesis by the combination of linear telechelics having moderately strained cyclic ammonium cations and another having appropriately nucleophilic carboxylate counteranions at the designated positions, as described in Chapter 4. Moreover, a dynamic control of the electrostatic polymer self-assembly is exploited to achieve the selective formation of a tadpole polymer through the de-assembling of the selected telechelic polymer aggregates by the dilution, to keep the balance of the charges between cations and anions. Thus, a tadpole polymer could be constructed by the combination of one difunctional and one monofunctional telechelic precursors carrying one trifunctional counteranion, as well as by the combination of an internally functionalized difunctional linear precursor and one difunctional counteranion. The subsequent covalent conversion by the ring-opening reaction of cyclic ammonium salt groups by carboxylate counteranions could provide a novel effective means to synthesize tadpole polymers.[5]

Thus in practice, an equimolar mixture of a bifunctional poly(THF) having an *N*-phenylpyrrolidinium salt end group carrying tricarboxylate

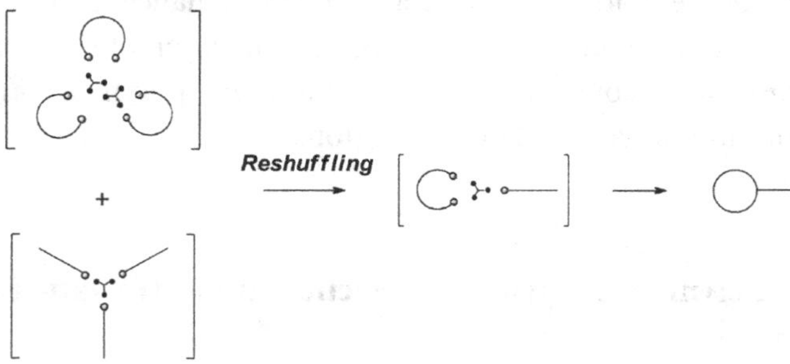

Scheme 2. Tadpole polymer formation by dynamic selection with a linear telechelic precursor in the ESA-CF process.

counteranions and a relevant monofunctional telechelic poly(THF) also carrying tricarboxylate counteranions was prepared, and subjected to covalent fixation under relevant dilution. The spontaneous reshuffling between the above two polymer self-assemblies could produce a unique tadpole polymer assembly in dilution, in which the polymer precursor and the carboxylate units tend to form a self-assembled product having the smallest number of components with the balance of the charges between cations and anions. The subsequent quantitative ring-opening reaction of the pyrrolidinium salt groups was confirmed to proceed even under applied dilution (Scheme 2).[5]

The SEC profile of the product was compared to the mixture of the products, obtained separately from the respective starting two types of the polymer self-assemblies; those are a dicyclic poly(THF) comprised of θ- and manacle-form polymeric isomers and a star poly(THF), respectively. SEC showed that an equimolar mixture of the dicyclic poly(THF) mixtures of θ- and manacle-forms and star poly(THF) gives a two-peak profile composed of the two fractions with equal peak areas. For the covalent conversion product obtained at various dilutions, the SEC profile approached toward unimodal as progressive dilution of the reaction solution. These results agree with the selective formation of a tadpole polymer product, which is composed of both cyclic and linear units by the covalent fixation of a polymer self-assembly through a reshuffling process, that is, the dissociation of multiple aggregates of polymer precursors and reassembly to form a self-assembled product comprised of the smallest number of the components with the balance of the charges between cations and anions.

Alternatively, a star-shaped trifunctional polymer precursor carrying a mixture of mono- and bifunctional carboxylate counteranions was employed for the selective synthesis of tadpole polymers through the dynamic selection from the electrostatic polymer self-assembly (Scheme 3).[5] Thus, as in the case of the linear polymer precursor system, a star telechelic poly(THF) having pyrrolidinium salt groups carrying a mixture of mono- and dicarboxylate (1/1 in mol/mol for counteranions) was prepared by reshuffling the two corresponding star precursors carrying the mono- and dicarboxylate counteranions. The subsequent heat treatment of the polymer self-assembly proceeded homogeneously under dilution while the gelation occurred in bulk or at a higher concentration. The quantitative

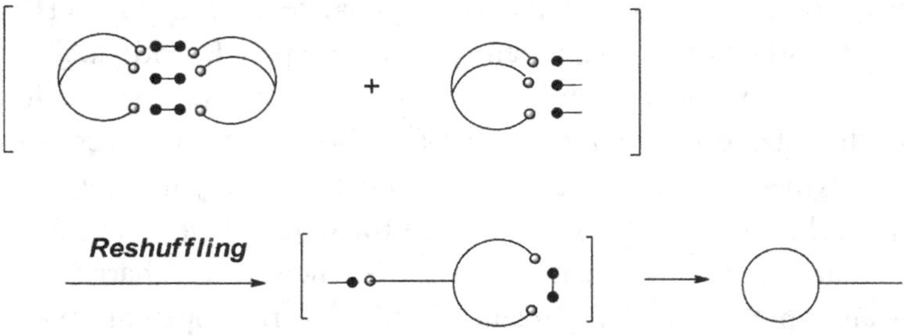

Scheme 3. Tadpole polymer formation by dynamic selection with a star telechelic precursor in the ESA-CF process.

ring-opening reaction of the pyrrolidinium salt end groups with benzoate or with the terephthalate anions was again confirmed to proceed under dilution.

   SEC showed the exclusive formation of a tadpole polymer from a unique assembly consisting of a single star polymer precursor carrying a pair of a benzoate and a terephthalate anion, in which cations and anions balance the charges. The obtained tadpole polymer was significantly smaller in the hydrodynamic volume than the relevant star polymer of equal molecular weight (thus, the total chain length), since the tadpole polymer is formed through the intramolecular cyclization of the star polymer precursor. It should be noted also that the size of a tadpole polymer is notably larger than that of the θ-shaped polymer, since the former is a *singly* cyclized but the latter is a *doubly* cyclized product from the star polymer precursor.

4. Polymeric Topological Isomers of θ- and Manacle-Forms

A pair of dicyclic, θ- and manacle-shaped polymers are a prototypical example of polymeric topological isomers, and are conceptually important in topological polymer chemistry, as discussed in Chapter 2.

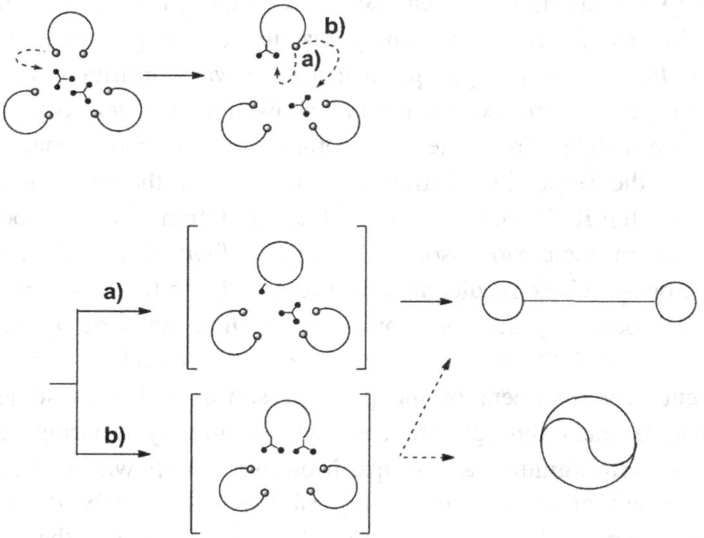

Scheme 4. Formation of dicyclic polymeric topological isomers by the ESA-CF process.

Thus, a polymer self-assembly consisting of three units of bifunctional telechelic poly(THF) accompanying two units of tricarboxylate (trimesate) was prepared and subjected to heat treatment in THF under dilution (0.2 g/L).[6,7] No gelation occurred under dilution. The covalent linkage between the two pyrrolidinium groups in telechelic poly(THF) and the three carboxylate groups in trimesate can produce the two topologically different, θ- and manacle-form polymer architectures, shown in Scheme 4. The random combination of cations and anions within the polymer self-assembly will produce θ- and manacle-form isomers in a ratio of 2:3. The size, that is, the hydrodynamic volume, of the manacle-form isomer is considered to be marginally larger than that of the θ-counterpart. In addition, the dielectric property of the two topological isomers should be distinct from each other because the spatial alignments of polar groups (*N*-phenyl groups) in the two forms are different. Indeed, the reversed phase chromatography (RPC) at the near critical conditions revealed the two components in the covalent fixation product, which were subsequently separated by means of fractionation.[6,7]

The two fractions were identical to each other in their ^{1}H NMR and IR, and the quantitative ring-opening reaction of the pyrrolidinium salt groups by the carboxylate groups in trimesate was confirmed. The SEC analysis of the covalent fixation product showed a consistent peak profile which is convoluted from the two components, θ- and manacle-form isomers, of the respective elution volumes and of the ratio of 22:78, estimated by the RPC method above. Thus, the intramolecular process to produce the manacle-form isomer is slightly favored in this covalent fixation process. These results are consistent with the formation of a self-assembled product by dilution, consisting of three units of bifunctional telechelic poly(THF) and two units of tricarboxylate (trimesate). Subsequent heat treatment of the polymer self-assembly could lead to topological isomers through the covalent fixation by the ring-opening reaction of pyrrolidinium salt groups. Moreover, as shown in Chapter 4, one component of these two topological isomers, namely the θ-form isomer, has been produced by an alternative process in which the covalent conversion of the assembled precursor obtained from a trifunctional star poly(THF) having N-phenylpyrrolidinium salt end groups and a trifunctional carboxylate.[7]

Alternatively, a pair of dicyclic, θ-shaped and manacle-shaped polymeric topological isomers was synthesized from a polymer self-assembly, comprised of three-armed star poly(THF) having N-

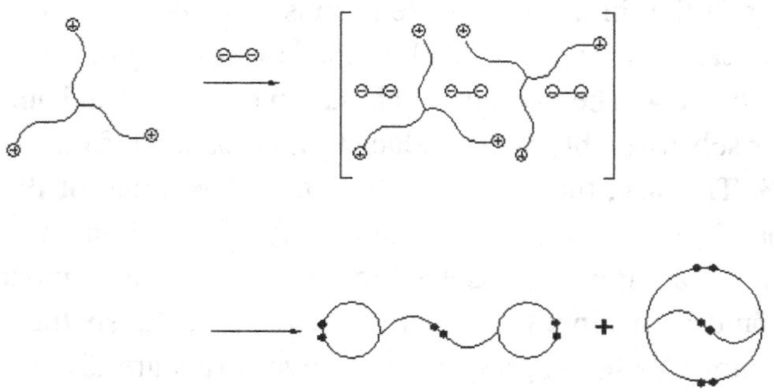

Scheme 5. Formation of dicyclic polymeric topological isomers with a star telechelic precursor by the ESA-CF process.

phenylpyrrolidinium salt end groups carrying dicarboxylate counteranions (Scheme 5).[8] The presence of the two constitutional polymeric isomers was again confirmed by means of the RPC technique. Moreover, SEC showed that a major component, presumably assignable as a manacle-shaped isomer, possesses notably larger hydrodynamic volume than the θ-counterpart.

As in the case with a polymer self-assembly using a linear telechelic precursor, the θ- and manacle-form isomers will be produced in the ratio of 2:3, when the random and stepwise linking takes place through the covalent conversion of the cationic precursor chain ends by terephthalate counteranions. In fact, the larger hydrodynamic volume component, corresponding to the manacle-form, was predominant (>70%, even higher than the statistical 60%) regardless of the molecular weight of the polymer precursor. The larger hydrodynamic volume component could be reasonably assigned as a manacle-shaped isomer and the other, having a smaller hydrodynamic volume, as a θ-counterpart, respectively. It was also conceived that the manacle-shaped product, obtainable through an intramolecular polymer cyclization, could be further favored by progressive dilution over the θ-shaped product formed through an intermolecular polymer combination.

Moreover, polymeric topological isomers having θ- and manacle-constructions including an olefinic group were prepared through the ESA-CF process by using a linear precursor having an inner olefinic group. And by the metathesis cleavage of the inner olefinic group included at the inner position of the two isomer frameworks, the unequivocal chemical assignment of each polymeric isomer has been achieved (Scheme 6).[9]

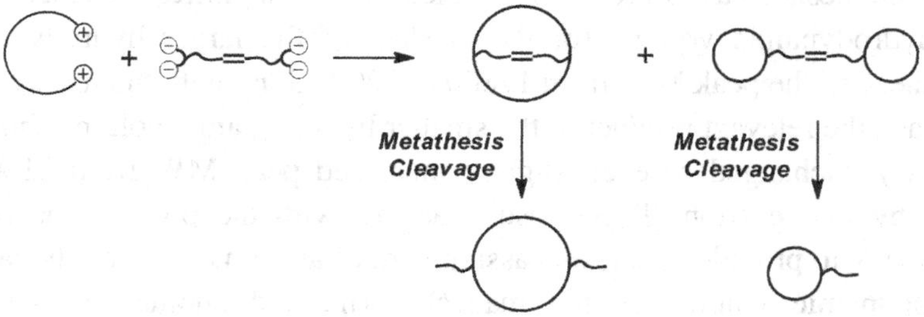

Scheme 6. Dicyclic polymeric topological isomers having an inner olefinic group for the assignment through the metathesis cleavage.

Thus by a high resolution SEC measurement, the presence of two fractions having the larger and the smaller hydrodynamic volume was confirmed in the ESA-CF product. Each fraction was subsequently separated by means of preparative SEC technique, to show the coincident ^{1}H NMR spectra with each other with the equivalent DP_n values, estimated based on the proton peak area ratio between N-phenyl groups of the linking units and poly(THF) main chain methylene units, despite the notable difference in their hydrodynamic volumes estimated by SEC. Accordingly, one component of the smaller hydrodynamic volume recovered as a major component (47% yield) was tentatively considered as a θ-shaped product, while another minor component (29% yield) as a manacle-shaped counterpart, respectively. Notably, the θ- and manacle-isomers could be formed in the ratio of 2/1, when the linking of the terminal groups took place randomly.

The two fractions having the larger and the smaller hydrodynamic volumes, respectively, were finally subjected to the cross-metathesis reaction with an excess of allyl benzene in the presence of the first generation Grubbs catalyst (Scheme 6).[10] By the selective chain cleavage of the olefinic group with a cross metathesis reaction, the manacle-isomer should produce two units of a tadpole-shape product having a significantly reduced SEC size with the half DP_n. On the other hand, the θ-isomer will produce a two-tail tadpole product having a comparable or even larger SEC size with the unchanged DP_n. The two reaction products were recovered in appreciable yields after purification, and ^{1}H NMR spectra of the two products coincided with each other, showing the signals attributed to the benzylmethylidene group. Remarkably, the SEC comparison of the two metathesis-cleaved products revealed that the significant reduction of the hydrodynamic volume for the product of the larger hydrodynamic volume, i.e., the peak MW from 17800 to 7200, along with the reaction. In contrast, the relevant product of the smaller hydrodynamic volume showed a nearly unchanged or even slightly increased peak MW from 8100 to 9900 by the reaction. These results accord with the process shown in Scheme 6, to provide conclusive assignment of an isomer having the larger hydrodynamic volume as the manacle-form, and another having the smaller hydrodynamic volume as a θ-form, respectively.

5. Co-Polymacromonomers by Reshuffling in Electrostatic Polymer Self-Assembly

By the ESA-CF method, an efficient ion-coupling of telechelic polymers having cyclic onium salt groups took place with poly(sodium acrylate) to prepare polymacromonomers of well-defined structures, as described in Chapter 3. A further structural control of polymacromonomers was achieved by taking advantage of the dynamic nature of electrostatic polymer self-assemblies (Scheme 7).[11,12] Thus first, poly(THF)s having an *N*-phenylpyrrolidinium salt end group and poly(sodium acrylate) were subjected to the ion-exchange reaction to produce an electrostatically self-assembled polymacromonomer precursor, which could be isolated and fully characterized by means of conventional spectroscopic techniques. The covalently linked polymacromonomer product was subsequently produced by the heat treatment, causing the ring-opening reaction of pyrrolidinium salt groups by a nucleophilic attack of acrylate anions to form amino-ester groups.

And by mixing the two separately formed polymer self-assemblies having different telechelic components having a short and long chain lengths, a dynamic reshuffling between the different branch segments was allowed to proceed between two electrostatically self-assembled poly(THF) polymacromonomer precursors as seen in Scheme 7.[12] The following heat treatment could produce a covalently linked poly(THF) polymacromonomer having mixed branch segments in high yields. This novel process will provide an efficient means for the synthesis of *co*-polymacromonomers, in contrast to the ineffective conventional copolymerization process involving different macromonomers.

6. Polymer Catenanes by Orthogonal Electrostatic and Hydrogen-Bonding Polymer Self-Assembly

Catenanes and knots are topologically appealing molecules and have continuously attracted broad academic interests.[13] Moreover, they have recently been highlighted as new platforms for breakthrough functions leading to future nanodevices and nanomachines.[14] The self-assembly synthesis of catenanes and knots having a shape-persistent and robust

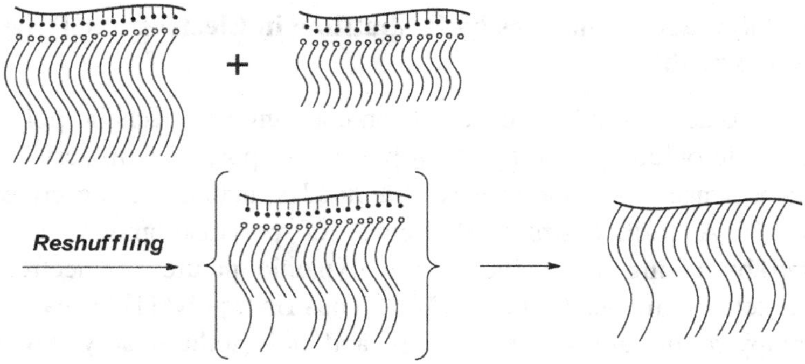

Scheme 7. Formation of co-polymacromonomers by reshuffling in the ESA-CF process.

structure has now been established by employing carefully designed components capable of forming a *just-fit-in-space* self-assembly through a variety of non-covalent interactions. Remarkable achievements by these protocols in recent years include simple to complex catenanes and knots, including Borromean rings.[15] The construction of polymer catenanes and knots by DNA molecules has also been achieved by exploiting the complementary base-pair units and the subsequent enzymatic ligation process specifically applicable to the DNA process.[16]

In contrast, polymer catenanes and knots with long and flexible synthetic polymer components have still been a formidable synthetic challenge. The end-linking polymer cyclization requires dilution, which promotes an intramolecular process over an intermolecular chain extension but circumvents the associative pairing of polymer precursors. In addition, a long and flexible polymer chain tends to form a randomly coiled, thus

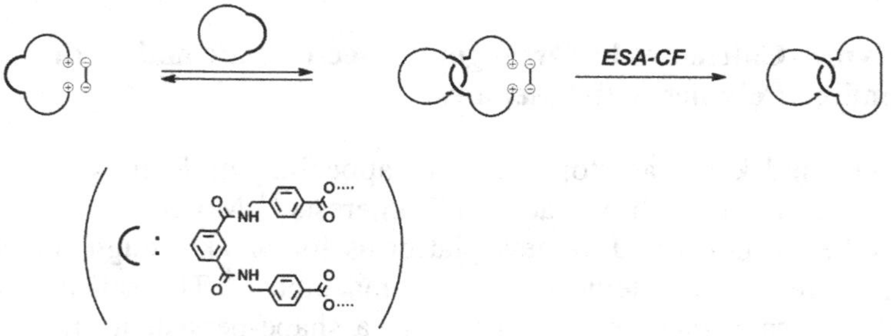

Scheme 8. Formation of a polymer catenanes by the orthogonal electrostatic and hydrogen-bonding polymer self-assembly.

contracted conformation, and the threading by another polymer chain appears ineffective, particularly in dilution. Therefore, previous attempts resulted in only less than 1% isolated yield of polymer catenanes, though they could provide a unique opportunity to elucidate the topology effect on polymer properties.[17]

The ESA-CF process could offer an alternative means to produce polymer catenanes (Scheme 8).[18] Thus, a hydrogen-bonding unit of an isophthaloylbenzylic amide group was incorporated in the ESA-CF process either by including in an initiator or in a counteranion. As the isophthaloylbenzylic amide group has been recognized as an effective structural motif to form self-complementary and intertwined hydrogen-bonding pair to produce various shape-persistent catenanes and knots,[19] two different cyclic polymers having the ring size of up to 150 atoms were prepared as components for a polymer [2]catenane. It is important to note that the combination of different cyclic polymer components provides the unequivocal mass spectroscopic support for the formation of the polymer [2]catenane, as the relevant polymer [2]catenane analogue comprised of the two identical cyclic polymer components is undistinguishable with a dimeric simple cyclic polymer by their molar masses.

Thus, a cyclic poly(THF), having a hydrogen-bonding, isophthaloylbenzylic amide group was prepared through the ESA-CF process with a telechelic poly(THF) having *N*-phenylpyrrolidinium salt groups carrying a dicarboxylate counteranion containing the hydrogen-bonding unit. Another telechelic poly(THF) having an isophthaloylbenzylic amide group at the center position and having *N*-phenylpyrrolidinium salt end groups carrying a biphenyldicarboxylate counteranion was subsequently prepared and subjected to a covalent conversion reaction in the presence of the preformed cyclic poly(THF) having a hydrogen-bonding unit. The acetone-insoluble fraction was isolated with notable yields of 5.7–7.1% in addition to the acetone-soluble fraction (40–60%). The control experiment by the relevant polymer cyclization in the presence of a simple cyclic poly(THF) without the hydrogen-bonding unit scarcely gave the acetone-insoluble fraction (0.1%).

The ^{1}H NMR of the acetone-soluble and acetone-insoluble fractions showed that both fractions are commonly comprised of the two cyclic poly(THF)s, and all signals were assignable to either of them. The composition ratios in the acetone-soluble and the acetone-insoluble fractions were then determined to show that the acetone-insoluble fraction contains the polymer [2] homocatenane product in addition to the polymer [2] hetero-catenane. The MALDI-TOF mass spectra of both the acetone-soluble and the acetone-insoluble fractions showed commonly a series of peaks corresponding to poly(THF) with peak interval of 72 mass units. And for the acetone-insoluble fraction, in particular, a series of peak molar mass was assignable to the polymer [2] hetero-catenane comprised of the two different cyclic poly(THF)s as the components. Moreover, a series of minor shoulder peaks assignable to the polymer [2] homo-catenane was detectable. The hydrodynamic volume of polymer [2]catenane was estimated by SEC comparison of the apparent molecular weights of cyclic poly(THF)s and of a polymer [2]catenane, to show the extent of the contraction of 3D size of the polymer catenane to be 0.68. And this value was comparable with those reported by the simulation study.[18]

Accordingly, the cooperative electrostatic and hydrogen-bonding self-assembly of polymer precursors and the subsequent covalent conversion was demonstrated as an effective means for the synthesis of polymer catenanes.

7. Ongoing Challenges and Future Perspectives

This Chapter showed a potential of the dynamic control in electrostatic polymer self-assemblies in the ESA-CF process, to produce a variety of complex topological polymers, including a *ring with branches* (tadpole) polymers, θ- and manacle-shaped polymeric topological isomers, as well as comb-type multicomponent co-polymacromonomers. The ESA-CF process was also applied to produce polymer catenanes by taking advantage of the orthogonal hydrogen-bonding interaction in the preformed polymer self-assemblies. A dynamic selection from an equilibrated self-assembly is certainly a promise of future polymer and material designs. Any inputs to achieve the selection of specific components from topological isomer mixtures will significantly contribute to the further development of the ESA-CF protocol. A combination of orthogonal non-covalent interaction with the ESA-CF process will become a powerful tool, as indicated a prototypical example of polymer catenane synthesis, to obtain other topologically intriguing polymers, including polymer rotaxanes and polymer knots.

References

1. S. J. Rowan, S. J. Cantrill, G. R. L. Cousins, J. K. M. Sanders and J. F. Stoddart, *Angew. Chem. Int. Ed.*, **41**, 898 (2002).
2. L. Brunsveld, B. J. B. Folmer, E. W. Meijer and R. P. Sijbesma, *Chem. Rev.*, **101**, 4071 (2001).
3. A. Harada, Ed., *Supromolecular Polymer Chemistry* (Wiley-VCH, Weinheim, 2012).
4. H. Oike, T. Mouri and Y. Tezuka, *Macromolecules*, **34**, 6592 (2001).
5. H. Oike, A. Uchibori, A. Tsuchitani, H.-K. Kim and Y. Tezuka, *Macromolecules*, **37**, 7595 (2004).
6. H. Oike, H. Imaizumi, T. Mouri, Y. Yshioka, A. Uchibori and Y. Tezuka, *J. Am. Chem. Soc.*, **122**, 9592 (2000).
7. Y. Tezuka, A. Tsuchitani and H. Oike, *Polym. Int.*, **52**, 1579 (2003).
8. Y. Tezuka, A. Tsuchitani and H. Oike, *Macromol. Rapid Commun.*, **25**, 1531 (2004).
9. Y. Tezuka, N. Takahashi, T. Satoh and K. Adachi, *Macromolecules*, **40**, 7910 (2007).
10. Y. Tezuka and F. Ohashi, *Macromol Rapid Commun.*, **26**, 608 (2005).
11. H. Oike, H. Imamura and Y. Tezuka, *Macromolecules*, **32**, 8816 (1999).
12. H. Oike, H. Imamura and Y. Tezuka, *Macromolecules*, **32**, 8666 (1999).
13. J.-P. Sauvage and C. Dietrich-Buchecker Eds., *Molecular Catenanes, Rotaxanes and Knots* (Wiley-VCH, Weinheim, 1999).
14. V. Balzani, A. Credi, F. M. Raymo and J. F. Stoddart, *Angew. Chem. Int. Ed.*, **39**, 3348 (2000).
15. S. J. Cantrill, K. S. Chichak, A. J. Peters and J. F. Stoddart, *Acc. Chem. Res.*, **38**, 1 (2005).
16. N. C. Seeman, *Nature*, **421**, 427 (2003).
17. K. Endo, *Adv Polym Sci.*, **217**, 121 (2008).
18. K. Ishikawa, T. Yamamoto, M. Asakawa and Y. Tezuka, *Macromolecules*, **43**, 168 (2010).
19. A. G. Johnston, D. A. Leigh, R. J. Pritchard and M. D. Deegan, *Angew. Chem. Int. Ed.*, **34**, 1209 (1995).

CHAPTER 6

CYCLIC AND MULTICYCLIC POLYMERS HAVING FUNCTIONAL GROUPS (*KYKLO*-TELECHELICS)

Yasuyuki Tezuka

Department of Organic and Polymeric Materials,
Tokyo Institute of Technology, Meguro-ku, Tokyo, Japan
E-mail: ytezuka@o.cc.titech.ac.jp

A variety of *kyklo*-telechelics are obtainable by the ESA-CF protocol. Those include single cyclic polymers having one speceific or plural (identical or different) functional groups, cyclic macromonomers having specifically a polymerizable group, tadpole polymers having either single or plural functional groups at the designated positions within *a ring with a branch* construction, as well as multicyclic polymers having single or plural functional groups at the prescribed positions within the multiple ring construction. Moreover, a functional cyclic polymer having a photonic probe for single molecule spectroscopy is presented.

1. Introduction

Cyclic and multicyclic polymer precursors having functional groups at the designated positions, termed *kyklo*-telechelics (a Greek, *kyklos*, means cyclic), will be of potential importance as a macromolecular building block to construct topologically unique and complex macromolecular architectures containing cyclic polymer units.[1-4] The ESA-CF process, described in the preceding Chapters, was successfully applied to produce such *kyklo*-telechelics through the efficient polymer cyclization process by employing the linear or branched telechelic precursors optionally possessing functional groups at the designated positions.

In this Chapter, a variety of *kyklo*-telechelics is presented.[5] Those include single cyclic polymers having one speceific or plural (identical or different) functional groups, cyclic macromonomers having specifically a polymerizable group, tadpole polymers having either single or plural functional groups at the designated positions within *a ring with a branch* construction, as well as multicyclic polymers having single or plural functional groups at the prescribed positions within the multiple ring construction. Moreover, a functional cyclic polymer having a photonic probe for single molecule spectroscopy is presented.

2. Single Cyclic Polymers Having Functional Groups

A variety of linear telechelic poly(THF)s having *N*-phenylpyrrolidinium salt groups, and optionally with an additional functional group at the center position of the polymer chain (termed *kentro*-telechelics) were employed in the ESA-CF process, to produce *kyklo*-telechelics having a single or two functional groups (Scheme 1).[5] Thus, a hydroxy-functionalized dicarboxylate, i.e., 5-hydroxyisophthalate, counteranion was introduced as a counteranion through the ion exchange reaction (Scheme 2). The obtained electrostatic polymer self-assembly was then subjected to the heat treatment under dilution. The complete ring-opening reaction was confirmed by the ^{1}H NMR analysis of the product recovered after simply evaporating the solvent. The presence of the hydroxyl group did not disturb the selective conversion of the electrostatic self-assembly into the covalently converted product. And the SEC showed that the product possesses a narrow size distribution and a notably smaller hydrodynamic volume than that of the linear analogue,

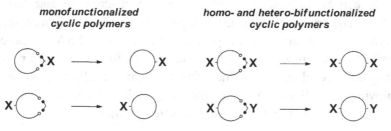

Scheme 1. Single cyclic polymers having functional groups (*kyklo*-telechelics).

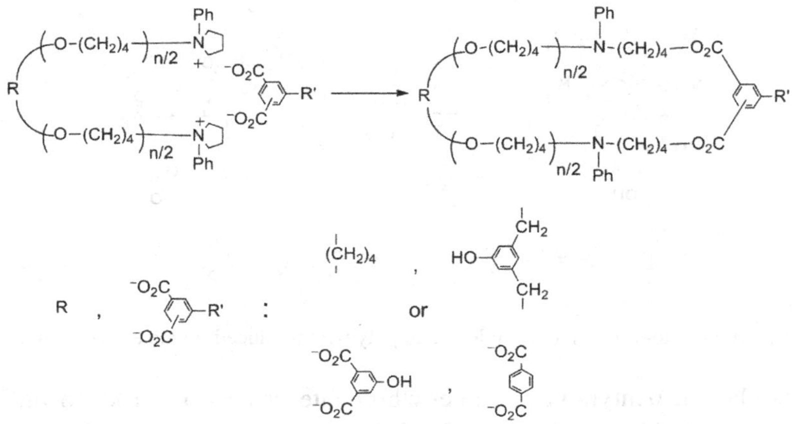

Scheme 2. Monofunctional single cyclic polymers by the ESA-CF process.

to support the polymer cyclization. The ratio of the hydrodynamic volumes between the cyclic and the linear products, estimated by the SEC with their apparent peak molecular weights was in good agreement with those previously reported. It was thus demonstrated that a cyclic poly(THF) having a hydroxy group, derived from the dicarboxylate counteranion, has been effectively produced.

Alternatively, the relevant hydroxy-functionalized cyclic poly(THF) was prepared by using a *kentro*-telechelic poly(THF), which carries not only *N*-phenylpyrrolidinium salt groups at both chain ends but also an additional hydroxy group at the center position of the polymer chain, obtainable by the living polymerization technique with a relevant protected functional initiator (Scheme 2).[6,7] Thus, the corresponding polymer precursor carrying a terephthalate counteranion was prepared and subjected to the heat treatment under dilution. The complete ring-opening reaction of the pyrrolidinium salt end groups again took place, as was confirmed by the ^{1}H NMR and the IR analysis of the product. Besides, the SEC showed that the obtained cyclic poly(THF) possesses a narrow size distribution, and a notably smaller hydrodynamic volume than that of the linear analogue.

The obtained cyclic precursor having a hydroxyl group was subsequently applied to prepare a series of cyclic poly(THF)s having

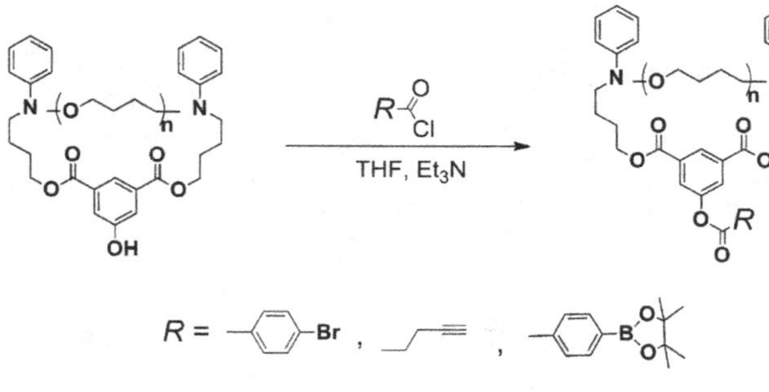

Scheme 3. Monofunctional single cyclic polymers produced by the esterification.

bromophenyl, pentynoyl and phenylboronate groups, in order to utilize them for Suzuki- and Sonogashira Pd coupling reactions (Scheme 3).[8] The esterification reaction of the hydroxyl group was performed with equimolar amount of the corresponding acid chlorides (4-bromobenzoyl chloride, 4-pentynoyl chloride and 4-pinacolboranyl benzoyl chloride, respectively) in THF in the presence of triethylamine. The esterification products were isolated in good yields after the silica-gel chromatography treatment.

The ^{1}H NMR spectra of the products showed the signals attributed to the bromophenyl, to the pentynoyl, and to the phenylboronate group, respectively, together with those due to the initiator fragment and the linking group signals of the cyclic polymer precursor. The MALDI-TOF mass spectra of the products showed a uniform series of peaks corresponding to poly(THF) (peak interval of 72 mass units), and each peak corresponds exactly to the molar mass summing up the initiator fragment and the linking group, respectively. Thus for the *kyklo*-telechelics having bromophenyl group, the peak (assumed to be the adduct with Na$^+$) was observed at 5063.6, corresponding to the expected cyclic product having a bromophenyl group, with the DP$_n$ of 60, (C$_4$H$_8$O) × 60 + C$_{39}$H$_{41}$N$_2$O$_6$Br, plus Na$^+$ as 5063.10.[8] The SEC showed that the esterification products retained the narrow size distributions as in the precursor having a hydroxyl group, to show the absence of polymer chain degradation during the esterification procedure.

Moreover, homo- and hetero-bifunctionalized cyclic poly(THF)s having either two identical or two different functional groups at the

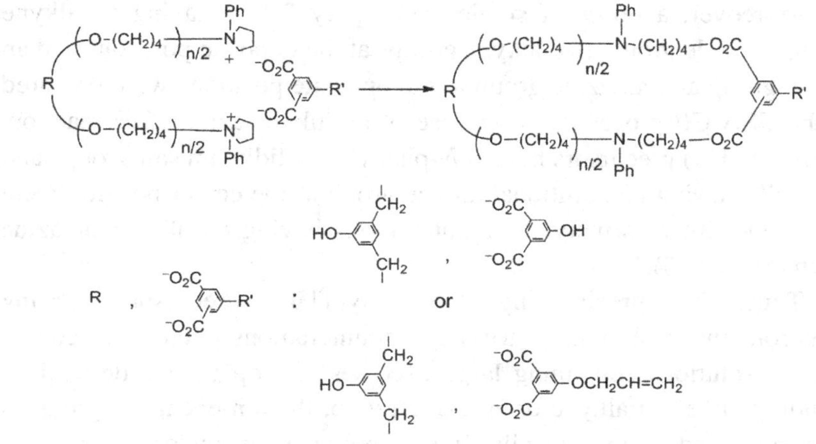

Scheme 4. Bifunctional single cyclic polymers by the ESA-CF process.

opposite positions, respectively, were synthesized (Scheme 4).[5] Thus, the *kentro*-telechelic polymer precursor carrying a 5-hydroxyisophthalate counteranion was prepared and subjected to the heat treatment under dilution. The ^{1}H NMR of the product showed the two singlets due to hydroxy protons not only from the 5-hydroxyisophthalate counteranion but also from the *kentro*-telechelic poly(THF) precursor. The intensities of these two signals were consistent with each other. Also, the SEC showed that the product possesses a narrow size distribution. It was thus confirmed that a homo-bifunctionalized cyclic poly(THF) having two hydroxy groups at the opposite positions was effectively produced.

In addition, the *kentro*-telechelic polymer precursor carrying a 5-allyloxyisophthalate counteranion was prepared and subjected to heat treatment under dilution. A hetero-bifunctionalized cyclic poly(THF) having a hydroxy and an allyloxy group at the opposite positions was subsequently produced in a quantitative yield (Scheme 4). The ^{1}H NMR showed again a signal due to the hydroxy proton derived from the *ketro*-telechelic poly(THF) precursor, in addition to signals due to allyloxy protons from the 5-allyloxyisophthalate counteranion. The proton intensity ratio of these signals agrees with that calculated from the expected molecular formula. The SEC showed the obtained hetero-bifunctionalized cyclic poly(THF) possesses a narrow size distribution.

Moreover, a series of single cyclic poly(THF)s having an alkyne group, an azide group, two alkyne groups at the opposite positions and an alkyne group and an azide group at the opposite positions were prepared by the ESA-CF process for their use of an alkyne-azido *click* reaction. The poly(THF) precursors having *N*-phenylpyrrolidinium salt groups and optionally having an additional alkyne group at the center position were employed with dicarboxylate counteranions having an alkyne or azide group (Scheme 5).[8,9]

Thus, by precipitating the poly(THF) precursor carrying trifluoromethanesulfonate (triflate) counteranions into ice-cooled aqueous solutions containing large excesses of isophthalate derivatives including either an alkyne or an azide group, the ion-exchange products were produced. The ionically linked polymer precursors were then subjected to the heat treatment by refluxing under dilution to cause the ring-opening reaction of the pyrrolidinium salt groups by the dicarboxylate counteranions, leading to the formation of single cyclic poly(THF)s having an alkyne group, an azide group, respectively. The completion of the reaction was confirmed by the ^{1}H NMR analysis, where the signals for the *N*-phenyl-adjacent methylene groups in the polymer precursor were replaced with that for the ester-adjacent

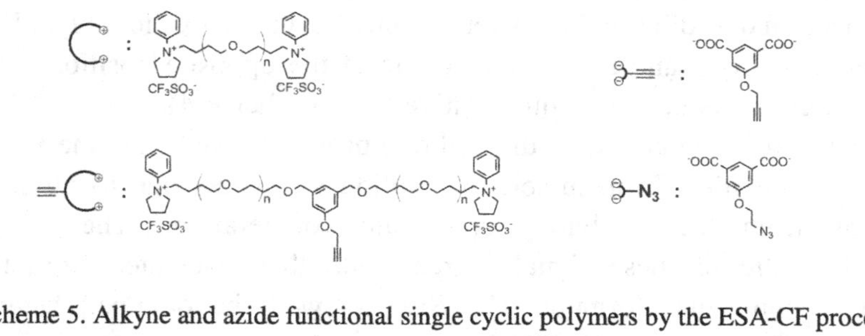

Scheme 5. Alkyne and azide functional single cyclic polymers by the ESA-CF process.

methylene group as a triplet in the covalent conversion products. The signals for the ethynyl and propargyl methylene groups as well as that for the azidomethylene group were also visible in the corresponding products.

Likewise, the *kentro*-telechelic poly(THF) having *N*-phenylpyrrolidinium salt groups and having an additional alkyne group at the center position was prepared by the living polymerization of THF using an alkyne-functionalized initiator. The electrostatic self-assemblies with isophthalate derivatives including either an alkyne or an azide group were subsequently produced by the ion-exchange reaction from the initial triflate. They were covalently converted to homo- and hetero-bifunctional *kyklo*-telechelic poly(THF)s having either two alkyne or alkyne and azido groups at the opposite positions, respectively. The ^{1}H NMR analysis of the former showed the signals for the two distinctive propargyl groups from the initiator and the counteranion, i.e., the ethynyl protons and the propargyl methylene protons, respectively. The signals for the ethynyl and azido-methylene protons were also visible in the latter.

The MALDI-TOF mass analysis showed uniform series of peaks with an interval of 72 mass units for the repeating THF units, and each peak exactly matches the molar mass summing up the initiator fragment, poly(THF) units and linking group. Thus, for an example, the peak at *m/z* = 3115.0, which is assumed to be the adduct with Na$^+$, corresponds to the single cyclic poly(THF) having an alkyne group possessing the expected chemical structure with DP$_n$ of 35; $(C_4H_8O) \times 35 + C_{35}H_{40}N_2O_5$ plus Na$^+$ as 3115.459.[9] Other *kyklo*-telechlics also showed representative peaks, which match the calculated molecular weights upon the expected linking structures. Furthermore, the SEC showed that the obtained *kyklo*-telechelics possess narrow size distributions and notably smaller hydrodynamic volumes than those of the linear counterparts, and their hydrodynamic volume ratios to the corresponding linear counterparts, estimated by the SEC peak molecular weights, were in agreement with those reported before.[9]

Finally, a series of homo- and heterobifunctional *kyklo*-telechelic poly(THF)s having an allyloxy group and an alkyne group, and another having an allyloxy group and an azide group were synthesized by the ESA-CF process using a set of *N*-phenylpyrrolidinium-terminated poly(THF) prepolymers having an alkene group at the center position of the chain, and carrying dicarboxylate counteranions having an alkyne or an azide group (Scheme 6).[10] They are subsequently applied for the metathesis condensation process as discussed in Chapters 7 and 8.

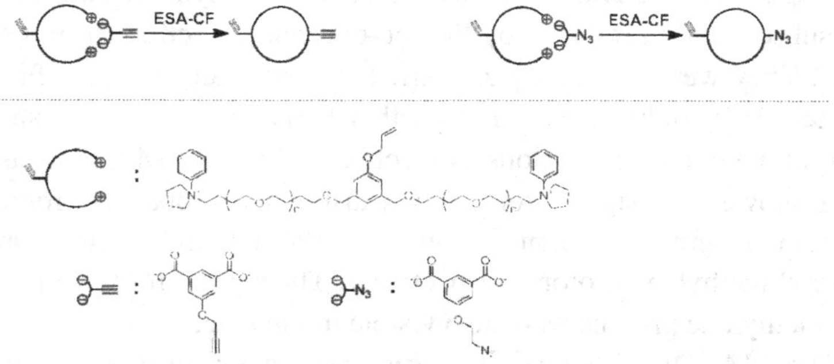

Scheme 6. Functional single cyclic polymers for *click* and *clip* (metathesis) chemistry by the ESA-CF process.

3. Cyclic Macromonomers

A cyclic macromonomer, a type of *kyklo*-telechelics having a specifically (co)polymerizable group, is useful to design an unusual polymer network structure having both covalent and physical linkages, as will be detailed in Chapter 11 in Part II. This network is formed by a chain threading of the propagating polymer segment through large cyclic polymer units attached to the backbone polymer segments. However, a tailored non-covalent linking process by threading through flexible and large size cyclic polymers has been a challenge despite its relevance to interpenetrating polymer networks (IPNs) of wide applications.[11] Since a long and flexible polymer chain tends to assume a randomly coiled and hence constrained three-dimensional structure, the chain threading by

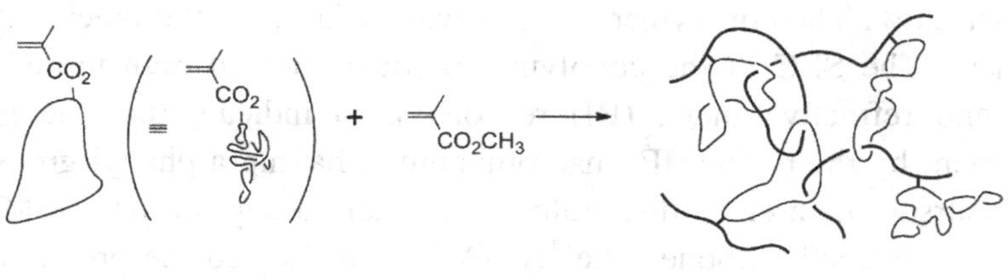

Scheme 7. Formation of non-covalent polymer networks with a cyclic macromonomer.

another polymer chain through such a large cyclic polymer unit is reluctant to proceed (Scheme 7). This is contrastive with a variety of medium size macrocyclic compounds up to around 100-membered ring, including 30~60 membered crown ethers, bipyridinium-based cyclophanes, cyclodextrins and macrocyclic amides.[12] They are regarded as configurationally stiff, and thus assuming an extended-cyclic conformation. The subsequent intriguing synthesis of (poly)catenanes and (poly)rotaxanes with such macrocycles are remarkable achievements, in which a variety of non-covalent interactions such as hydrogen bonding, π-π stacking, metal-coordinating and van der Waals interactions direct the chain threading events.[12]

Thus, the hydroxy-functionalized *kyklo*-telechelic poly(THF), described in the preceding section, was subjected to the esterification by methacryloyl chloride in the presence of triethylamine (Scheme 8).[13] The SEC analysis of the product showed a narrow size distribution, and the ^{1}H NMR showed the signals due to the methacrylate group. The degree of polymerization of the obtained cyclic macromonomer was typically 70, corresponding to be as large as 280-membered ring (Mn = 5200).

A free radical copolymerization of the obtained poly(THF) cyclic macromonomer with MMA was then carried out in solution. First, the polymerization was interrupted to keep a limited conversion of MMA at

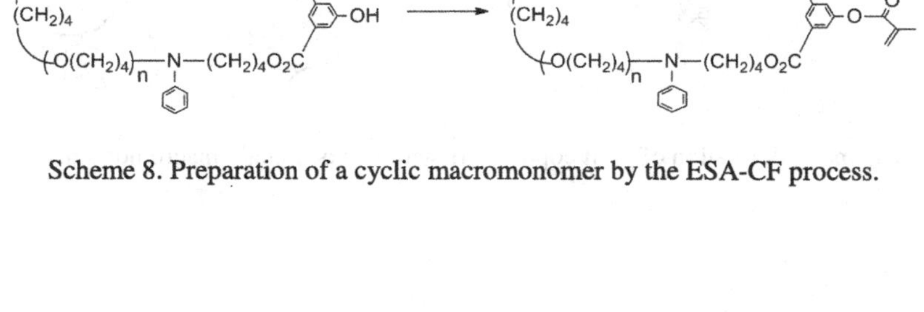

Scheme 8. Preparation of a cyclic macromonomer by the ESA-CF process.

less than 20%. The copolymer fraction was isolated by the precipitation treatment. The SEC of the copolymer product showed both ultraviolet (UV) and refractive index (RI) responses, to indicate that the graft component by the poly(THF) macromonomer, having a phenyl group at the precursor chain ends, distributes statistically along the poly(MMA)-based copolymer backbone. The ^{1}H NMR also showed the presence of the poly(THF) component in the copolymer, while the methacrylate group signals are completely disappeared. No side reactions such as a decomposition of the cyclic structure were detectable. The relative molar ratio of the graft components in the copolymer was estimated to be very close to the feed ratio against MMA, in accord with the random copolymerization to occur.

Subsequently, a complete-conversion copolymerization was carried out. A nearly whole portion of the recovered product became insoluble but swollen in various solvents. No gelation, on the other hand, took place in the relevant quantitative-conversion copolymerization of MMA with a relevant open-chain polymer precursor, obtained from a monofunctional poly(THF) with a pyrrolidinium salt end group (Scheme 9). The recovered product was totally soluble, and the ^{1}H NMR and IR showed no sign of side reactions, responsible for the covalent cross-linking by chain transfer reactions during the copolymerization process. These results indicate that the gelation in the copolymerization of MMA with the cyclic macromonomer took place through the physical cross-linking, i.e., the threading by the propagating chain through the pendant cyclic branches attached to the polymer backbone as shown in Scheme 7. And more importantly, the chain threading through large polymer loops occurred only when they are covalently attached to the polymer backbone.

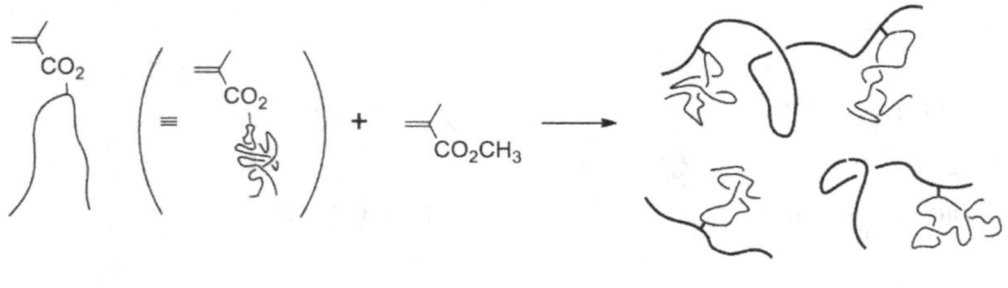

Scheme 9. Formation of graft copolymers with an open-chain macromonomer.

4. Tadpole Polymers Having Functional Groups

Tadpole polymers having functional groups at the prescribed positions, either at the tail-end or at the top-head positions of the tadpole construction, were produced through the dynamic structural control in the ESA-CF process, discussed in Chapter 5.[14]

Thus, a star poly(THF) precursor having *N*-phenylpyrrolidinium salt end groups carrying a mixture of 4-hydroxybenzoate and terephthalate anions (1/1 in mol/mol for counteranion) was prepared in the similar manner to produce a simple tadpole polymer described in Chapter 5. Another star polymer precursor carrying a mixture of benzoate and 5-hydroxyisophthalate anions (1/1 in mol/mol for counteranion) was also prepared (Scheme 10). These electrostatic polymer self-assemblies were subsequently subjected to heat treatment under dilution to cause covalent conversion of ionic pairs. The SEC of the recovered products after the heat treatment of the two polymer assemblies above showed unimodal peak profiles with narrow size distributions at the elution volumes corresponding to the tadpole polymers, smaller than the star precursor. The quantitative formation of each type of the hydroxyl-functionalized tadpole polymers was further confirmed by the ^{1}H NMR analysis of the product, in which signals due to 4-hydroxybenzoate and 5-hydroxyisophthalate moieties were visible, respectively.

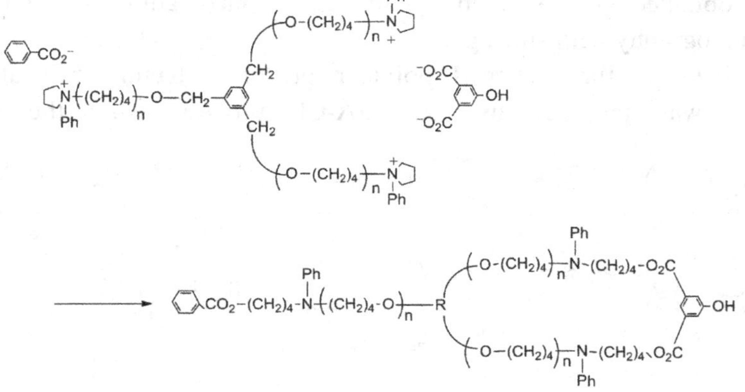

Scheme 10. Preparation of a tadpole polymer having a functional group.

5.   Multiyclic Polymers Having Functional Groups at the Prescribed Positions

A class of dicyclic polymer constructions includes three types; 8-shaped (*spiro*-form), manacle-shaped (*bridged*-form) and θ-shaped (*fused*-form) structures. The ESA-CF process could produce 8-shaped poly(THF)s having either two allyl or two alkyne groups at the opposite positions of the two ring units, by the relevant *kentro*-telechelic poly(THF)s (Scheme 11).[9,15] Thus, a linear poly(THF) precursor having *N*-phenylpyrrolidinium salt end groups and an allyloxy group at the center of the chain, was prepared, and the subsequent ion-exchange reaction of the pyrrolidinium salt end groups was conducted by the precipitation of the poly(THF) precursor, initially carrying triflate counteranions, into aqueous solution containing an excess of tetracarboxylate anions as a sodium salt form. A symmetrical tetracarboxylate was purposely chosen to avoid the formation of isomeric 8-shaped products.

The recovered ion-exchanged product was then dissolved in THF under various dilutions. At the concentration of 0.125 g/L, an assembly consisting of two units of the poly(THF) precursor and one unit of a tetracarboxylate counteranion was produced by the de-assembly of the ionically aggregated form. The subsequent covalent conversion was performed by heating to reflux in THF to cause the ring-opening reaction of the pyrrolidinium salt groups by each carboxylate group in the counteranion. The 8-shaped polymer precursor having two allyl groups was obtained in a high yield after purification by column chromatography with silica gel.[15]

Likewise, the 8-shaped polymer precursor having two alkyne groups was prepared by the ESA-CF process using the same

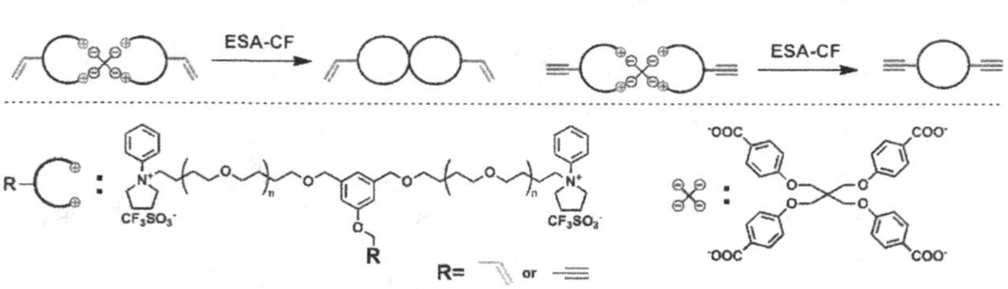

Scheme 11. Eight-shaped polymers having functional groups at the opposite positions of the two ring units.

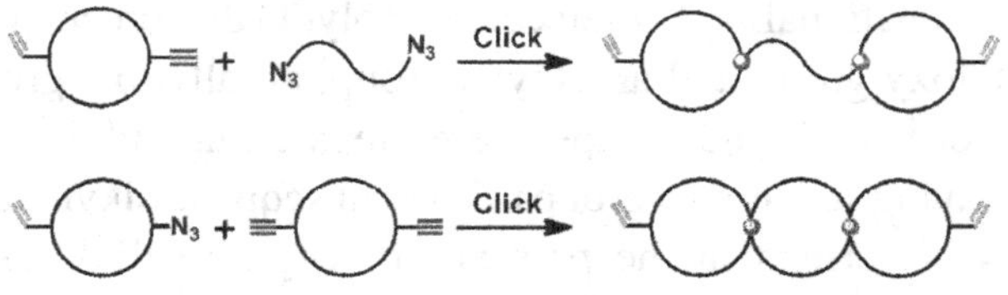

Scheme 12. Dicyclic and tricyclic polymers having functional groups at the opposite positions of the ring units.

tetracarboxylate counteranion, together with an $\alpha,\omega,kentro$-telechelic poly(THF) having N-phenylpyrrolidinium salt end groups and an alkyne group at the center position.[9] The ^{1}H NMR spectroscopic analysis of the product showed the characteristic signals for the propargyl groups, together with those for the tetracarboxylate group. The MALDI-TOF mass spectra of the product showed uniform series of peaks with an interval of 72 mass units for the repeating THF units, and each peak exactly matches the molar mass summing up the initiator fragment, poly(THF) units and linking group. Thus, the peak at m/z = 6012.0, which was assumed to be the adduct with Na^+, corresponding to the expected chemical structure with the DP_n of 60; $(C_4H_8O) \times 60 + C_{103}H_{112}N_4O_{16}$ plus Na^+ as 6011.471.[9] The SEC analysis showed a narrow size distribution of the product, and apparent molecular weight significantly higher than that of the starting linear precursor, estimated after the reaction with sodium benzoate, but lower than twice of that. The contraction of the 3D size is estimated as 0.80 in good agreement with the preceding studies.[9]

Furthermore, a manacle-shaped dicyclic polymer precursor having two allyl groups at the opposite positions of the two ring units, and a spiro-tricyclic precursor having two allyl groups at the opposite positions of the three ring units were successfully obtained by employing a series of the relevant *kyklo*-telechelics of symmetric and asymmetric functionalities at the opposite positions of the ring unit, which were obtained by the ESA-CF process with a set of α, ω, *kentro*-telechelic poly(THF)s having N-phenylpyrrolidinium groups and an alkene or an alkyne group at the center position of the chain, and carrying dicarboxylate counteranions having an alkyne or an azide group (Scheme 12).[10]

Thus, bifunctional *kyklo*-telechelic poly(THF) precursors having either an allyloxy group and an alkyne group, an allyloxy group and an azide group, or two alkyne groups, were prepared in addition to a linear poly(THF) having azide end groups. The subsequent alkyne-azide *click* reaction was performed in the presence of copper sulfate and sodium ascorbate by employing complementary sets of polymer precursors, namely cyclic/linear and cyclic/cyclic combinations to produce a *bridged*-dicyclic (manacle) polymer precursor and a tandem *spiro*-tricyclic counterpart, respectively, both having two allyloxy groups at the opposite positions of the ring units. These products were isolated by preparative SEC in satisfactory yields. The progress of the reaction and the subsequent purification process were monitored by SEC, which showed the peak of the products shifted toward the higher molecular weight region. These products were finally purified by preparative SEC fractionation. The extent of the contraction of the 3D size for dicyclic and tricyclic products was estimated as 0.82 and 0.63, respectively, by their apparent peak molecular weights by SEC, with reasonable agreements with the previously reported values for the relevant polymers.

The ^{1}H NMR showed that the signals for the ethynyl protons and that of the azidomethylene protons of the cyclic/linear precursors were replaced by the triazole proton signals, respectively, to confirm the effective click reaction. The signals for the allyloxy units were visible intact during the click process. The MALDI-TOF mass spectra of the products showed uniform series of peaks with an interval of 72 mass units corresponding to the repeating THF units, and moreover each peak exactly matched the molar mass calculated from chemical structure of the product. Thus, for the tandem *spiro*-tricyclic product having two allyl groups, the peak at $m/z = 9023.5$, assumed to be the adduct with Na$^+$, corresponds to the product possessing the expected chemical structure with a DP$_n$ of 90; $(C_4H_8O) \times 90 + C_{148}H_{180}N_{12}O_{24}$, plus Na$^+$ equals 9023.524.[10]

In addition, a pair of dicyclic topological isomers having θ-shaped and manacle-shaped constructions, having two allyl groups at the prescribed positions were prepared (Scheme 13).[16] Thus, a series of cationic polymer precursors, i.e., a *kentro*-telechelic poly(THF) having *N*-phenylpyrrolidinium salt groups and an additional pendant allyloxy group at the center position, as well as anionic counterparts, i.e., a telechelic poly(THF) having dicarboxylate end groups were employed for the ESA-CF process. The subsequent ion-exchange reaction between the polymer precursor, initially carrying triflate counteranions, and another initially carrying tetrabutylammonium countercations, was conducted by a simple coprecipitation procedure. In practice, a THF solution containing an equimolar amount of the polymer pairs, monitoring with the ^{1}H NMR, was precipitated into ice-cooled water to remove tetrabutylammonium triflate. An electrostatic polymer-polymer assembly product was recovered by simple filtration, and was subjected to the heat treatment under appropriate dilution, in order to maintain the self-assembly consisting of the smallest number of cationic and anionic units, i.e., two units of the polymer cation and one unit of the polymer anion, in which the cations and anions balance the charges.

Covalently linked products were subsequently recovered in good yields. The ^{1}H NMR showed that the signals attributed to the N-phenylpyrrolidinium salt group in the ionic self-assemblies were replaced by a triplet signal arising from ester methylene protons in the covalently converted product. Pendant allyl protons were, on the other hand, visible intact during the reaction. The SEC analysis of the product showed a unimodal peak, and distinctively higher apparent molecular weight than each of the precursor analogues. Nevertheless, the hydrodynamic volume of the product, estimated from the apparent peak molecular weight by SEC, was notably smaller than that of the relevant linear polymer having the same molecular weight. The product having the contracted hydrodynamic volume, compared with the linear analogue, was consistent with its multicyclic structure. Indeed, the ratio of the apparent molecular weight, i.e., the measure of the hydrodynamic volume, of the product against the linear polymer analogue possessing the same

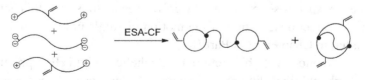

Scheme 13. Dicyclic polymeric topological isomers having allyl groups at the opposite positions.

molecular weight was 0.69, and these values were close to the reported ones (0.69-0.77).[16]

The presence of the two constitutional isomers in the obtained product was then confirmed through the RPC and the MALDI TOF mass techniques. Thus, the recovered product was first subjected to the RPC analysis, and the presence of two components was confirmed under an optimized eluent condition. The two components were then fractionated and subjected to the MALDI TOF mass analysis. In both fractions, a series of peaks corresponding to poly(THF) (peak interval of 72 mass units) were detected, and significantly, corresponding exactly to the expected chemical structures. Thus, for one fraction, the peak (the mass/charge assumed to be the adduct with Na^+) at 4049.0 was detected, corresponding to the expected product having either a θ- or manacle-constructions, with the DP_n of 50, $(C_4H_8O) \times 50 + C_{78}H_{88}N_6O_{10}$, plus Na^+ as 4050.03. And for another fraction, the peak (the mass/charge assumed to be the adduct with Na^+) at 6393.2 corresponds to another dicyclic polymeric isomer product, with the DP_n of 70, $(C_4H_8O) \times 70 + C_{78}H_{88}N_6O_{10}$, plus Na^+ as 6392.17.[16] This is indicative of the presence of the two components having identical series of molar masses, i.e., topological isomers, in the polymer products.

6. Cyclic Polymers Having Non-Reactive Functional Groups

A cyclic polymer having a fluorescent chromophore unit was also prepared by the ESA-CF process for the single-molecule spectroscopy to elucidate the topology effect on the diffusion behavior by cyclic polymer molecules. Cyclic polymers are unique by their absence of chain ends, and their diffusion by non-reptation dynamics has been envisaged both theory/simulation and experimental studies.[17] A single-molecule fluorescence spectroscopy is a powerful means to study this subject through the direct observation of the motion of single polymer molecules, as discussed in Chapter 13 in Part II.

Thus, mono- and bifunctional telechelic poly(THF) precursors having *N*-phenylpiperidinium (6-membered ring) groups possessing a perylene dicarboxylate counteranion was employed for the ESA-CF process, to produce ring and linear poly(THF)s having a perylene unit (Scheme 14).[18] It is remarkable that the cyclic and linear poly(THF)s must be produced through the ring-emitting esterification reaction. In contrast, the relevant cyclic and linear products prepared with the telechelic poly(THF)s having *N*-phenylpyrrolidinium (5-membered ring) groups were not suitable for fluorescent measurements, as the spontaneous quenching took place due to the aniline moiety included in the linking group of the cyclic polymer structure, formed by the ring-opening reaction of cyclic ammonium salt end groups.

7. Ongoing Challenges and Future Perspectives

This Chapter presented a variety of *kyklo*-telechelics obtained by the ESA-CF protocol. Those include single cyclic polymers having one speceific or plural (identical or different) functional groups, cyclic macromonomers having specifically a polymerizable group, tadpole polymers having either single or plural functional groups at the designated positions within *a ring and a branch* construction, as well as multicyclic polymers having single or plural functional groups at the prescribed positions within the multiple ring construction. These newly developed *kyklo*-telechelics are critically important to realize

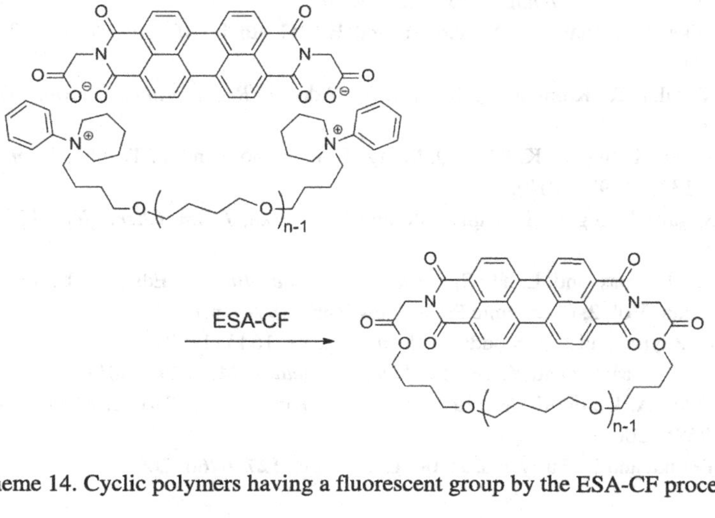

Scheme 14. Cyclic polymers having a fluorescent group by the ESA-CF process.

unprecedented complex polymer topologies, by conjunction with effective covalent chemistries, as discussed in the following Chapter 8. Further elaboration of the structural design of *kyklo*-telechelics, such as the introduction of more than two functional groups at the prescribed positions will extend the scope of topological polymer chemistry. A further choice of reactive groups for effective linking chemistries, such as a thiol-ene *click* process, should also be explored. Moreover, a cyclic polymer having a photonic probe for single molecule spectroscopy was presented. A variety of cyclic and multicyclic polymers having specific photonic, electronic, as well as bioactive components are to be designed by the ESA-CF protocol, to examine and eventually create any topology effects by cyclic and multicyclic polymers, so far unknown with conventional linear or branched counterparts.

References

1. N. Hadjichristidis, A. Hirao, Y. Tezuka and F. Du Prez, Eds., *Complex Macromolecular Architectures* (Wiley, Singapore, 2011).
2. Y. Tezuka, *Polym. J.*, in press (DOI: 10.1038/pj.2012.92)
3. T. Yamamoto and Y. Tezuka, *Polym Chem.*, **2**, 1930 (2011).
4. T. Yamamoto and Y. Tezuka, *Eur. Polym. J.*, **47**, 535 (2011).
5. H. Oike, S. Kobayashi, T. Mouri and Y. Tezuka, *Macromolecules*, **34**, 2742 (2001).
6. H. Oike, Y. Yoshioka, S. Kobayashi, M. Nakashima, Y. Tezuka and E.J. Goethals, *Macromol. Rapid Commun.*, **21**, 1185 (2000).
7. H. Oike, S. Kobayashi, Y. Tezuka and E.J. Goethals, *Macromolecules*, **33**, 8898 (2000).
8. Y. Tezuka, R. Komiya, Y. Ido and K. Adachi, *React. Funct. Polym.*, **67**, 1233 (2007).
9. N. Sugai, H. Heguri, K. Ohta, Q. Meng, T. Yamamoto and Y. Tezuka, *J. Am. Chem. Soc.*, **132**, 14790 (2010).
10. N. Sugai, H. Heguri, T. Yamamoto and Y. Tezuka, *J. Am. Chem. Soc.*, **133**, 19694 (2011).
11. D. A. Thomas and L. H. Sperling, in *Polymer Blends*, Eds., D. R. Pauland S. Newman, Vol. 2, (Academic Press, New York, 1978), p.1.
12. F. M. Raymo and J. F. Stoddart, *Chem. Rev.*, **99**, 1643 (1999).
13. H. Oike, T. Mouri and Y. Tezuka, *Macromolecules*, **34**, 6229 (2001).
14. H. Oike, A. Uchibori, A. Tsuchitani, H.-K. Kim and Y. Tezuka, *Macromolecules*, **37**, 7595 (2004).
15. Y. Tezuka and K. Fujiyama, *J. Am. Chem. Soc.*, **127**, 6266 (2005).

16. K. Adachi, H. Irie, T. Sato, A. Uchibori, M. Shiozawa and Y. Tezuka, *Macromolecules*, **38**, 10210 (2005).
17. T. C. B. McLeish, *Adv. Phys.*, **51**, 1379 (2002).
18. S. Habuchi, N. Satoh, T. Yamamoto, Y. Tezuka, M. Vacha, *Angew. Chem. Int. Ed.*, **49**, 1418 (2010).

CHAPTER 7

METATHESIS POLYMER CYCLIZATION

Yasuyuki Tezuka

Department of Organic and Polymeric Materials,
Tokyo Institute of Technology, Meguro-ku, Tokyo, Japan
E-mail: ytezuka@o.cc.titech.ac.jp

A metathesis polymer cyclization (MPC) process using telechelic precursors having olefinic groups is presented. A telechelic poly(THF) having allyl groups was subjected to the MPC process, and the subsequent hydrogenation could afford a *defect-free* ring poly(THF) as a unique model ring polymer. The MPC process was applied to telechelic precursors obtained by the ATRP technique, to afford not only cyclic homopolymers but also cyclic block copolymers. Furthermore, the double MPC process by using tetrafunctional branched precursors and by using cyclic precursors having olefinic groups was performed to produce dicyclic polymer products. The transformation of polymer topologies by the cross-metathesis process was also achieved with a dicyclic polymer precursor having an olefinic group introduced at the specific position.

1. Introduction

A strictly unimolecular polymer cyclization process has now been introduced in addition to the ESA-CF process, where asymmetric telechelic precursors having complementarily reactive end groups are employed (Chapter 10 in Part II). The reaction under dilution gives a cyclic polymer product in good yields, since the stoichiometric balance is inherently maintained between complementarily reactive groups located within the same polymer molecule. Nevertheless, the synthesis of the heterobifunctional polymer precursors often requires multistep processes involving protection/deprotection of reactive end groups, limiting its synthetic potentials.

In this Chapter, a metathesis polymer cyclization (MPC) process is presented by using readily accessible symmetric telechelic precursors having *identical* and reactive, more specifically olefinic end groups. A telechelic poly(THF) having allyl groups was subjected to the MPC process, in the presence of a commercially available metathesis catalyst, to prepare a ring poly(THF), and moreover, a *defect-free* ring poly(THF) comprised exclusively of monomer units therefrom. The MPC method was combined with the ATRP process, by which block copolymer precursors are conveniently obtainable. Furthermore, the double MPC process by using tetrafunctional branched precursors and by using cyclic precursors having olefinic groups was performed to produce dicyclic polymer products. The transformation of polymer topologies by the cross-metathesis process was also achieved with a dicyclic polymer precursor having an olefinic group introduced at the specific position.

2. Metathesis Polymer Cyclization

2.1. *Ring (simple cyclic) polymers and defect-free ring polymer*

The olefin-metathesis reaction has been successfully applied for the cyclization of small- to medium-sized substrates having terminal alkene groups, and for the synthesis of a variety of topologically unique molecules like catenanes and knots, as well as for the polycondensation of acyclic dienes (ADMET) to produce a variety of functional polymer materials.[1] A linear polymer precursor having allyl groups was subsequently subjected to a metathesis condensation, also known as a ring-closing metathesis (RCM), under dilution in the presence of a Grubbs catalyst, ruthenium(II) dichloride phenylmethylene bis(tricyclohexylphosphine) $[RuCl_2(PCy_3)_2(=CHPh)]$.[2]

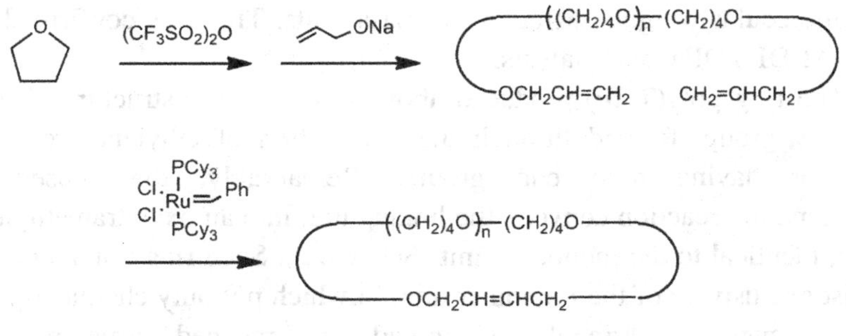

Scheme 1. Synthesis of a ring poly(THF) by the MPC process.

Thus, a telechelic poly(THF) having allyl end groups was prepared by the end-capping reaction of a bifunctionally living poly(THF) with sodium allyloxide.[3] The metathesis condensation was subsequently performed under reflux in methylene chloride at various polymer concentrations (0.2–2.0 g/L) in the presence of a Grubbs catalyst. The catalyst was charged in the comparable molar quantity to allyl groups at the polymer chain ends in order to achieve quantitative conversion of allyl groups in the precursor. The ^{1}H NMR showed characteristic signals due to allyl protons in the precursor were replaced by those due to inner alkene protons (both cis and trans signals are visible). The SEC showed that the product having a nearly symmetrical profile (>95% purity from SEC area ratio) was isolated by the reaction at 0.2 g/L dilution. Moreover, the hydrodynamic volume ratio of the product against the precursor, estimated by SEC, was 0.81 in good agreement with the previous cases to confirm the effective cyclization (Scheme 1).

The cyclic poly(THF) and its linear precursor were also examined by the MALDI-TOF mass spectroscopy. Both polymers showed a uniform series of peaks corresponding to poly(THF) (peak interval of 72 mass units); and each peak of the cyclic product corresponds exactly to the molar mass summing up the linking structure produced by the metathesis condensation reaction of allyl end groups in the linear precursor. As an example, the peak (assumed to be the adduct with Na$^+$) at 4419.8 corresponds to the product with the DP$_n$ of 60, $(C_4H_8O) \times 60 + C_4H_6O$, plus Na$^+$ as 4419.517.[3] As the cyclic poly(THF) products are produced from the precursor by the elimination of an ethylene molecule, their molecular weights differ by 28 mass units. This was confirmed by the MALDI-TOF mass analysis.

The ring poly(THF) produced above has a linking structure of a 2-butenoxy group, formed through the elimination of ethylene from the precursor having allyl end groups. Remarkably, the subsequent hydrogenation reaction converts the linking unit into an oxytetramethylene group, identical to the monomer unit (Scheme 2). Such ring polymers that consist exclusively of the monomer unit, in which not only chemically but also geometrically irregular chain-end or branched structures are completely eliminated, are considered to be defect-free and are valuable for the study of the fundamental properties of randomly coiled long-chain polymers both in solution and in bulk. The practical synthesis of ring polymers of a defined, defect-free structure, however, has not been achieved until recently even through a ring-enlarging chain polymerization, where an initiator or a catalyst fragment within the growing cyclic polymer intermediate is inherently included as an irregular structure, and the ring topology of the polymer product scarcely remains intact upon removing the initiator or catalyst fragment.[4]

Thus, a defect-free ring poly(THF) was synthesized together with its relevant linear counterpart with inert end groups, whose structure corresponds to one formed through the bond-breaking at the middle position of the butane (tetramethylene) unit of the ring polymer and the

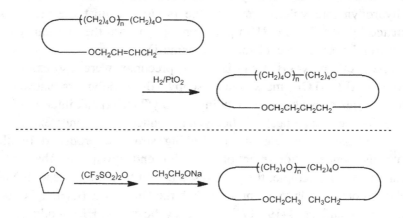

Scheme 2. Synthesis of a defect-free ring poly(THF) and its linear analogue.

subsequent addition of two hydrogen atoms (Scheme 2).[5] The hydrogenation reaction was successfully conducted in THF under a H_2 atmosphere in the presence of Adams' catalyst (PtO_2), which could be removed by a simple filtration with a celite-packed column to give almost colorless products in high yields. A relevant linear poly(THF) counterpart having ethoxy end groups was prepared by a simple end-capping reaction of a bifunctional living poly(THF) with sodium ethoxide.

The ^1H NMR of the defect-free ring polymer showed no signals except for those arising from the main-chain methylene units, while the spectrum of the linear counterpart showed a triplet signal at 1.20 ppm due to methyl protons of ethoxy end groups.

The MALDI-TOF mass showed a uniform series of peaks that correspond to poly(THF) (peak interval of 72 mass units), and each peak corresponds exactly to the molar mass summing up solely the monomer units for the defect-free ring polymer, and additional ethoxy end groups for the linear counterpart, respectively. As an example, the peak (assumed to be the adduct with Na^+) at 5142.6 corresponds to the ring product, with the DP_n of 71 [(C_4H_8O) $\times$ 70 + C_4H_8O, plus Na^+ of 5142.605].[5] The molecular weight of the linear analogue that has ethoxy end groups should differ by two mass units from the defect free ring poly(THF), and this was indeed confirmed by comparing the MALDI-TOF mass spectra.

The isothermal crystallization experiments were carried out with these model polymers. A distinctive polymer topology effect was anticipated in the dynamics of the ring polymer as opposed to the linear counterpart as a result of the absence of chain ends as well as the entropic restriction of chain conformations.[5,6] It was shown indeed the *slower* crystallization kinetics for ring poly(THF)s, ascribed to their topologically constrained conformation. Moreover, a banded structure in the spherulites was produced exclusively for the ring poly(THF) but not for the linear counterpart. The topology effects upon the crystallization are discussed in detail in Chapter 15 in Part II.

2.2. *ATRP-MPC process for cyclic block copolymers*

An atom transfer radical polymerization (ATRP) process has allowed the remarkable control in radical polymerization process, and has been widely applied for the preparation of a variety of polymers having functional groups, telechelic polymers, and star polymers as well as block and graft copolymers.[7] By the combination of the ATRP with the MPC process, a wider variety of functional cyclic polymers, which are not readily accessible through cationic or anionic processes, could become obtainable. Thus, the ATRP of methyl acrylate (MA) was conducted with a difunctional initiator, dimethyl-2,6-dibromoheptanedioate, followed by a radical addition (Keck) reaction with allyltributyltin to afford a linear telechelic poly(MA) having olefinic groups. The subsequent MPC reaction was performed in the presence of the Grubbs catalyst 1st generation under dilution in CH_2Cl_2. The product was recovered upon alumina column chromatography, and finally purified by SEC fractionation.[8]

The ^1H NMR showed that, along with the MPC reaction, the signals due to allyl end groups in the precursor were replaced by those due to the inner olefinic unit, to confirm the efficient RCM reaction to take place even under applied dilution. The cyclized product and its linear precursor were compared by the MALDI-TOF mass technique to reveal that the cyclic product with a uniform series of peaks corresponding to poly(MA) (peak interval of 86 mass units); each peak corresponds exactly to the molar mass summing up the linking structure produced by the MPC of the linear precursor having allyl groups. As an example, the peak (assumed to be the adduct with Na^+) at 3534.6 corresponds to the product with the DP_n of 40, $(C_4H_6O_2) \times 40 + C_5H_8$, plus Na^+ as 3534.73.[8] As the cyclic product is produced from the linear precursor by the elimination of an ethylene molecule, their molecular weights differ 28 mass units. This was confirmed by the MALDI-TOF mass spectra.

The effective synthesis of cyclic diblock copolymers was subsequently achieved by employing the ATRP-MPC process using either A-B or A-B-A type allyl-telechelic block copolymer precursors, obtained via the ATRP technique (Scheme 3 and 4, respectively).[9] Thus first, the ATRP of MA was conducted using allyl bromide as an initiator in the presence of Cu(I)Br and 2,2'-bipyridyl. After adding the second

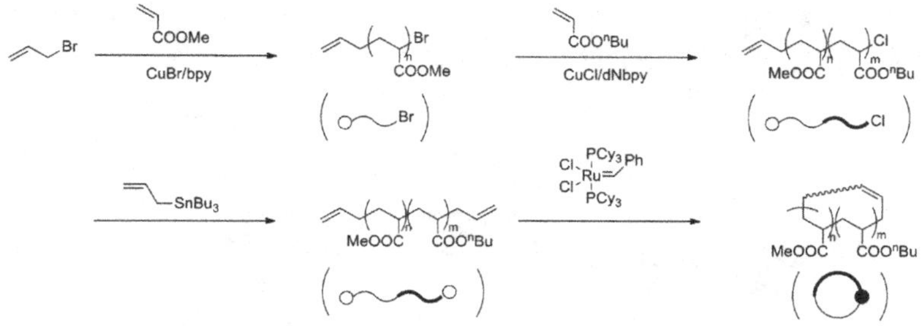

Scheme 3. Synthesis of a cyclic diblock copolymer by the MPC process with an AB diblock copolymer precursor.

monomer, *n*-butyl acrylate (BA) to produce the block copolymer, the end-capping reaction was conducted with allyltributylstannane to cause the allylation of terminal bromoalkyl groups. The ^{1}H NMR of the allyl-telechelic poly(MA)-*b*-poly(BA) showed the signals attributed to terminal allyl groups, and the signal of methyne proton adjacent to the bromine atom was completely eliminated after the allylation. The SEC traces of the poly(MA) precursor having a bromoalkyl end group and the subsequent A-B type allyl-telechelic block copolymer, poly(MA)-*b*-poly(BA), were compared to show the shift of the unimodal profile of the former toward the higher molecular weight region with retaining unimodal peak profile. These results indicated the quantitative conversion of the poly(MA) precursor to initiate the ATRP of the second monomer, BA (Scheme 3).

Moreover, an A-B-A type allyl-telechelic block copolymer precursor, i.e., poly(BA)-*b*-poly(ethylene oxide, EO)-*b*-poly(BA), was prepared through the ATRP of BA using a bifunctional poly(EO) macroinitiator, followed by the endcapping reaction with allyltributylstannane (Scheme 4).[9] Thus first, the poly(EO) macroinitiator was prepared by the esterification of hydoroxy-telechelic poly(EO) (*M*n of 2000) with 2-bromoisobutyryl bromide. The block copolymerization of the second monomer, BA, was conducted in the presence of Cu(I)Br and a 2,2′-bipyridyl derivative and the subsequent end-capping reaction was performed with allyltributylstannane to cause the allylation of terminal bromoalkyl groups in the similar manner as described in the production of A-B block copolymers.

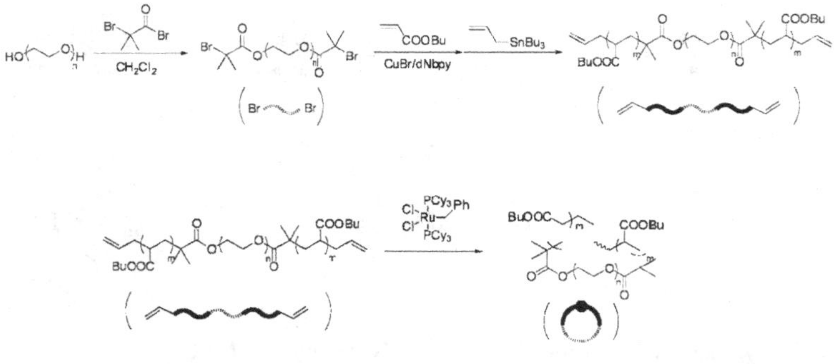

Scheme 4. Synthesis of a cyclic diblock copolymer by the MPC process with an ABA triblock copolymer precursor.

The ^{1}H NMR spectrum of the block copolymer product showed the signals attributed to the terminal allyl groups, in addition to the overflowed main-chain methylene signals from poly(EO) segment. The SEC traces of the poly(EO) precursor having 2-bromoisobutyryl end groups and the subsequent block copolymer, poly(BA)-*b*-poly(EO)-*b*-poly(BA), were compared to show the shift of the unimodal profile toward the higher molecular weight region with retaining unimodal peak profile.

The obtained A-B and A-B-A type allyl-telechelic block copolymers were subsequently subjected to the MPC process under dilution in the presence of the Grubbs catalyst (Scheme 3 and 4). The ^{1}H NMR showed the signals attributed to the inner olefinic group in the products, replacing those due to the terminal allyl groups visible before the reaction. The SEC showed a unimodal peak in the absence of noticeable shoulder traces at the higher MW region, indicating the exclusive intramolecular reaction under the applied dilution. The SEC also showed that the apparent molecular weights (3D size) of the cyclic block copolymers were reduced by comparing with the starting allyl-telechelic linear precursors. The smaller 3D size, i.e., the hydrodynamic volume, against their linear precursors is consistent with the formation of the cyclic polymer products.

It is notable that a micelle formed from the obtained cyclic amphiphile, i.e., poly(EO)-*b*-poly(BA), exhibits a significantly enhanced thermal stability in comparison with the one from the linear A-B-A block copolymer counterpart. This finding is regarded as the first example of an amplified topology effect by a synthetic cyclic polymer upon self-assembly.[10] The topology effects upon the micelle and other self-assemblies are discussed in detail in Chapter 16 in Part II.

Furthermore, the ATRP–RCM polymer cyclization process was successfully applied to produce another amphiphilic block copolymer, i.e., cyclic polystyrene-*b*-poly(EO) (cyclic PS-*b*-PEO).[11] Thus, a bromobenzyl-terminated PS-*b*-PEO-*b*-PS was prepared by the ATRP of styrene using a PEO macroinitiator having 2-bromoisobutyryl groups. The subsequent introduction of allyl end groups was achieved with allyltrimethylsilane (ATMS) in the presence of TiCl₄, in place of Keck allylation employed for the end-capping of polyacrylates. The obtained allyl-telechelic triblock copolymer PS-*b*-PEO-*b*-PS was then subjected to the MPC to produce amphiphilic cyclic PS-*b*-PEO.

Another class of cyclic block copolymers, namely a pair of orientationally isomeric cyclic stereoblock polylactides (PLAs) possessing head-to-head (HH) and head-to-tail (HT) linkages between the poly(L-lactide) (PLLA) and poly(D-lactide) (PDLA) segments was also prepared through the MPC process by using of three asymmetrically functionalized telechelic precursors. Thus, α-ethenyl-ω-azido-PLLA was reacted with α-ethenyl-ω-ethynyl-PDLA and with α-ethynyl-ω-ethenyl-PDLA by means of the alkyne-azide addition reaction to form ethenyl telechelic stereoblock PLAs with HH and HT orientations, respectively. The subsequent MPC produced cyclic stereoblock PLAs with the corresponding linking manners to examine the topology effects on the melting properties of isomeric linear and cyclic PLAs having the contrastive linking orientations.[12]

2.3. *Dicyclic polymers by branched or cyclic polymer precursors*

Dicyclic 8-shaped polymers are obtainable through the ESA-CF process, involving two units of telechelic precursor having cyclic ammonium salt groups carrying one unit of tetracarboxylate counter anion, as shown in Chapter 5. The MPC process could offer an alternative means to produce an 8-shaped polymer from a 4-armed star polymer precursor (Scheme 5).[13] Thus, a star poly(MA) having four allyl end groups was prepared by the ATRP of MA using pentaerythritol tetrakis(2-bromoisobutyrate) as a tetrafunctional initiator, and the subsequent Keck allylation with allyltributyltin. The double metathesis condensation reaction of the star polymer precursor was then performed under dilution in the presence of the Grubbs catalyst. The 8-shaped polymer product was isolated after the purification through alumina column followed by precipitation into hexane.

The ^{1}H NMR showed that the signals due to the allyl groups in the precursor were totally replaced by those due to the inner olefinic unit after the reaction. The MALDI-TOF mass showed that all peaks by the product corresponded exactly to the molar mass for the expected structure produced by the double metathesis condensation of four allyl groups in the 4-armed star polymer precursor. Thus, the peak (assumed to be the adduct with Na^+) at 3986.6 corresponds to the product with the DP_n of 40, $(C_4H_6O_2) \times 40 + C_{29}H_{44}O_8$, plus Na^+ as 3987.28.[13] As the 8-shaped PMA product is produced from the 4-armed PMA precursor by the elimination of two ethylene molecules, their molecular weights differ by 56. And this was indeed confirmed by comparing the MALDI-TOF mass spectra. Remarkably also, neither a twin-tail tadpole PMA formed by single metathesis condensation of the 4-armed star PMA precursor, nor the unreacted 4-armed star PMA were observed by the MALDI-TOF mass analysis.

The SEC showed that the 4-armed star PMA precursor and the 8-shaped PMA retained unimodal distribution to indicate negligible chain

Scheme 5. Synthesis of an eight shaped polymer by the MPC process with a four-armed star polymer precursor.

degradation during the metathesis process. The SEC also showed that the the 8-shaped polymers were smaller in their 3D sizes than either linear or simple ring counterparts.

Moreover, an H-shaped polymer precursor having allyl end groups was prepared to perform the double MPC to produce dicyclic polymers, consisting of polymeric topological isomers of θ- and manacle-forms (Scheme 6).[14] Notably, a pair of θ-shaped and manacle-shaped polymers was also formed simultaneously in the ESA-CF process, discussed in Chapter 5.

Thus, the H-shaped polymer precursor having four terminal allyl groups was prepared by an asymmetric telechelic poly(THF) having an *N*-phenylpyrrolidinium salt carrying triflate counteranions, and having an allyl end group and another symmetric telechelic precursor having two dicarboxylate end groups carrying tetrabutylammonium countercations. An ionically linked H-shaped polymer product was prepared therefrom, comprised of four units of the former and one unit of the latter through the coprecipitation of the mixture of the both prepolymers into aqueous solution. The product was subsequently heated to reflux in THF to cause the ring-opening reaction of the pyrrolidinium salt groups to form the covalently converted H-shaped polymer product having allyl end groups.

The subsequent double MPC was performed under dilution in the presence of the Grubbs catalyst. The effective metathesis condensation was confirmed to proceed even under applied dilution, as the ^{1}H NMR showed that the signals attributed to inner alkene groups were visible after the reaction, replacing those attributable to the allyl group of the H-shaped precursor.

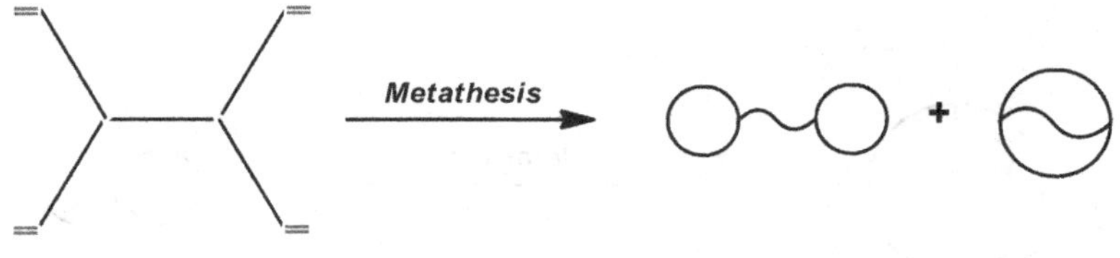

Scheme 6. Synthesis of a pair of dicyclic topological isomers of manacle- and θ-shaped topologies by the MPC process with an H-shaped polymer precursor.

The two polymeric topological isomers, having manacle- and θ-constructions, could be obtained by this intramolecular metathesis condensation as seen in Scheme 6. When the linking of the four terminal groups takes place randomly, the θ- and manacle-isomers could be formed in the ratio of 2/1. The RPC of the metathesis product showed the two fractions corresponding to the relevant polymeric topological isomers, and the ^{1}H NMR spectra of both fractions coincided with each other. The SEC comparison of the two fractions showed that one component was noticeably larger in its hydrodynamic volume, and was assignable as the manacle-isomer rather than the θ-isomer.

Moreover, inter- and intramolecular MPC processes were applied to produce 8-shaped, dicyclic polymers (Scheme 7).[15] Thus, a bimolecular reaction of a cyclic poly(THF) prepolymer having an allyloxy group, as well as unimolecular reactions with a twin-tailed tadpole polymer precursor having allyoxy groups at the tail-ends, and with a cyclic polymer precursor having two allyloxy groups at the opposite positions of the ring unit, were introduced respectively.

Though the bimolecular reaction in the presence of the Grubbs catalyst was found to proceed with noticeable side reactions, alternative unimolecular MPC processes either with a twin-tailed tadpole or with a

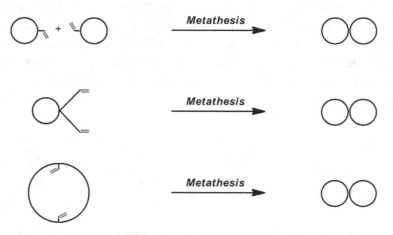

Scheme 7. Synthesis of an 8-haped polymer by the MPC process with cyclic polymer precursors.

kyklo-telechelic counterpart having two allyl groups at opposite positions proceeded effectively even under dilution to produce the corresponding 8-shaped poly(THF)s. Further extension of metathesis condensation with multicyclic polymer precursors is discussed in the following Chapter 8.

3. Metathesis Transformation of Cyclic Polymer Topologies

A pair of polymeric topological isomers of dicyclic, θ-shaped and manacle-shaped constructions, including a metathesis-cleavable olefinic group at the prescribed inner position in the specific segment component was described in Chapter 5. The subsequent topological conversion by an olefin-cleavage reaction through the cross-metathesis reaction was performed to give a tadpole and two-tail tadpole polymer products, to allow the unequivocal assignment of the two isomeric topological forms of the starting θ-shaped and manacle-shaped polymers.[16]

Another example of the topological conversion was performed through the metathesis cleavage process by employing a dicyclic 8-shaped polymer having an olefinic unit at the focal position (Scheme 8).[17] Thus, a pair of 8-shaped poly(THF)s having a metathesis-cleavable olefinic unit at the junction position were synthesized through the ESA-CF process using a telechelic poly(THF) having *N*-phenylpyrrolidinium salt groups carrying a tetracarboxylate counteranion containing a trans-3-hexenyl group. The subsequent metathesis cleavage reaction of the olefinic group was conducted through the cross-metathesis with

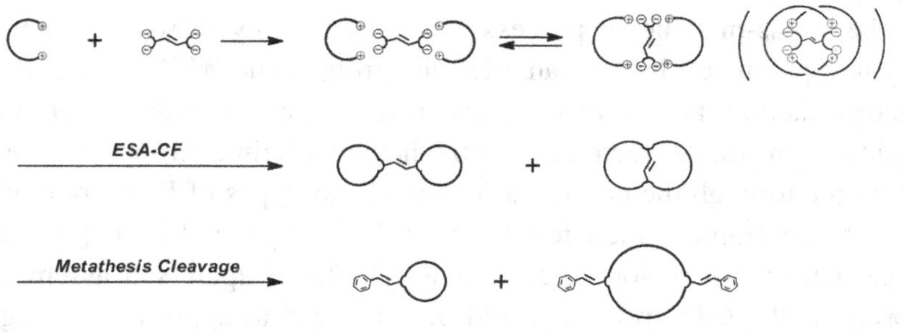

Scheme 8. Metathesis transformation of isomeric 8-shaped polymer pairs obtained by the ESA-CF process.

allylbenzene, to convert the polymer topology from a dicyclic 8-shape into two simple loops having the two distinctive sizes consisting of either one or two prepolymer units. Those correspond to the isomeric linking mode of the two prepolymer segments on the pair of carboxylate groups at either the neighboring or the remote sites of the tetracarboxylate. The MALDI-TOF mass analysis for a series of the SEC-fractionated metathesis cleavage products showed that the absence of any fraction corresponding to polymeric [2]catenane products, to imply that the entanglement of the two prepolymer segments does not occur even though they are placed spatially close to each other. An alternative means for the construction of polymer catenanes was presented in Chapter 5, by making use of the hydrogen-bond interaction with a tailored telechelic precursor and subsequent polymer cyclization by the ESA-CF process.

4. Ongoing Challenges and Future Perspectives

This Chapter presented the MPC process by telechelic precursors having olefinic groups. A telechelic poly(THF) having allyl groups was subjected to the MPC process to produce a ring poly(THF), and the subsequent hydrogenation could afford a *defect-free* ring poly(THF) as a unique model polymer. The MPC method was also applied with the precursors obtainable by the ATRP technique, to afford, in particular, various cyclic block copolymers. Furthermore, the double MPC process using branched or cyclic precursors having olefinic groups was applied to produce dicyclic polymers. The transformation of polymer topologies by the cross-metathesis process was also achieved with a dicyclic polymer precursor having an olefinic group. The MPC process was demonstrated as an versatile means to make various types of complex cyclic polymers, as their precursors having olefinic groups are ready accessible through the end-modification of any types of living polymers. Further opportunities include different types of cyclic block copolymers, in particular, to explore their topology effects upon self-assemblies. Moreover, the MPC process could be improved to apply for the higher molecular weight precursors along with the ongoing progress toward more effective metathesis catalysts. A class of defect-free ring polymers consisting of exclusively monomer units has now become obtainable also through the ring-expansion polymerization technique. Thus, a variety of new model ring polymers will open new opportunities to rigorously elucidate any topology effects arisen from cyclic constructions.

References

1. K. J. Ivin and I. C. Mol, Eds., *Olefin Metathesis and Metathesis Polymerization* (Academic Press, London, 1997).
2. R. H. Grubbs Ed., *Handbook of Metathesis* (Wiley-VCH, Weinheim, 2003).
3. Y. Tezuka and R. Komiya, *Macromolecules*, **35**, 8667 (2002).
4. Y. Tezuka, *Polym. J.*, in press (DOI: 10.1038/pj.2012.92)
5. Y. Tezuka, T. Ohtsuka, K. Adachi, R. Komiya, N. Ohno and N. Okui, *Macromol Rapid Commun.*, **29**, 1237 (2008).
6. N. Okui, N. Ohno, S. Umemoto and Y. Tezuka, *Bussei Kenkyu*, **92**, 51 (2009).
7. K. Matyjaszewski Ed., *Controlled/Living Radical polymerization: Progress in ATRP*, ACS Symp. Ser. 1023, (ACS, Washington, D. C. 2009).
8. S. Hayashi, K. Adachi and Y. Tezuka, *Chem. Lett.*, **36**, 982 (2007).
9. K. Adachi, S. Honda, S. Hayashi and Y. Tezuka, *Macromolecules*, **41**, 7898 (2008).
10. S. Honda, T. Yamamoto and Y. Tezuka, *J. Am. Chem. Soc.*, **132**, 10251 (2010).
11. E. Baba, S. Honda, T. Yamamoto and Y. Tezuka, *Polym. Chem.*, **3**, 1903 (2012).
12. N. Sugai, T. Yamamoto and Y. Tezuka, *ACS MacroLetters.*, **1**, 902 (2012).
13. S. Hayashi, K. Adachi and Y. Tezuka, *Polym. J.*, **40**, 572 (2008).
14. Y. Tezuka and F. Ohashi, *Macromol Rapid Commun.*, **26**, 608 (2005).
15. Y. Tezuka, R. Komiya, M. Washizuka, Macromolecules, 36, 12-17 (2003)
16. K. Adachi, H. Irie, T. Sato, A. Uchibori, M. Shiozawa, Y. Tezuka, Macromolecules, 38, 10210-10219 (2005).
17. K. Ishikawa, T. Yamamoto, H. Harada, Y. Tezuka, Macromolecules, 43, 7062-7067 (2010).

CHAPTER 8

MULTICYCLIC POLYMER TOPOLOGIES BY THE ESA-CF WITH *KYKLO*-TELECHELIC PRECURSORS

Yasuyuki Tezuka

Department of Organic and Polymeric Materials,
Tokyo Institute of Technology, Meguro-ku, Tokyo, Japan
E-mail: ytezuka@o.cc.titech.ac.jp

A variety of multicyclic polymer topologies of either *spiro, bridged* or *fused* forms, was effectively constructed by the ESA-CF technique in conjunction with the effective covalent linking chemistries. An alkyne-azide *cross-coupling*, i.e., *click*, reaction and a metathesis *homo-coupling*, i.e., *clip*, reaction were thus applied by employing tailored single cyclic and multicyclic polymer precursors (*kyklo*-telechelics) obtained by the ESA-CF protocol.

1. Introduction

The multicyclic polymer topologies are divided into two classes, namely *catenated* (mechanically linked) and *bonded* (covalently linked) forms, upon the connecting mode of the ring units. And for the *bonded multicyclic* topologies, the three subtypes are included; *spiro, bridged* and *fused* foms shown in Scheme 1, according to the linking mode of the ring units. The ESA-CF protocol could produce all the three bicyclic constructions of 8 (*spiro*), manacle (*bridged*) and θ (fused) forms, as well as a trefoil (*spiro* tricyclic) topology, as described in Chapter 4. Moreover, the ESA-CF process could produce *kyklo*-telechelics, cyclic and multicyclic polymers having pre-determined functional groups at the designated positions, as listed in Chapter 6.

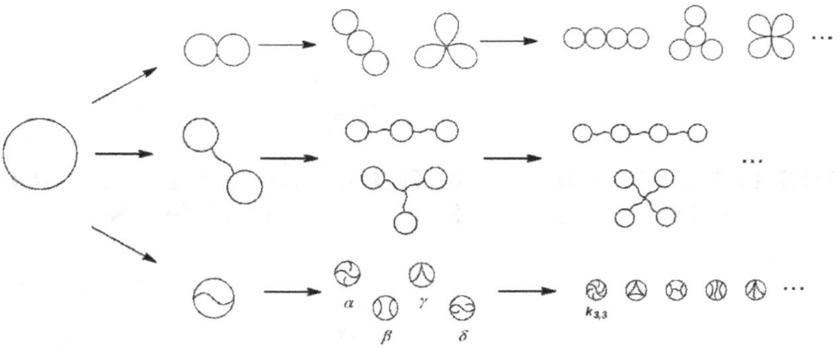

Scheme 1. *Spiro-*, *bridged-* and *fused-*multicyclic polymer topologies.

In this Chapter, the effective construction of a variety of multicyclic polymer topologies of either *spiro*, *bridged* or *fused* forms is presented. The effective covalent linking chemistries, in particular an alkyne-azide *cross-coupling*, i.e., *click*, reaction and a metathesis *homo-coupling*, i.e., *clip*, reaction were thus applied by employing tailored single cyclic and multcyclic polymer precursors (*kyklo*-telechelics) as large as 300-membered rings, obtainable by the ESA-CF protocol. The *click* and *clip* chemistries have now been widely employed to prepare a wide variety of functional polymers.[1–3] The metathesis polymer cyclization (MPC) process was already discussed in Chapter 7. It is notable also that the click cross-coupling process offers a rational design to selectively combine polymer precursors having different topologies, producing complex constructions of lower geometrical symmetries that are difficult to be obtained by one-step processes. The combined *click* and *clip* protocol is shown, moreover, as a powerful means to produce multiply-*fused* complex polymer topologies.

2. *Spiro*-Multicyclic Polymers

A *spiro*-dicyclic, 8-shaped polymer was directly obtainable by the ESA-CF process with a self-assembly containing two units of a linear telechelic precursor carrying one unit of tetracarboxylate counteranion (Chapter 4). The MPC process was also applied by employing star- and *kyklo*-telechelic precursors, including a four-armed star telechelic prepolymer having allyloxy end groups, a twin-tailed tadpole polymer precursor having two allyloxy groups at the tail-end positions, and a ring polymer precursor having two allyloxy groups at opposite positions (Chapter 7).

A *spiro*-tricyclic polymer having a trefoil topology was also produced by the ESA-CF process with the self-assembly consisting of three units of a linear telechelic precursor carrying a hexafunctional carboxylate counteranion as shown in Chapter 4 (Scheme 2). Another type of *spiro*-tricyclic polymer having a tandem form was obtained by the *click* linking of two cyclic prepolymers having two alkyne groups at the opposite positions of the ring unit and another having an azide group, both obtainable by the ESA-CF technique shown in Chapter 6 (Scheme 2).[4] Moreover, a *spiro*-tetracyclic polymer having a tandem construction was produced by the click coupling between an 8-shaped dicyclic prepolymer having two alkyne groups at the opposite positions of the two ring unit and another single cyclic precursor having an azide group (Scheme 2).[4]

The *click* reaction was conveniently monitored by means of a SEC technique, where the peak molecular weight (Mp) was shifted toward the higher molecular weight region. The observed Mp for the tricyclic product was almost identical to the sum of those for the precursors. On the other hand, the Mp observed for tetracyclic product was marginally lower than the individually added Mp values of the prepolymers. The subsequent isolation of these products was performed by means of a

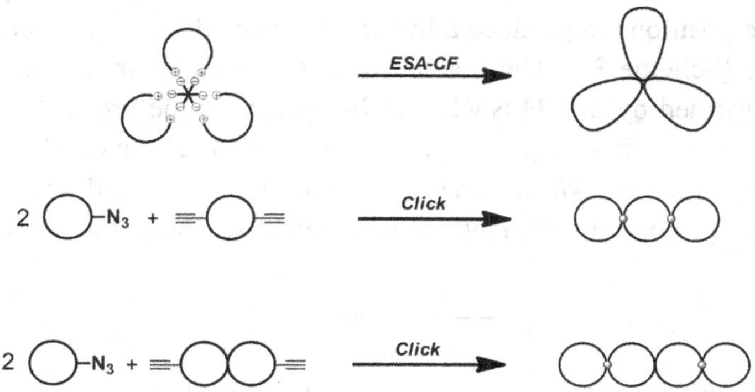

Scheme 2. Construction of *spiro*-tricyclic and *spiro*-tetracyclic polymer topologies.

preparative SEC fractionation technique. The SEC showed also that the tricyclic and tetracyclic products were comparably contracted in their 3D sizes.

The ^{1}H NMR confirmed the selective reaction between the alkyne and azide groups of the prepolymers. Thus, the signals for the ethynyl protons and those for the azidomethylene protons in the prepolymers completely disappeared after the reaction. Alternatively, triazole proton signals emerged in both the tricyclic and tetracyclic products. Moreover, the IR absorbance of the azide groups at 2104 cm^{-1} observed in precursor was scarcely visible in the products, indicating that the reaction proceeded effectively. The MALDI-TOF mass showed uniform series of peaks with an interval of 72 mass units for the tricyclic and tetracyclic products corresponding to the repeating THF units, and each peak exactly matched the total molar mass of the complementary precursors. Thus for an example, the peak at 11538.4, which was assumed to be the adduct with Na$^+$, corresponds to the tetracyclic product possessing the expected chemical structure with the DP$_n$ of 120; (C$_4$H$_8$O) $\times$ 120 + C$_{171}$H$_{194}$N$_{14}$O$_{26}$ plus Na$^+$ equals 11537.372. The sum of twice the molar mass of the respective precursors, one with the DP$_n$ of 30 (2 $\times$ (2786.5 $-$[Na$^+$])) of 5527.0 and another with the DP$_n$ of 60 (6012.0 $-$ [Na+]) of 5989.0 was 11516.0, in agreement with the molar mass of the tetracyclic product with the DP$_n$ of 120 (11538.4 $-$ [Na$^+$]) of 11515.4 given above.[4]

The *click* reaction was further applied to the polyaddition of a single cyclic precursor having an alkyne group and an azide group at the opposite positions to produce a linearly connected *spiro*-type multicyclic polymer (Scheme 3).[4] The reaction between the alkyne and azide units was confirmed by the ^{1}H NMR and IR spectra of the product. Thus, the signals for the ethynyl, propynyl methylene, and azidomethylene groups in the asymmetric *kyklo*-telechelic precursor were replaced with those for a triazole proton and the methylene group adjacent to the 4-position of

Scheme 3. Construction of a *spiro*-multicyclic polymer topology.

the triazole ring in the product. The IR absorbance of the azide groups observed in the precursors was scarcely visible in the product, indicating that the reaction proceeded effectively.

Finally, SEC showed, upon reaction, notable peak shifts toward the higher molecular weight region with multimodal distributions. From the estimated molecular weights by SEC, the hexamer of a linearly arranged *spiro*-type construction was formed in average (Scheme 3). In addition, the SEC traces indicated that concurrent intramolecular reactions took place. Thus, an 8-shaped product was formed as a side product by the intramolecular reaction of the asymmetric *kyklo*-telechelic precursor having complementary alkyne and azide groups.

3. *Bridged*-Multicyclic Polymers

A *bridged*-dicyclic, manacle-shaped polymer was obtainable together with a *fused*, θ-shaped polymeric isomer, as described in Chapter 5, through the ESA-CF process with an assembly composed of three units of a linear bifunctional precursor carrying two units of trifunctional carboxylate counteranions, or with an assembly composed of two units of a star-shaped trifunctional precursor carrying three units of bifunctional carboxylate counteranions. A pair of dicyclic polymeric topological isomers having θ- and manacle-constructions were also formed through the MPC process using an H-shaped precursor having an allyloxy group at each chain end (Chapter 7).

The selective construction of a dicyclic, manacle-shaped polymer topology was achieved through the click-coupling of a bifunctional linear telechelic precursor having azide groups with a cyclic polymer precursor having an alkyne groups obtainable through the ESA-CF protocol (Scheme 4).[4] In addition, a bridged-tricyclic, three-way paddle-shaped polymer was obtained by the relevant click-coupling reaction between a star-shaped trifunctional precursor having azide groups with a cyclic polymer precursor having an alkyne groups as seen in Scheme 4.[4]

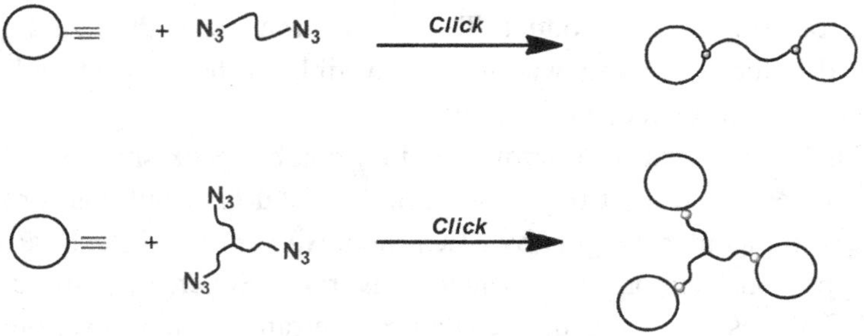

Scheme 4. Construction of *bridged*-tricyclic and *bridged*-tetracyclic polymer topologies.

Thus, linear and three-armed star telechelic poly(THF)s having azide end groups were prepared by the end-capping reaction of bifunctional and trifunctional living poly(THF)s, respectively. In addition, a *kyklo*-telechelic precursor having an alkyne group was prepared by the ESA-CF protocol (Chapter 6), and was subjected to the click reaction with linear and three-armed star telechelic precursors having azide end groups. To ensure complete reaction, a slight excess of the *kyklo*-telechelic precursor was charged relative to the azido-functional counterparts. The SEC showed a noticeable peak shift toward the higher molecular weight region for the products. The Mp for both the dicyclic and tricyclic products were nearly equal to the sum of those of the precursors. The products were subsequently isolated by means of a preparative SEC fractionation technique. The SEC also showed that the extent of the contraction in the 3D sizes of the dicyclic and tricyclic products were 0.68 for both di- and tricyclic products, and was comparable with those of previously reported various multicyclic topologies.

The ^{1}H NMR and IR analysis of the dicyclic and tricyclic products together with the respective precursors confirmed the selective reaction between the alkyne and azide groups of the prepolymers. Thus, the signal for the ethynyl proton in the *kyklo*-telechelic precursor and that for the azidomethylene protons in the azido-functional counterpart completely disappeared after the reaction, and a triazole proton signal emerged in the products. The IR absorbance of the azide groups observed in the linear and star precursors was scarcely visible in the products, indicating again that the reaction proceeded effectively.

The MALDI-TOF mass showed a uniform series of peaks with an interval of 72 mass units for the dicyclic and tricyclic products corresponding to repeating THF units, and each peak exactly matched the total molar mass of the complementary precursors. Thus for an example, the peak at m/z of 13902.5, which was assumed to be the adduct with Na^+, corresponds to the tricyclic paddle-shaped product possessing the expected chemical structure with the DP_n of 160; (C_4H_8O) × 160 + $C_{129}H_{156}N_{18}O_{24}$ plus Na^+ equals 13902.924. The sum of 3-fold the molar mass of he *kyklo*-telechelic precursor with the DP_n of 35 (3 × (3115.0 – $[Na^+]$) of 9276.0 and the molar mass of the star precursor with the DP_n of 55 (4625.2 – $[Na^+]$) of 4602.2 is 13878.2, was in good agreement with the molar mass of the product with the DP_n of 160 (13902.5 – $[Na^+]$) of 13879.5 given above.[4]

Moreover, *bridged*-multicyclic polymers consisting of cyclic and linear or branched polymer units were produced by the *click* polycondensation of a cyclic precursor having two alkyne groups at the opposite positions of the ring unit (Chapter 6) with the respective linear/branched telechelic precursors having azide groups (Scheme 5).[4] These topologically unique polymers are regarded as topological block copolymers including alternative cyclic and linear (or branched) segments. In the case of reaction between bifunctional and trifunctional precursors, the conditions were chosen to avoid uncontrolled gelation.

The ^1H NMR showed again that the signals for the ethynyl and for propynyl methylene groups of the *kyklo*-telechelic precursor, as well as that for the azidomethylene groups in the linear and star counterparts completely disappeared after the reaction. Instead, the signals for triazole protons and the methylene groups adjacent to the 4-position of the

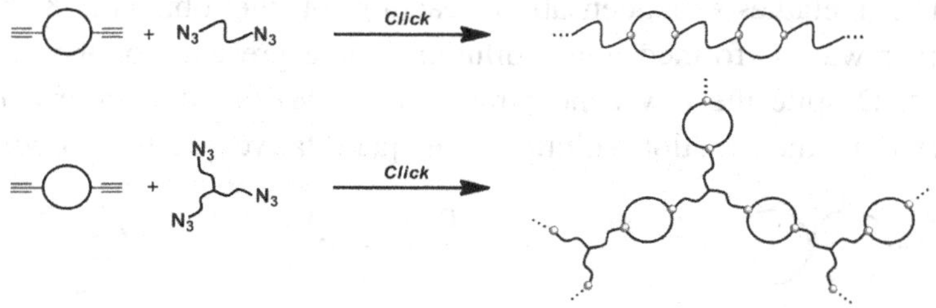

Scheme 5. Construction of *bridged*-multicyclic polymer topologies.

triazole rings emerged in the products. The IR absorbance of the azide groups observed in the precursors was again scarcely visible in the products, indicating that the reaction proceeded effectively.

Finally, the SEC implied that the product of the cyclic/linear units comprised of, in average, four rings connected with three linear chains in an alternating fashion. Likewise, the product of the cyclic/star branched components consisted of, in average, four ring units and three star branched units, respectively.

4. *Fused*-Multicyclic Polymers

A *fused*-dicyclic, θ-shaped polymer was obtainable through the ESA-CF process with a self-assembly containing a three-armed, star telechelic precursor carrying a tricarboxylate counteranion (Chapter 4). Alternatively, the θ-shaped polymer was formed together with a manacle-shaped polymeric isomer either through the ESA-CF or the metathesis clip process using the relevant polymer precursors (Chapters 5 and 7).

A doubly-*fused* tricyclic polymer topology includes four different forms of α-, β-, γ, and δ-graph constructions (Scheme 1). And a δ-graph polymer was first produced by the combination of the ESA-CF with the metathesis *clip* process (Scheme 6).[5] Thus, an 8-shaped polymer precursor having two allyl groups placed at the opposite ends of the two ring units in the 8-shaped structure was prepared through the covalent conversion of the electrostatic self-assembly composed of two units of a *kentro*-telechelic polymer precursor having *N*-phenylpyrrolidinium salt end groups and an allyl group at the center of the chain, accompanying one unit of tetrafunctional carboxylate counteranion, as detailed in Chapter 6.

The metathesis condensation reaction of the obtained 8-shaped precursor was performed under dilution in the presence of the Grubbs catalyst. Despite the low concentration (10^{-5} M) of allyl groups of the precursor in the reaction solution, the quantitative reaction proceeded

Scheme 6. Construction of a *fused*, δ-graph polymer topology.

within 48 h. The δ-graph polymer product was isolated by precipitation into precooled hexane.[5]

The ^{1}H NMR analysis of the product together with the 8-shaped precursor showed that the signals due to the allyl groups in the precursor are totally removed, and those due to the inner olefinic unit appeared instead. The δ-graph poly(THF) and its 8-shaped precursor were subsequently compared by MALDI-TOF mass spectroscopy. Both the δ-graph product and the precursor showed a uniform series of peaks corresponding to poly(THF) (peak interval of 72 mass units). And the 8-shaped polymer precursor showed a major series of the peaks corresponding to the Na$^+$ adduct. Thus, the peak (assumed to be the adduct with Na$^+$) at 7312.69 corresponds to the product with the DP$_n$ of 80, $(C_4H_8O) \times 80 + C_{95}H_{100}N_4O_{14}$, plus Na$^+$ as 7312.328. After the reaction, each peak corresponds exactly to the molar mass summing up the linking structure produced by the metathesis condensation reaction of allyl groups in the 8-shaped precursor. Thus, the peak (assumed to be the adduct with Na$^+$) at 7284.74 corresponds to the product with the DP$_n$ of 80, $(C_4H_8O) \times 80 + C_{93}H_{96}N_4O_{14}$, plus Na$^+$ as 7284.297. Since the δ-graph poly(THF) product is produced from the 8-shaped precursor by the elimination of an ethylene molecule, their molecular weights differ by 28 mass units. This was confirmed by comparing above two MALDI-TOF mass spectra.[5]

The SEC showed the topology effect on the 3D size of the obtained polymer products. The δ-graph product retained a nearly symmetrical SEC profile, with its hydrodynamic volume marginally smaller than that of the 8-shaped precursor. The degree of the contraction of the 3D size for the δ-graph polymer against the linear counterpart was subsequently estimated to be 0.63. It is remarkable that the observed degree of the contraction for the δ-graph polymer is in the same range of the θ-shape polymer having a singly *fused*, double cyclic topology (0.61–0.63). This implies the repulsive interaction between densely emanating branches, i.e., four arms from two junctions, or by the exclusion volume effect by the presence of the four polymer chains within the restricted loop structure of the δ-graph polymer.

Moreover, another doubly-*fused* tricyclic, γ-graph polymer, and a triply-fused tetracyclic polymer possessing an unfolded tetrahedron graph constructions were obtained by a tandem alkyne-azide *click* and an olefin metathesis *clip* reaction process upon the ESA-CF protocol (Scheme 7).[6]

Thus, a *bridged*-dicyclic (manacle) polymer precursor and a tandem *spiro*-tricyclic counterpart, respectively, both having two allyloxy groups at the opposite positions of the ring units were prepared as detailed in Chapter 6, and subjected to the intramolecular olefin metathesis reaction under dilution by repeated addition of the Grubbs catalyst into the reaction solution. As the crude reaction products contained a noticeable portion of the intermolecular condensation products, the product was first recovered by a column chromatography with silica gel, and finally isolated by the purification with the preparative SEC fractionation technique.

The ^{1}H NMR revealed that the signals for the allyloxy units in the precursors were completely replaced by those of the inner olefinic units in the products. It was indicative of the effective metathesis condensation reaction in both cases even under applied dilution. The MALDI-TOF mass analysis of the isolated tricyclic and tetracyclic products together with their precursors showed the successful construction of a doubly-*fused* tricyclic and a triply-*fused* tetracyclic polymer topologies. Both of the products showed a uniform series of peaks as in the precursors. Thus

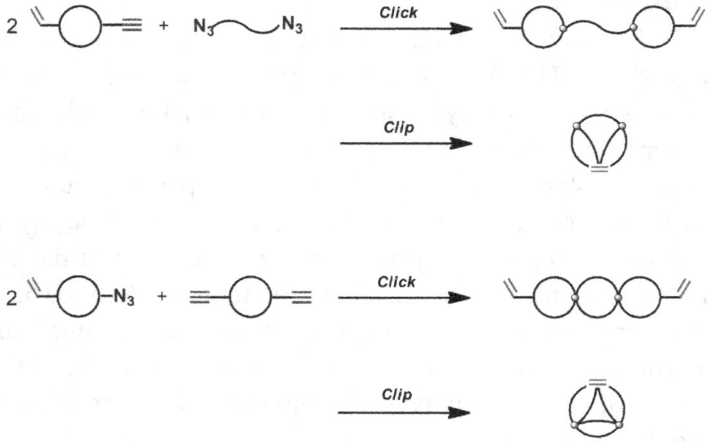

Scheme 7. Construction of *fused*-tricyclic (γ-graph) and *fused*-tetracyclic (unfolded tetrahedron) polymer topologies.

for an example, the triply-*fused* tetracyclic polymer exhibited the peak at 8996.3, which is assumed to be the adduct with Na^+, corresponds to the product possessing the expected chemical structure with the DP_n of 90; $(C_4H_8O) \times 90 + C_{146}H_{176}N_{12}O_{24}$, plus Na^+ equals 8995.476.[6] As this product is produced from the precursor by the elimination of an ethylene molecule, their molecular weights should differ by 28 mass units. This was confirmed by comparing the two mass spectra above.

The extent of the contraction of the 3D size for the tricyclic product from the dicyclic precursor, and for the tetracyclic one from the tricyclic counterpart was estimated by means of the SEC measurements. The noticeable reduction was observed from 0.82 to 0.60 for the former and from 0.63 to 0.52 for the latter, respectively. Accordingly, the programmed polymer folding could produce unusually compact polymer conformation in their 3D structures.

5. Ongoing Challenges and Future Perspectives

This Chapter presented a variety of complex multicyclic polymer topologies of either *spiro*, *bridged* or *fused* forms, obtained through the effective covalent linking chemistries, in particular an alkyne-azide *cross-coupling*, i.e., *click*, reaction and a metathesis *homo-coupling*, i.e., *clip*, reactions, by employing tailored single cyclic and dicyclic polymer precursors (*kyklo*-telechelics) as large as 300-membered rings obtained also by the ESA-CF protocol. Ongoing synthetic challenges in order to extend the current frontier of the synthetic polymer chemistry include such topologically significant but still elusive polymers constructions as an α-graph (doubly fused tricyclic), a $K_{3,3}$ graph (triply fused tetracyclic), and a prisman graph (triply fused tetracyclic). These complex polymers are also regarded as important model polymers to elucidate systematically any topology effects of basic polymer properties upon the precise structural parameters, like terminus and junctions. These could also benefit to create new properties and functions of polymers based on their topology effects, and will stimulate modeling and simulation studies now comparable with experimental results.

References

1. H. C. Kolb, M. G. Finn and K. B. Sharpless, *Angew. Chem. Int. Ed.*, **40**, 2004 (2001)
2. J.-F. Lutz, *Angew. Chem. Int. Ed.*, **46**, 1018 (2007).
3. B. S. Sumerlin and A. P. Vogt, *Macromolecules*, **43**, 1 (2010).
4. N. Sugai, H. Heguri, K. Ohta, Q. Meng, T. Yamamoto and Y. Tezuka, *J. Am. Chem. Soc.*, **132**, 14790 (2010).
5. Y. Tezuka and K. Fujiyama, *J. Am. Chem. Soc.*, **127**, 6266 (2005).
6. N. Sugai, H. Heguri, T. Yamamoto and Y. Tezuka, *J. Am. Chem. Soc.*, **133**, 19694 (2011).

PART II

Cyclic Polymers - Developments in the New Century -

CHAPTER 9

OVERVIEW ON PHYSICAL PROPERTIES OF CYCLIC POLYMERS

Jacques Roovers

Formerly at the National Research Council of Canada
Ottawa, Ontario. K1A 0R6
E-mail: JNRoovers@rogers.com

The random walk of a cyclic chain has fewer possible configurations than the linear chain. The cyclic chain has therefore the lower entropy. Ring formation induces a reduction in the size of the chain and affects the dynamical properties that depend on size. Interactions between two rings suffer constraints not found in the interaction between two linear chains. These constraints appear as an "exclude volume" effect. The constraints are most important in the dense melt and lead to a further reduction of the size of cyclic chains. Diffusion and relaxation processes are thereby also affected. Cyclic diblock copolymers prefer the disordered state and have a lesser tendency to form microphase separated morphologies than linear diblock copolymers. Micellization of cyclic diblocks is retarded and leads to weaker micelles.

1. Introduction

Cyclic polymers have once been proposed to explain the undetectable end-groups of high MW polymers. They have also been identified as low MW impurities in polycondensation products, but never played an important role. Flory did not consider them in his review of the subject.[1] The formation of ring polymers in polycondensation reactions was elucidated in 1950.[2,3] It was shown that a polycondensation with perfect stoichiometry and no side reactions could lead to 100% cyclic product. However, since the molar concentration of the cyclics depends on the −5/2 power of N, the number of repeat units in the ring, the product would be very polydisperse and contain mostly low molecular weight species N = 3, 4, etc. The method was, therefore, not considered efficient for preparing high MW ring polymers.

The discovery in the 1960s that viral particles contained a single cyclic DNA[4–6] elevated the study of cyclic polymers from a curiosity to a subject of great general interest. It is interesting to note that some of the first evidence for circularity was based on ultracentrifugation sedimentation of the DNA and comparison of the sedimentation coefficient with that of the once-cut, i.e. the linear form.[5–7] Shortly after, electron microscopy visually revealed the circular form.[8] The number of circular DNA molecules found from different sources in nature, including mitochondria, has since grown extensively.[9] Reviews about the special features of circular DNA have appeared.[10,11]

The end of the 1970s and beginning of the 1980s saw the first synthesis of ring polymers. Dodgson and Semlyen[12] used equilibrium polycondensation for the synthesis of cyclic polydimethylsiloxane under calculated ideal conditions. Geiser and Höcker[13] worked out the practical conditions for the cyclization of living dianionic polystyrenes. The method was improved and higher molecular weight samples could be produced.[14,15] All cyclic polymer samples needed purification and fractionation. Analysis was limited to SEC-DRI, with very poor resolution of cyclic and linear chains. Nevertheless, despite the lack of data on the cyclic purity of the samples, they form the basis from which our present knowledge about the dilute solution properties of cyclic polymers is derived.[16] A proper method to purify and to assay the cyclic purity has only been developed more recently by Cho *et al.*[17] The method combined with the new MALDI-TOF analysis of ring/linear polymer pairs holds great promise for advancing the synthesis, characterization and the study of the properties of cyclic polymers.

This Chapter forms a bridge between Part I and Part II. Several important properties of cyclic polymers are further developed in Chapters 13 to 16.

2. Topological Features of Ring Polymers

A ring is mathematically in a different topological class than linear and branched (without cyclic segments) polymers. As a result cyclic polymers can have two permanent features not found in linear polymers: they may be concatenated and they may have knots along the chain. Both have been found and created in DNA chains.[10] The cyclization of linear chains in the presence of overlapping rings can lead to concatenation. A statistical theory as a function of concentration and chain length and applied to DNA chains has been developed.[18,19] Static and dynamic structure factors for multiple m-membered rings are a prototype for catenated rings albeit with locked tie points.[20] Perhaps catenated rings have been formed undetected. One early publication aimed for it specifically but analysis of the reaction products remained difficult at that time.[21] More recently a c-PS-catena-c-PI was synthesized and well characterized.[22]

Permanently knotted ring polymers are a consequence of intramolecular entanglements of the linear chain at the moment of cyclization.[15,23] As a first order process it is independent of concentration but may become relevant for high molecular weight polymers with a small entanglement molecular weight. Frank-Kamenetskii[24] was the first to model the knotting probability of a ring polymer. Expansion to larger rings $N \leq 1600$ yields for the probability of the unknotted (trivial) ring:

$$P_N^0 = C_0 \exp(-\alpha N) \tag{1}$$

with $\alpha = 7.6 \ 10^{-6}$ and $C_0 \approx 1$.[25] Similar study confirmed that such simulations model the athermal SAW condition $<Rg^2>_r = N^{2v}$.[26] For N of the order of $1/\alpha$ the most abundant trefoil (31) knotted ring occurs with a probability:

$$P_{31}(N) \sim \alpha N \tag{2}$$

When $N \gg 1/\alpha$ the probability for the trefoil knot goes through a maximum and other knots appear. The probability of knots in phantom rings, where a lattice site may have multiple occupancies, is much higher of the order of $(2.5–5) \ 10^{-3}$.[27,28] The value of α has been recently established over the whole range from SAW to the collapsed state ($<Rg^2> \sim N^{2/3}$) as a function of the nearest neighbor contact potential and ranges from 10^{-6} (SAW) to $2 \cdot 10^{-3}$. In particular, for the θ-state SAW (α = 1–$2)10^{-4}$ with a steep dependence around the θ point.[29]

Static and dynamic structure factors of daisy rings, multiple rings with a central tie have been derived.[20,30] Daisy rings can be considered a model for knotted polymers with a tight knot.

The equilibrium dimensions $\langle R_g^2 \rangle$ of 10 different specific rings have been studied by numerical simulation.[31] There remains a difficult subject of the underlying problem of the knot localization. Two extremes are the tight knot, requiring the minimum amount of chain length and depending on the chain thickness, and the completely delocalized knot which is spread over the whole ring polymer. Simulations have established that SAW knotted rings contain weakly delocalized knots. The knotted part of the ring grows as $\langle l \rangle \sim N^{3/4}$.[32,33]

Knots in the θ-state are delocalized over the whole ring polymer.[34] The expansion of a knotted ring from the θ-state to the athermal good solvent may be increased because of simultaneous transition of the knot from delocalized to partly localized. A special technique (electrostatic approximation) was used to calculate [f] and [η] of specific types of knotted ring polymers, in swollen and under theta conditions.[35]

To date only one paper has dealt explicitly with identifying knotting in a synthetic polymer.[36] While the results remain qualitative the experiments were performed with extreme care. A small fraction of knotted rings were produced from a linear difunctional polystyrene (MW = 380k) in cyclohexane at 30 °C (i.e. 4.5 degrees below the θ temperature) and compared with the product of cyclization when carried out on the same polymer in a good solvent. The analysis is by high resolution SEC such that linear and cyclic homologues are clearly separated as well as both dimeric products. When combined with in-line multi-angle light scattering the ring polymer produced in cyclohexane shows a slightly wider elution curve with a constant MW across the whole elution band but a decreasing $\langle R_g^2 \rangle_r$ on the tailing edge. Assuming a 10% formation of knotted rings at $N = 730$ yields α = $1.4 \cdot 10^{-4}$ in fair agreement with Mansfield.[29] That no better separation is obtained seems to be in agreement with the small ratio of the intrinsic viscosity of the expected trefoil 3_1 ring and its unknotted counterpart.[35]

3. Dilute Solution Properties

The radius of gyration of the unperturbed phantom ring was calculated by Kramers[37] and Zimm.[38] The result is extremely simple: at constant MW

$$<Rg^2>_r = (1/2) <Rg^2>_l \qquad (3)$$

Of course, the relation does not apply to small rings that do not satisfy the statistical nature of the calculation. This rule generally applies to all comparisons of rings and linear polymer properties.

The 1960s was a period in which dilute solution properties of ring polymers were studied theoretically and compared to the properties of the linear homologues. These developments remain a guideline for the characterization of ring polymers. The expansion of the unperturbed ring polymer expressed as a series expansion in terms of $z = (3/2\pi <b^2>)^{3/2} N^{1/2} \beta$ where b^3 is the segment volume and β is the mutually excluded volume per pair of segments. In the case of cyclic polymer

$$\alpha^2 = <Rg^2>_r / <Rg^2>_{r,0} = 1 + (\pi/2) z + \dots \qquad (4)$$

compared to $1 + (134/105) z$ for the linear chain, suggesting that rings expand faster from their θ dimension than linear polymers.[39] The dependence of the scattered intensity on the scattering vector, $q = (4\pi/\lambda)\sin(\theta/2)$ has also been determined. It is observed that both the expansion coefficient and the scattering function of cyclic and regular 5-arm stars are similar as are their coil segment density.

Theoretical calculations for the dilute solution translational friction coefficient, [f], measured in sedimentation or diffusion experiments,[40,41] and the rotational friction coefficient as measured by the intrinsic viscosity $[\eta]$[42,43] were also developed. Such calculations refer to the unperturbed dimension and the non-free draining case. The respective ratios to the linear chain are

$$[s_0]_l / [s_0]_r = [f]_r / [f]_l = 8/3\pi = 0.849 \qquad (5)$$

$$[\eta]_r / [\eta]_l = 0.662 \qquad (6)$$

The introduction of excluded volume into the theory has led occasionally to opposite signs for whether rings expand hydrodynamically more or less than the linear chain.[44-46] This controversy has more to do with theoretical assumptions and approximations and conclusive experimental results remain to be obtained to settle the question. The experimental data at the θ condition as well as in good solvent mostly confirm the ratios given in Eqs. (3), (5) and (6).[16]

The topology of the ring has consequences in dilute solution: two ring polymers have more restricted possibility of approaching and interpenetrating each other. Des Cloiseaux and Mehta[28] conjectured that this would not change the exponent in the $<Rg^2>_r$ *versus* M^v relation of the isolated rings (zero concentration) neither in the unpertured state ($v = 1$) nor under the athermal condition ($v = 2 \cdot 0.588$). The topological constraints would have an effect similar to the excluded volume,[47] and does appear at finite concentration in the second virial coefficient, A_2, *etc*. This was confirmed by a lattice model treatment of rings in solution.[48]

Older experimental data found positive A_2 at the Flory θ temperature of ring polystyrenes and that $θ_r - θ = -6.0$ °C slightly increasing with molecular weight.[15,49] Newer experimental data on samples purified by liquid chromatography at the critical condition to better than 97% purity yield a constant $θ_r - θ = -6.8$ °C for polystyrene samples from 16,000 to 570,000.[50] Because the topological shift in the θ temperature is entropic in origin it is expected to be independent of solvent. This was indeed observed for in deuterated cyclohexane cyclic PS has a 6 °C lower $θ_r$ than in hydrogenous cyclohexane.[51] The value of A_2 at θ and $θ_r$ were calculated for ring PS in cyclohexane, the latter was found to have a shallow minimum at θ -7.5 °C at Mn = 60,000 and not to vary more than 2.5 °C between Mn = 10^4 and 10^6.[52] The θ point is not far removed from the upper critical solution temperature (UCST) and the $θ_r$ data implicitly indicate that the ring polymers are more soluble than the linear homologues. This is the reason why ring/linear mixtures can be fractionated at all.

At the critical condition, which is generally close to the θ condition, there is a balance between the volume excluded in a pore to a linear chain and weak adsorption of the chain on the pore surface. As a result the partition coefficient of linear polymers, $K_d = 1$, is independent of molecular weight and in a chromatographic experiment all linear chains elude at constant volume. At the same condition the partition coefficient of ring polymer is given by

$$K_d = 1 + \pi^{1/2} (R / D) \qquad (7)$$

where R/D is the ratio of radius of gyration over pore diameter.[53,54] The elution of the rings is retarded, the smaller rings eluding first. This has been experimentally verified for polystyrene rings.[55,56]

4. Ring Polymer Constraints in the Melt

4.1. *Constrained dimensions in the melt*

As the concentration of the rings increases the number of contacts and the topological constraints between rings grows. In the melt rings have a self-concentration in their pervaded volume, $<Rg^2>_r^{3/2}$, that decreases with MW because $<Rg^2>_r \sim M$. Simultaneously, the number of hetero-contacts must increase but these are subjected to topological constraints: concatenation is impossible and ring-ring perforation is required to be by a double chain, *etc.* Relative to linear chains, numerous conformations are thus excluded. As a result, with increasing MW the pervaded volume and the radius of gyration grow less rapidly for rings than for linear polymers in the melt.

Cates and Deutsch[57] have estimate the entropic deficit of a ring in its own melt and concluded that $<Rg^2>_r \sim M^{4/5}$ rather than $\sim M$ for linear polymers. They did not hold fast on the exact exponent as they were fully aware of the uncertainties in the underlying assumptions. The dimension of a ring in its own melt is expected to be less than in dilute solution at the θ-condition. A somewhat different approach by Suzuki[58] reached the conclusion that $<Rg^2>_r \sim M^{2/3}$ in the limit of high molecular weight. A long list of computer simulations has been devoted to the MW dependence of $<Rg^2>_r$ in the melt. Improved programming and, especially, increased computing power have since led to a more complete picture in which the $<Rg^2>_r$ on M dependence changes from the classical $<Rg^2>_r \sim M$ at low molecular weight to $<Rg^2>_r \sim M^{2/3}$ at high molecular weight.

We like to mention the atomistic molecular dynamics simulations for polyethylene chains by two different groups.[59–61] They reproduce the experimental radii of gyration of linear PE correctly. At low MW,

$<Rg^2>_r \sim M$ and equal to $\frac{1}{2} <Rg^2>_l$. However, even for the relatively low MW = 21,000 (200 Kuhn steps) downward curvature of $<Rg^2>_r$ is observed. Recent molecular dynamics simulations on four very high MW chains at $\Phi = 0.8$, where screening is complete, confirm that $<Rg^2>_r \sim M^{2/3}$ is reached for a ring equivalent to PE with MW = 114,000.[62] The authors point out that because $<Rg^2>_r \sim M^{2/3}$ does not mean that the rings are collapsed. They are somewhat crumpled but have plenty of volume that is invaded by neighboring rings. The rings remain entangled albeit less than linear chains.

Only one group of researchers has managed to measure experimentally the size of ring polydimethylsiloxane (deuterated) in ring polydimethylsiloxane (hydrogenous) by SANS and compare the results to identical measurements on linear mixtures.[63] The range of Kuhn steps varied from 17 to 52 and the observed $<Rg^2>_r \sim M^{0.42\pm0.05}$ agrees rather well with the computer results for these chain sizes. Furthermore, $<Rg^2>_r < 0.5 <Rg^2>_l$ and the exponent of the scattered intensity in the high q-range (>2) for rings qualitatively support the result.[64]

4.2. Blends of linear and cyclic polymers

At constant enthalpy, the entropic deficit of the ring has consequences in a number of transitions. In going from the melt to a more disordered state the entropy of the ring increases more than of the equivalent linear polymer. This has been discussed for the θ temperature and the underlying lower demixing temperature. This is also the case when a ring is diluted with linear chains. There is no topological constraint between a ring and a linear chain. Replacing rings with linear chains increases the entropy of the remaining rings. In the process the $<Rg^2>_r$ of the ring swells from its melt size to its unperturbed dimension.[65]

In blending a ring polymer with another linear polymer the ring polymer will appear "more soluble".[66] This was experimentally confirmed for the well-studied system polyvinylmethylether-PS. Ring PS has a 7–8 °C higher LCST than the comparable linear PS blend.[67] Similar observations were made on cyclic bisphenol-A-polycarbonate blends with PS.[68] The better solubility of the ring was assimilated in a lower interaction constant. A number of ring/linear blends have been compared to the linear/linear equivalents but the low molecular weights of samples make it most likely that other factors like end-groups of the linear, closure chemistry of the ring polymer, density differences, and enthalpic differences affect the results too much for solid conclusions. A recent study of blends of linear PS with low MW ring poly(oxyethylene) (PEO) used several techniques to probe the phase transformation in more detail.[69]

4.3. *Dynamic properties of ring polymer melts*

Dynamic processes like diffusion, relaxation and viscosity are expected to be affected by the architecture and topology of ring polymers. Presently the dynamic properties of linear and branched low molecular weight polymers melts conform to the Rouse model. There is general agreement that the Rouse model also applies to low molecular rings.[61,70,71] Accordingly, the diffusion coefficient depends on M and the melt viscosity depends on $<Rg^2>$ so that $(\eta_0)_r = \frac{1}{2} (\eta_0)_l$ in agreement with Eq. (3). The dynamic properties of high molecular weight entangled polymers are successfully modeled by combination of reptation and constraint release mechanisms. However, because both mechanisms rely on the presence of chain ends they can not readily be transposed to cyclic polymers. The earliest zero-shear melt viscosity data on rings were contradictory. From one set the melt viscosity of ring and linear polymers is found similar depending on $M^{3.5}$.[71] The other set of data suggests that the melt viscosity of rings is lower with weaker molecular weight dependence.[70] The observed strong increase of the $(\eta_0)_r$ with small additions of linear "impurities"[72,73] implicitly favors $(\eta_0)_r < (\eta_0)_l$. Theoretical work proposes now a model that represents ring polymers as "lattice animal" with a longest relaxation time and melt viscosity dependent on $M^{5/2}$.[74-76] Limited recent results confirm this newer theoretical result.[77] Nevertheless, the theory still needs to explore mutual ring effects and better experimental data on certified-pure higher molecular weight rings are required. The concept and the role of "entanglements" between rings need to be clarified. Newer development of the dynamic properties of ring polymers are reviewed in detail in Chapters 13 and 14.

4.4. *Crystallization of cyclic polymers*

Molecularly homogeneous cyclic polyethyleneoxides (PEO) containing 18 and 27 monomer units (Mn = 792 and 1188 resp.) crystallize as loops with the same structure as the corresponding linear oligomers that crystallize in the extended chain form. The differences in enthalpy and entropy of melting reflect on the difference between two loops and two end groups per stem.[78,79] Lower molecular weight cyclic PEOs have altogether different crystalline structures.[79]

Linear polyethyleneglycols (PEG) which have a finite molecular weight distribution crystallize in the extended chain form. The lamellar thickness is equal to the length of the 7/2 helix in which each atom contributes 0.95 Å. When Mn is larger than 4000 the lamellae can be obtained in different thicknesses. Annealing at progressively higher temperatures decreases the integral number of chain folds and increases the lamellar thickness. The cyclic PEGs produce in all cases lamellae of looped chains. Their thickness is equal to that of the linear PEG with half the molecular weight.[80–83] Larger cyclics have melting temperatures that are equal to those of linear PEGs with the same lamellar thickness but their enthalpy of melting is much lower. This indicates that they form less ordered crystals and for this reason have the lower entropy of melting than the linear.

Extrapolation of the heat of melting, corrected to 70 °C and to infinite lamellar thickness, indicates an identical equilibrium heat of fusion for linear and cyclic PEGs of low molecular weight (Mn < 3000). Since the cyclics melt at a higher temperature than the linear at equal lamellar thickness it follows that according to

$$Tm^0 = \Delta Hm^0 / \Delta Sm^0 \qquad (8)$$

the entropy of melting is less for the cyclic than of the linear form. Assuming that both have the same entropy in the crystal, the entropy of the ring is lower than for the linear form in the melt.[83]

The crystallization rates of low Mw linear and cyclic poly(dimethylsiloxane)s (PDMS) were studied by non-isothermal DSC. Linear PDMS with Mn > 2200 and cyclic PDMS with Mn > 3100 cold crystallize after quenching when scanned at 10 °C/min. At intermediate MW, PDMS with 1630<Mn<2190 and cyclic with 2410<Mn<2744 show only cold crystallization when slowly scanned at 2 °C/min. Linear PDMS between 530<Mn<1630 and cyclics with 880<Mn<1730 could not be crystallized.[84,85] The results suggest that cyclic PDMS crystallize more difficultly than linear chains. The cold crystallization studied is strongly affected by chain mobility. The lower rate of crystallization of some cyclic PDMS is most likely due to the higher glass transition temperature (Tg) for rings than linear fractions of the same molecular weight. The Tg of linear PDMS decreases with decreasing MW — the normal "free volume" effect – but the Tg of cyclics increases with decreasing MW. The Tg of cyclics is therefore always higher than for linear low MW fractions.[84]

A cyclic polyethylene (Mw = 200,000) is found to crystallize and melt at 115 °C/132 °C compared to a linear PE at 113 °C/130 °C. The crystallinity was the same in both samples.[86]

Cyclic and linear poly(ε-caprolactone)s, P(ε–CL), with molecular weight between 75 000 and 140 000 (DPI between 1.3 and 2.0) were prepared by zwitterionic polymerization and characterized by SEC-[η] and SEC-LS as essentially the expected architecture.[87] WAXS identified that the unit cell of the two forms are identical and time resolved SAXS during isothermal crystallization proved that the lamellar spacing and the lamellar thickness (20 nm and 9 nm respectively) were similar indicating that the cyclic polymer is also incorporated in the lamellae with multiple folds and can span multiple lamellae. The time resolution data established that the cyclic polymer crystallizes faster than the linear polymer. This was further confirmed by DSC isothermal crystallizations and the determination of the $t_{1/2}$ for different molecular weights. Extrapolation of the Tm after isothermal crystallization as a function of crystallization temperature (Weeks plot) established that the equilibrium Tm^0 of the cyclic polymer is about 2 degrees higher than that of the linear polymer. Attempts to extrapolate the melting temperature as a function of the inverse lamellar thickness to obtain Tm^0 failed in the case of the cyclic polymer. As in the case of the cyclic PEOs the higher Tm^0 at constant enthalpy of melting suggests that the entropy of melting is less for the cyclic than for the linear polymer in agreement with a lower entropy of a cyclic polymer in the melt.

The faster crystallization of cyclic P(ϵ–CL)s than linear polymers, observed also by several other authors,[88,89] may be due to a stronger undercooling and/or lower melt viscosity. The lower viscosity is possibly due to partial collapse of the ring conformation that contributes to lesser entanglement.[87]

Recently, an interesting difference was observed between the spherulites of linear polytetrahydrofuran (PTHF) and the perfectly cyclic polymer. While the former crystallizes into classic negative birefringent spherulites the latter forms banded spherulites with a radial pitch. The spherulites of the cyclic PTHF grow more slowly and are less perfect as suggested by their 5 °C lower Tm.[90] A complete review of crystallization of ring polymers is presented in Chapter 15.

5. Cyclic Block Copolymers

5.1. *Microphase separation*

The order-disorder temperature (ODT) of a block copolymer indicates the endothermic transition from the microphase separated state to the melt state. The transition is governed by the interaction parameter χ of the A and B segments. χ is considered independent of the polymer architecture, composition and, to a first approximation, of the molecular weight of the block copolymer, but is temperature dependent. At the critical (spinodal) temperature, T_c:

$$\chi = A^* / T_c + B^* = C^* / N. \qquad (9)$$

$N = nA + nB$ is the sum of the number of A and B segments expressed in terms of a reference volume. A*, B* and C* are constants. C* depends on the architecture: is equal to 2 for homopolymers, 10.5 for diblocks,[91] 17.9 for symmetric triblocks[92] and 17.5 for cyclic diblocks.[93] From Eq. (9) it can be seen that T_c is lower for a cyclic than for a diblock but similar to a symmetric triblock copolymer. A lower T_c indicates that the cyclic diblock transfers more easily to the disordered melt. Morozov[94] established that the complete microphase diagram of the cyclic diblock is very similar to that of the linear diblock being displaced by about 7 units along the χN axis from the phase diagram of the linear diblock. A simulation by the dissipative particle dynamics method of a cyclic diblock with $N = 20$ over a wide range of the segment ratio A/B allowed the observation of lamellae, hexagonal cylinders and body centered cubic regions as well as transition zones with perforated lamellae, random networks, liquid rods and micelle-like dispersions.[95]

Two cyclic poly(ethyleneoxide-b-butyleneoxide) copolymers, c-P(EO-b-BO), ($\Phi_{EO} \approx 0.5$) have been submitted to ODT/DOT measurement near the spinodal. Compared to the linear diblock the ODT is 100 and 120 °C lower than the linear diblock and slightly lower than the parent triblock copolymers.[96] The PEO fraction of two other c-P(EO-b-BO) of about one half the MW crystallized before the ODT could be observed but it is expected that they would remain disordered because their χN is less than 18. The ODT of an asymmetric cyclic poly(styrene-b-isoprene), c-P(S-b-I), with $\Phi_{PS} \approx 0.78$ was found to be 157 °C, that of the linear precursor diblock 177 °C, but it is uncertain that the latter represents the true ODT.[97]

The lamellar domain spacing of the cyclic block is 0.64 ($1\sqrt{2}$) times smaller than that of the linear diblock of same MW and composition[93] and, in the limit of high N, the so called strong segregation limit, equal to that of the symmetric triblock.[98] Experimental results for domain spacings are most abundant for near symmetric cyclic diblock copolymers and compared to the parent linear triblock copolymers. The ratio D_c/D_l is mostly between 0.86 and 1. At least three different models have been proposed to explain the results. The ratio D_c/D_l increases to 1 with increasing molecular weight as the cyclic diblock and linear triblock move from the weak and intermediate to the strong segregation limit.[98] Alternatively, D_c/D_l is less for the cyclic diblock than the triblock because the two hetero-links of the diblock are bound to be on the same interface, the cyclic diblock forming two loops, while it is known that 40% of linear triblocks are more extended because they span the micro-phase when the two hetero-junctions are located on opposite interfaces.[99,100] The latter effect however is quite independent of MW. There is a more recent proposal that, supported by experimental data over a wide MW range (including the weak segregation region), suggests that the chain extension in linear triblock $D \sim N^{2/3}$ is reduced in the case of cyclic diblocks to $D \sim N^{3/5}$ because in the melt the entropic penalty of the cyclic diblock leads from $R \sim N^{1/2}$ to $R \sim N^{3/5}$.[101] The present experimental evidence does not clearly favor a particular model.

Because the phase boundaries of the different microphase structures differ between the various architectures it is occasionally possible to observe a different microstructure for the asymmetric ($\Phi_A > \Phi_B$) cyclic and the parent symmetric triblock[100] or diblock.[97] Ohta and coworkers[22] managed to prepare a small amount of c-PS-catenated-c-PI with 50 w% PS and obtained a lamellar morphology with a 20 nm repeat length. This length is intermediate between that for the linear diblock (25 nm) and the cyclic diblock (16 nm). This result clearly indicates that the catenated double ring topology tries to limit the styrene-isoprene contacts into a narrower region than the cyclic diblock but wider than the single contact in the linear diblock.

5.2. *Micellation of cyclic block copolymers*

Block copolymers when placed in a selective solvent for one of the blocks associate and form, most often spherical, micelles. A model predicts that the critical micelle concentration (CMC) of the cyclic polymer c-(A_n-b-B_m) is slightly lower than that of the parent $A_{n/2}$-b-B_m-b-$A_{n/2}$ triblock.[102] Both are much higher than for the linear diblock. This confirms that the cyclic diblock prefers the disordered conformation. The small difference with the triblock is due to the smaller entropic penalty to form a micelle from a ring polymer: the conformation of the ring polymer in solution resembles more the conformation in the micelle. The average association number in the micelle is in the order diblock > c-diblock > triblock, $P_D > P_C > P_T$. The cyclic diblocks form the larger micelles.

The micellation of a cyclic diblock c-P(S-b-Bd) with 54 w% PS-d_8 was studied in n-decane, a selective solvent for PBd (PBd forming the swollen corona), and in DMF, a selective solvent for PS (PS forming the corona).[103] SANS with contrast matching was used to obtain separately the core radius R_c and the overall static radius R_0 of the micelles. The thickness of the corona is given by $R_0 - R_c = H$. The corona of the cyclic diblock is very nearly equal to that of the half block and slightly smaller than the corona of the end-block collapsed triblock copolymers in which a majority of the center block is also expected to loop in the corona. Moreover, the ratio of H of the cyclic diblock to the linear diblock is 0.62 which is slightly less than the ratio of radii of gyration of a cyclic over a linear polymer $(1/\sqrt{2})$. In all cases H is larger than Rg indicating that the chains in the corona are stretched. The density profile in the corona of the cyclic diblock is constant, $(\rho(r) \sim \text{constant})$, rather than star-like $(\rho \sim r^{-4/3})$ in the linear diblock.

It was found that, in contrast to the very stable micelles of linear diblock P(S-b-I) with 80% styrene in the core, the micelles of the cyclic diblock easily further coagulate from spherical into wormlike cylindrical threads.[104,105] From the results by Iatrou[103] it is know that the looped chains of the cyclic copolymer form a thinner corona than in the diblock of the same MW and the low soluble fraction (isoprene in this case) in the corona is only borderline able to stabilize the spherical micelles. Cohesive collisions will convert the spheres into cylinders. A comparison of the c-P(S-b-I) with that of the triblock copolymer P(S-b-I-b-S) would be interesting in this regard.

Amphiphilic block copolymers are of great technological importance. The micellation in water of cyclic block copolymers has been compared with those of the linear diblock and triblocks. Most detailed data are from the in-depth study of cyclic P(EO_{42}-b-BO_8) (42 ethylene oxyde and 8 butylene oxide units) in comparison with P(EO_{41}-b-BO_8) and P(EO_{21}-b-BO_8-b-EO_{21}).[106] The CMC of the cyclic and triblock are very similar, but ten times higher than for the diblock. The temperature dependence of the CMC allows the calculation of the endothermic enthalpy of micellation which is smaller for the cyclic than for the triblock, probably due to the reduced hydrophobic effect on BO_8 in the cyclic form.[106] Further improved results were obtained on a larger pair of c-P(EO_{144}-b-BO_{27}) and P(EO_{72}-b-BO_{27}-b-EO_{72}) which have a narrower distribution in the PEO block length and form larger micelles and, therefore, show better the difference in CMC (the CMC of the cyclic is about half that of the linear). The cyclic diblocks have larger radii and higher association numbers than the triblock micelles.[107] The trend is confirmed with a pair of c-P(EO$_{104}$-b-PO$_{34}$) and P(EO$_{52}$-b-PO$_{34}$-b-EO$_{52}$). The CMC is about 100 times higher for propylene oxide than butylenes oxide block copolymers, e.g. at 55 °C the CMC of the cyclic is 1.1 g/L and 1.2 g/L for the linear triblock. At 45 °C the CMCs are 4.5 and 9 g/L respectively.[108] All these experimental results support the theoretical model.[102] Full detailed on these systems have been reviewed.[109]

The micelles of cyclic poly(butylacrylate-b-ethylene oxide), c-P(BA$_{12}$-b-EO$_{59}$), and P(BA$_6$-b-EO$_{59}$-b-BA$_6$) have a CMC of about 0.1 g/L and a LCST at 24 °C. On further increasing the temperature the micelles from the linear triblock form linear aggregates as indicated by the cloud point at 30 °C but the micelles from the cyclic diblock remain stable till 70 °C.[110]

Two different laboratories obtained cyclic poly(N-isopropylacrylamide), P(NIPAM). P(NIPAM) has internal amphiphilicity and is soluble in water at low temperature but has a LCST at 32 °C or above depending on concentration, molecular weight and end-groups. The LCST, measured at 1% turbidity and extrapolated to zero concentration, is slightly higher for the cyclic than for the linear block copolymer.[111] At finite concentration the dependences of the LCST can be reversed.[112] Similar phenomena are also observed for the cloud point temperature as measure by the 50% turbidity temperature. Calorimetric measurements of the cloud point clearly show that the phase transition is less endothermic for the cyclic than for the linear polymer. Ye et al.[113] further compared the formation and size of the particles formed by cyclic and linear P(NIPAM)s. All aspects of micellation are further reviewed in Chapter 16.

References

1. P. J. Flory, *Chem. Rev.*, **39**, 137 (1940).
2. H. Jacobson and W. H. Stockmayer, *J. Chem. Phys.*, **18**, 1600 (1950).
3. H. Jacobson and W. H. Stockmayer, *J. Chem. Phys.*, **18**, 1607 (1950).
4. W. Fiers and R. L. Sinsheimer, R. L. *J. Mol. Biol.*, **5**, 408 (1962).
5. R. Dulbecco and M. Vogt, *Proc. Natl. Acad. Sci., USA*, **50**, 236 (1963).
6. R. Weil and J. Vinograd, *Proc. Natl. Acad. Sci., N. Y.* **50**, 730 (1963).
7. W. Fiers and R. L. Sinsheimer, *J. Mol. Biol.*, **5**, 424 (1962).
8. D. Freifelder, A. K. Kleinschmidt and R. L. Sinsheimer, *Science*, 146, 254 (1964).

9. W. R. Bauer and J. Vinograd, in (1974) *Basic Principles in Nucleic Acid Chemistry*, 2, ed. Ts'o, P. O. P., Chapter 4 "Circular DNA," (Academic Press, New York, 1974) p. 265.

10. J. C. Wang, in *Cyclic Polymers*, ed. Semlyen, J. A., Chapter 7 "Circular DNA," (Elsevier Applied Science Publisher, London, 1986) p. 225.

11. A. V. Vologodskii, *Cyclic Polymers*, 2nd Ed., ed. Semlyen, J. A., Chapter 2 "Circular DNA," (Kluwer Academic Publishers, Dordrecht, The Netherlands, 2000) pp. 47.

12. K. Dodgson and J. A. Semlyen, *Polymer*, **18**, 1265 (1977).

13. D. Geiser and H. Höcker, *Macromolecules*, **13**, 653 (1980).

14. G. Hild, A. Kohler and P. Rempp, *Eur. Polym. J.*, **16**, 525 (1980).

15. J. Roovers and P. M. Toporowski, *Macromolecules*, **16**, 843 (1983).

16. J. Roovers, in *Cyclic Polymers*, 2nd Ed., ed. Semlyen J.A. (Kluwer Academic Publishers Dordrecht. The Netherlands, 2000) Chapter 10, p. 347.

17. D. Cho, S. Park, K. Kwon, T. Chang and J. Roovers, *Macromolecules*, **34**, 7570 (2001).

18. H. Jacobson, *Macromolecules*, **17**, 705 (1984).

19. H. Jacobson, *Macromolecules*, **21**, 2842 (1988).

20. W. Burchard, E. Michel and V. Trappe, *Macromolecules*, **29**, 5934 (1996).

21. B. Vollmert and J.-X. Huang, *Makromol. Chem., Rapid Commun.*, **2**, 467 (1981).

22. Y. Ohta, Y. Kushida, D. Kawaguchi, Y. Matsushita and A. Takano, *Macromolecules*, **41**, 3957 (2008).

23. F. Brochard and P. G. de Gennes, *Macromolecules*, **10**, 1157. (1977).

24. M. D. Frank-Kamenetskii, A. V. Lukashin and A. V. Vologodskii, *Nature*, **258**, 398 (1975).

25. Janse van Rensburg, E .J. and S.G. Whittington, *J. Phys. A: Math. Gen.*, **23**, 3573 (1990).

26. A. Yao, H. Matsuda, H. Tsukahara, M. K. Shimamura and T. Deguchi, *J. Phys. A: Math. Gen.*, **34**, 7563 (2001).

27. A. V. Vologodskii, A. V. Lukshin, M. D. Frank-Kamenetskii and V. V. Anshelelevich, *Sov. Phys., JETP*, **39**, 1059 (1974).

28. J. des Cloiseaux, *J. de Phys.*, **42**, L-433 (1981).

29. M. L. Mansfield, *J. Chem. Phys.*, **127**, 244902 (2007).

30. D. W. Matuschek and A. Blumen, *Macromolecules*, **22**, 1490 (1989).

31. M. L. Mansfield and J. F. Douglas, *J. Chem. Phys.*, **133**, 044903 (2010).

32. B. Marcone, E. Orlandini, A. L. Stella and F. Zonta, *J. Phys., A: Math. Gen.*, **38**, L15 (2005).

33. B. Marcone, E. Orlandini, A. L. Stella and F. Zonta, *Phys. Rev. E*, **75**, 041105. (2007).

34. E. Orlandini, A. L. Stella and C. Vanderzande, *Phys. Rev., E*, **68**, 031804 (2004).

35. M. L. Mansfield and J. F. Douglas, *J. Chem. Phys.*, **133**, 044904 (2010).

36. Y. Ohta, Y. Kushida, Y. Matsushita and A. Takano, *Polymer*, **50**, 1297 (2009).

37. H. A. Kramers, *J. Chem. Phys.*, **14**, 415 (1946).
38. B. H. Zimm and W. H. Stockmayer, *J. Chem. Phys.*, **14**, 1301 (1949).
39. E. F. Casassa, *J. Polym. Sci.: Part A*, **3**, 605 (1965).
40. O. B. Ptitsyn, *Sov. Phys., Techn. Papers*, **4**, 65 (1959).
41. M. Fukatsu and M. Kurata, *J. Chem. Phys.*, **41**, 4539 (1966).
42. V. Bloomfield and B. H. Zimm, *J. Chem. Phys.*, **44**, 315 (1966).
43. G. Tanaka and H. Yamakawa, *Polym. J.*, **4**, 446 (1972).
44. H. Yamakawa, in *Modern Theory of Polymer Solutions*. (Harper and Row, N. Y., 1971) p. 321.
45. T. Norisuye and H. Fujita, *J. Polym. Sci.: Polym. Phys. Ed.*, **16**, 999 (1978).
46. J. Shimada and H. Yamakawa, *J. Polym. Sci.: Polym. Phys. Ed.*, **16**, 1927 (1978).
47. J. des Cloiseaux, *J. de Phys.*, **42**, L-433 (1981).
48. J. Léonard, *J. Phys. Chem.*, **93**, 4346 (1989).
49. J. Roovers and P. Toporowski, *J. Polym. Sci.: Poly. Phys. Ed.*, **23**, 1117 (1985).
50. A. Takano, Y. Kushida, Y. Ohta, K. Masuoka and Y. Matsushita, *Polymer*, **50**, 1300 (2009).
51. P. Lutz, G. B. McKenna, P. Rempp and C. Strazielle, *Macromol. Chem., Rapid Commun.*, **7**, 599 (1986).
52. K. Iwata, *Macromolecules*, **22**, 3702 (1989).
53. A. A. Gorbunov and A. M. Skvortsov, *Polym. Sci., U.S.S.R.*, **26**, 2305 (1984).
54. A. A. Gorbunov and A. M. Skvortsov, *Adv. Colloid Interface Sci.*, **62**, 31 (1995).
55. H. C. Lee, H. Lee, W. Lee, T. Chang and J. Roovers, *Macromolecules*, **33**, 8119 (2000).
56. W. Lee, H. Lee, H. C. Lee, D. Cho, T. Chang, A. A. Gorbunov and J. Roovers, *Macromolecules*, **35**, 529 (2002).
57. M. E. Cates and J. M. Deutsch, *J. de Phys.*, **47**, 2121 (1986).
58. J. Suzuki, A. Takano, T. Deguchi and Y. Matsushita, *J. Chem. Phys.*, **131**, 144902. (2009).
59. K. Hur, R. G. Winkler and D. Y. Yoon, *Macromolecules,* **39**, 3975 (2006).
60. K. Hur, C. Jeong, R. G. Winkler, N. Lacevic, R. H. Gee and D. Y. Yoon, *Macromolecules*, **44**, 2311 (2011).
61. G. Tsolou, N. Stratikis, C. Baig, P. S. Stephanou and V. G. Mavrantzas, *Macromolecules*, **43**, 10692 (2010).
62. J. D. Halverson, W. B. Lee, G. S. Grest, A. Y. Grosberg and K. Kremer, *J. Chem. Phys.*, **134**, 204904 (2011).
63. V. Arrighi, G. Gagliardi, A. D. Dagger, J. A. Semlyen, J. S. Higgins and M. J. Shenton, *Macromolecules*, **37**, 8057 (2004).
64. G. Gagliardi, V. Arrighi, R. Ferguson, A. D. Dagger, J. A. Semlyen and J. S. Higgins, *J. Chem. Phys.*, **122**, 064904 (2005).
65. B.V.S. Iyer, A.K. Lele and S. Shanbhag, *Macromolecules*, **40**, 5995 (2007).
66. A. R. Khokhlov and S. K. Necheav, *J. de Phys., II France*, **6**, 1547 (1996).
67. M. Santore, C. C. Han and G. M. McKenna, *Macromolecules*, **25**, 3416 (1992).

68. W. L. Nachlis, J. T. Bendler, R. P. Kambour and W. J. McKnight, *Macromolecules*, **28**, 7869 (1995).
69. S. Singla and H. W. Beckham, *Macromolecules*, **41**, 9784 (2008).
70. J. Roovers, *Macromolecules*, **18**, 1359. (1985).
71. G. B. McKenna, B. J. Hostetter, N. Hadjichristidis, L J. Fetters and D. J. Plazek, *Macromolecules*, **22**, 1834 (1989).
72. G. B. McKenna and D. J. Plazek, *Polym. Commun.*, **27**, 304 (1986).
73. J. Roovers, *Macromolecules*, **21**, 1517 (1988).
74. M. Rubinstein, *Phys. Rev. Lett.*, **57**, 3023 (1986).
75. S. P. Obukhov, M. Rubinstein and T. Duke, *Phys. Rev. Lett.*, **73**, 1263 (1994).
76. S. T. Milner and J. D. Newhall, *Phys. Rev. Lett.*, **105**, 208302 (2010).
77. M. Kapnistos, M. Lang, D. Vlassopoulos, W. Pyckhout-Hintzen, D. Richter, D. Cho. T. Chang and M. Rubinstein, *Nat. Mater.*, **7**, 997 (2008).
78. Z. Yang, G.-E. Yu, J. Cooke, K. Viras, H. Matsuura, A. J. Ryan and C. Booth, *J. Chem. Soc., Faraday Trans.*, **92**, 3173 (1996).
79. Z. Yang, J. Cooke, K. Viras, P. A. Gorry and C. Booth, *J. Chem. Soc., Faraday Trans.*, **93**, 4033 (1997).
80. T. Sun, G.-E. Yu, C. Price, C. Booth, J. Cooke and A. J. Ryan, *Polym. Commun.*, **36**, 3775 (1995).
81. K. Viras, Z.-G. Yan, C. Price, C. Booth and A. J. Ryan, *Macromolecules*, **28**, 104 (1995).
82. G.-E. Yu, T. Sun, Z.-G. Yan, C. Price, C. Booth, J. Cooke, A. J. Ryan and K. Viras, *Polymer*, **38**, 35 (1997).
83. J. Cooke, K. Viras, G.-E.. Yu, T. Sun, T. Yonemitsu, A. J. Ryan, C. Price and C. Booth, *Macromolecules*, **31**, 3030 (1998).
84. S. J. Clarson, K. Dodgson and J. A. Semlyen, *Polymer*, **26**, 930 (1985).
85. D. J. Orrah, J. A. Semlyen, K. Dodgson and S. B. Ross-Murphy, *Polymer*, **28**, 985 (1987).
86. C. W. Bielawski, D. Benitez and R. H. Grubbs, *Science*, **297**, 2041 (2002).
87. E. J. Shin, W. Jeong, H. A. Brown, B. J. Koo, J. L. Hedrick and R. M. Waymouth, *Macromolecules*, **44**, 2773 (2011).
88. M. E. Córdova, A. T. Lorenzo, A. Müller, J. N. Hoskins and S. M. Grayson, *Macromolecules*, **44**, 1742 (2011).
89. K. Schäler, E. Ostas, K. Schröder, T. Thurm-Albrecht, W. H. Binder and K. Saalwächter, *Macromolecules*, **44**, 2743 (2011).
90. Y. Tezuka, T. Ohtsuka, K. Adachi, R. Komiya, N. Ohno and N. Okui, *Macromol. Rapid Commun.*, **29**, 1237 (2008).
91. L. M. Leibler, *Macromolecules*, **13**, 1602 (1980).
92. A. M. Mayes and M. Olvera de la Cruz, *J. Chem. Phys.*, **91**, 7227 (1989).
93. J. F. Marko, *Macromolecules*, **26**, 1442 (1993).
94. A. N. Morozov and J. G. E. M. Fraaije, *Macromolecules*, **34**, 1526 (2001).

95. H.-J. Qian, Z.-Y. Lu, L.-J. Chen, Z.-S. Li and C.-C. Sun, *Macromolecules*, **38**, 1395 (2005).
96. A. J. Ryan, S.-M. Mai, J. Patrick, A. Fairclough, I. W. Hamley and C. Booth, *Phys. Chem. Chem. Phys.*, **3**, 2961 (2001).
97. S. Lecommandoux, R. Borsali, M. Schappacher, A. Deffieux, T. Narayanan and C. Rochas, *Macromolecules*, **37**, 1843 (2004).
98. R. L. Lescanec, D. A. Hajduk, G. Y. Kim, Y. Gan, R. Yin, S. M. Gruner, T. E. Hogen-Esch and E. L. Thomas, *Macromolecules*, **28**, 3485 (1995).
99. M. W. Matsen and M. Schick, *Macromolecules*, **27**, 187 (1994).
100. Y. Zhu, S. P. Gido, H. Iatrou, N. Hadjichristidis and J. W. Mays, *Macromolecules*, **36**, 148 (2003).
101. Y. Matsushita, H. Iwata, T. Asari, T. Uchida, G. ten Brinke and A. Takano, *J. Chem. Phys.*, **121**, 1129 (2004).
102. K. H. Kim, J. Huh and W. H. Jo, *J. Chem. Phys.*, **118**, 8468 (2003).
103. H. Iatrou, N. Hadjichristidis, G. Meier, H. Frielinghaus and M. Monkenbusch, *Macromolecules*, **35**, 5426 (2002).
104. E. Minatti, R. Borsali, M. Schappacher, A. Deffieux, V. Soldi, T. Narayanam and J.-L. Puteaux, *Macromol. Rapid Commun.*, **23**, 978 (2002).
105. E. Minatti, P. Viville, R. Borsali, M. Schappacher, A. Deffieux and R. Lazzaroni, *Macromolecules*, **36**, 4125 (2003).
106. G.-E. Yu, Z. Yang, D. Attwood, C. Price and C. Booth, *Macromolecules*, **29**, 8479 (1996).
107. G.-E. Yu, Z. K. Zhou, D. Attwood, C. Price, C. Booth, P. C. Griffiths and P. Stilbs, *J. Chem. Soc., Faraday Trans.*, **92**, 5021 (1996).
108. G.-E. Yu, C. A. Garrett, S.-M. Mai, H. Altinok, D. Attwood, C. Price and C. Booth, *Langmuir*, **14**, 2278 (1998).
109. C. Booth and D. Attwood, *Macromol. Rapid Commun.*, **21**, 501 (2000).
110. S. Honda, T. Yamamoto and Y. Tezuka, *J. Am. Chem. Soc.*, **132**, 10251 (2010).
111. J. Xu, J. Ye and S. Liu, *Macromolecules*, **40**, 7069 (2007).
112. X.-P. Qiu, F. Tanaka and F. M. Winnik, *Macromolecules*, **40**, 7069 (2007).
113. J. Ye, J. Xu, J. Hu, X. Wang, G. Zhang, S. Liu and C. Wu, *Macromolecules*, **41**, 4416 (2008).

CHAPTER 10

THE RING-CLOSURE APPROACH FOR SYNTHESIZING CYCLIC POLYMERS

Boyu Zhang and Scott M. Grayson

Department of Chemistry, Tulane University
New Orleans, LA 70118, USA
E-mail: sgrayson@tulane.edu

The ring-closure approach remains the most commonly used approach for the preparation of cyclic polymers. In the last decade, a number of developments within the field of polymer chemistry have resulted in a number of innovations within the subfield of cyclic polymers, particularly in regards to improved methods of their synthesis. This chapter aims to highlight the advances in the ring closure approach, within the context of previous techniques, and with respect to alternative techniques for preparing cyclic polymers. While the ring expansion technique exhibits advantages to prepare extremely large cyclic polymers, the ring closure approach using highly activated coupling reactions remains the most versatile technique for preparing smaller, well-defined polymer macrocycles with diverse chemical composition.

1. Introduction

Precision control over polymer architecture has been a longstanding goal for polymer chemists due to the fact that the physical properties of macromolecules are inherently dependent on their structure and connectivity on the nanoscale. The wide diversity of polymer architecture, including linear polymers, polymer brushes, star polymers, ladder polymers, dendrimers, hyperbranched polymers, and network polymers has been studied to elucidate the dependence of their physical properties (e.g. glass transition temperatures (T_g), melt transition temperatures (T_m), intrinsic and melt viscosities, thermostabilities, solubilities, rheological properties, viscoelastic properties, etc.) and chemical properties (e.g. both reactivity and stability) on their covalent structure. For all of the abovementioned polymer architectures, their end groups play a substantial role in their observed physical properties, highlighting the uniqueness of cyclic polymers, which inherently lack end groups. However, synthetic challenges in preparing polymer macrocycles have hampered the detailed investigation of their physical properties. Historically, synthetic techniques for preparing cyclic polymers have suffered from low yields of the desired cyclic macromolecule, a lack of functional group compatibility, and difficulty in scaling up to larger amounts. The most significant challenge is the removal of linear impurities often unavoidable in the cyclic synthesis. Linear impurities skew the observed physical properties of a given sample, obscuring the fundamental architecture-based trends. Hence, better synthetic techniques that yield high-purity cyclic materials have remained an important target within the macromolecular synthesis community.

Paul Ruggli[1] and Karl Ziegler[2] first demonstrated that high dilution could be used to favor intramolecular cyclization events for small organic molecules with complementary functionalities. This arises from the fact that when such molecules are diluted, the effective concentration of reactive functionalities on the same molecule remains high since they are covalently tethered within reactive proximity. The thermodynamic trends are well known: intermediate rings with 5 or 6 atoms can be prepared efficiently due to the minimization of ring strain while the cyclization of small rings consisting of 3–4 covalent bonds are disfavored due to Baeyer strain, and the cyclization of rings comprising of 7–13 covalent bonds are discouraged by Pitzer and transannular strains. However, the entropy for such cyclizations becomes a more important consideration than strain for larger rings. Ruzicka predicted in the 1930s that the synthesis of larger cyclic molecules would be subject to high entropic penalties even though the conformational flexibility of much larger rings also results in negligible strain energies.[3] However, decades later, the discovery and structural determination of cyclic biomacromolecules such as gramicidin S, a cyclic peptide,[4] and cyclic DNA,[5] confirmed that the entropic penalties could be overcome, and affirmed the feasibility of synthesizing cyclic macromolecules.

The first examples of synthetic cyclic polymers were prepared via the ring-chain equilibrium approach. Early studies on condensation polymerization yielded trace amounts of cyclic oligomers as byproducts. A representative example of this was demonstrated by Ross *et al.* during their polymerization of poly(polyethylene terephthalate), which yielded a few percent cyclic oligomer, with very high polydispersity (Fig. 1).[6]

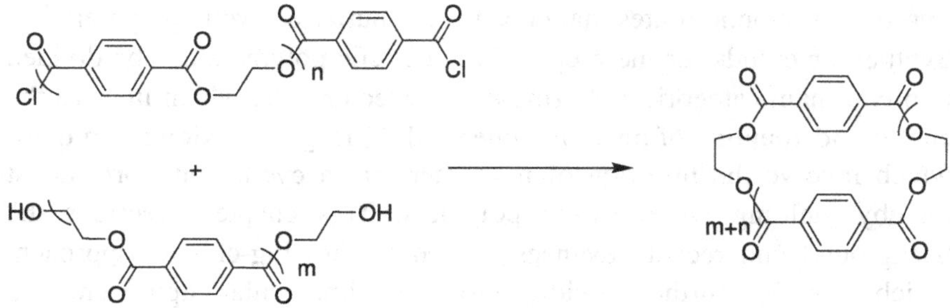

Fig. 1. Synthesis of cyclic polyester by polycondensation method.[6]

By tuning reaction conditions, the generation of cyclic polymers could be favored, and has been explored for making cyclic poly(dimethylsiloxanes) and polyesters.[7] However, this approach inevitably yields significant amounts of linear impurities and products exhibiting broad polydispersities. Extensive purification is required to obtain cyclic polymer of sufficient purity for architecture-related studies. For example, higher molecular weight linear polymer impurities resulting from intermolecular couplings can be selectively removed through repeated fractional precipitations to enable the isolation of the desired cyclic materials.[8] In addition, because the cyclic polymers exhibit a reduced hydrodynamic volume resulting from their more compact conformation, preparative gel permeation chromatography (GPC) can be employed on small scale for the isolation of polymer macrocycles from linear polymers of similar molecular weight.[9] Liquid chromatography at the critical condition also provides an alternative method for the isolation cyclic macromolecules as large as 200 kDa because under these specific conditions, their separation depends on architecture and end group functionality rather than size or molecular weight.[10] Alternatively, for specific backbones, other methods of purification can also be employed. For example, Beckham and coworkers removed the linear poly(ethylene glycol) (PEG) impurities from a predominantly cyclic sample by taking advantage of the propensity of α-cyclodextrin to thread onto linear PEG, triggering their selective precipitation.[11]

While these and other purification technologies can be applied to isolate cyclic polymers, a more efficient strategy would rely upon improved synthetic routes that enhance the purity of cyclic polymers.[12,13] Synthetic methods for the preparation of cyclic polymers can be divided into two main categories: (1) ring-closure techniques, which involve the end-to-end coupling of linear polymer and (2) ring-expansion techniques, which involve the insertion of monomer into a cyclic initiator/catalyst thereby yielding larger cyclic polymers. This chapter describes the background and recent advances related to the ring-closure approach, which can be further divide into the bimolecular approach, the homofunctional unimolecular approach and the heterofunctional unimolecular approach. The advantages and complications for each approach will be compared with respect to the other ring closure approaches, as well as the ring expansion techniques.

2. Bimolecular Ring Closure Approach

The bimolecular ring closure approach entails the reaction of a difunctional polymer with a difunctional coupling agent. It involves two successive reaction steps, the first step is intermolecular between the polymer and the coupling reagent, and the second step is intramolecular between the remaining complementary end groups now on the same polymer chain. The bimolecular approach was the first method developed to address the prevalence of linear byproducts associated with the ring-chain equilibrium; however, this technical is also susceptible to the formation of linear byproducts because of its bimolecular nature. It can be difficult to measure the exact mole ratio of the end groups (in part because of imprecise methods for determining M_n) yet the bimolecular approach requires that a perfect stoichiometric ratio of both reagents is used. Another complication is the incompatibility of the coupling conditions required during the two successive reactions. Specifically, the rate of initial bimolecular coupling is favored under high concentration, while clean cyclization in the second step requires high dilution to minimize the generation of linear oligomeric byproducts. As a result, either the reaction is carried out under dilute conditions with very low throughput, or, for faster, less dilute conditions, tedious purification techniques are often required to remove the linear byproduct and isolate pure cyclic product. In a few specific examples, however, clever synthetic designs have been employed to overcome this tendency to generate linear byproducts.

The bimolecular approach was first demonstrated for well-defined (narrow polydispersity) polymers by the research groups of Höcker,[14] Rempp,[15] and Vollmert[16] in 1980. In these early studies living anionic polymerization was utilized to produce polymer anions that cyclized upon addition of a bifunctional linker (Fig. 2).

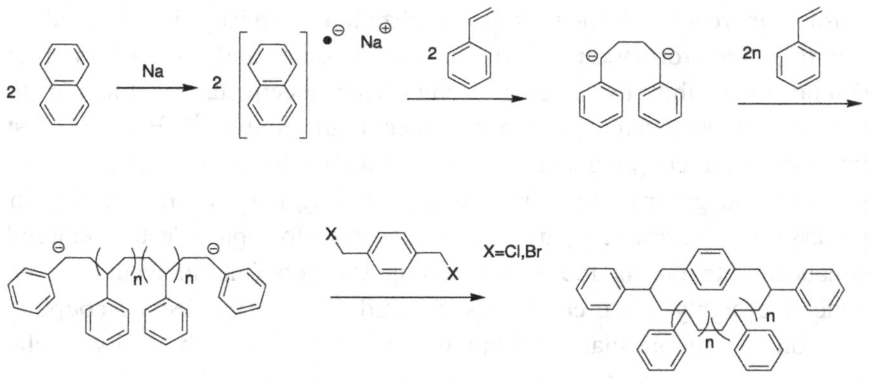

Fig. 2. Synthesis of cyclic polystyrene by bimolecular coupling of bis-anionic linear polystyrene with dihalo-*p*-xylene.[14]

For these initial synthetic reports of cyclic polystyrene (PS), sodium naphthalenide was first reacted with styrene at a molar ratio 1:1 to generate a bis-anionic initiator. A bis-anionic PS chain was then obtained by addition of styrene monomer to the initiator. While maintaining stringent impurity-free conditions, 1,4-dibromo-*p*-xylene was added to the active bisanionic styrene chain as coupling reagent to afford cyclic PS. However, the styryl anion was still present based on its characteristic red color, even after adding exactly 1 equivalent of linker, verifying the generation of acyclic byproducts. Excess linker was then added to ensure that any linear byproducts that maintained active chain ends would oligomerize into much larger polymers. Fractionation could then be employed to isolate the relatively low molecular weight cyclic polymers from the much larger oligomeric byproducts. The reported yields of cyclic polymers produced using this method were often very low (near 50%) because of the losses during purification. Since these initial reports, cyclic PS with molecular weights as high as 450 kDa has been reported as well as cyclic diblock or triblock polymers via to the sequential addition of different monomers during the living polymerization.[17] However for all of these investigations, the isolation of clean cyclic polymer requires rigorous purification resulting in a more tedious sample preparation and relative low yields.

Higher purity products can be achieve using the bimolecular ring closure approach, as long as the cyclization chemistry is designed to suppress the formation of oligomers. Tezuka and coworkers first demonstrated this by elegantly employing electrostatic attraction to template bimolecular cyclization under high dilution.[18] For the first intramolecular coupling reaction, the attractive forces between cationic polymer end groups and a bis-anionic coupling reagent encourage their pre-assembly into a salt pair (Fig. 3) and therefore provide the required enhanced rate during the first coupling reaction (intermolecular) even under the highly dilute conditions required to favor the second coupling reaction (intramolecular cyclization). As a result, high purity cyclic polymers can be made without extensive purification. This technology, termed "electrostatic self-assembly and covalent fixation" (ESA-CF) utilizes a linear precursor having two cyclic ammonium salt end groups that couple with a bisfunctional carboxylate counterion. Tezuka and coworkers first reported this approach in 2000 by using *N*-phenylpyrrolidinium end groups on poly(tetrahydrofuran) (polyTHF) and a bisfunctional carboxylate coupling reagent. At polymer concentrations of 4.6×10^{-5} M in THF, electrostatic pre-assembly yielded predominantly the cyclic salt, while the moderately strained cyclic ammonium end groups encouraged a rapid and nearly quantitative covalent coupling with the carboxylate nucleophiles upon heating to 66 °C yielding the covalent macrocycle.

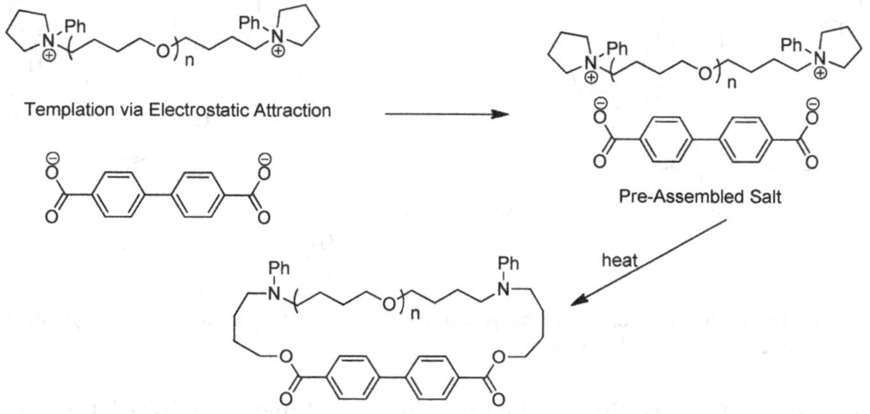

Fig. 3. The preparation of cyclic polymers by electrostatic templating approach.[18]

Since these initial reports, cyclic PEG,[19] PS,[20] and polydimethylsiloxane[21] were prepared by same ESA-CF methodology using relevant prepolymers obtained by living polymerization. Furthermore, cyclic polymers bearing an addressable functional group for additional coupling have also been prepared by using telechelic biscationic polymer precursors and dicarboxylic acid linkers in which one or both component bear a pendant functionality for coupling (e.g. alkyne or azide groups) (Fig. 4).[22] These reactive cyclic polymers can then be coupled to prepare more complicated multicyclic polymer topologies.

Another approach for preparing high purity cyclic polymers using the bimolecular approach is utilizing extremely rapid coupling chemistries, which enable the initial bimolecular coupling despite the low reactant concentration. Using the highly efficient thiol-ene coupling reaction,[23,24] Dove and coworkers report the synthesis of stereoregular cyclic polylactide (PLA).[25] The major advantages of this reaction are the tremendously rapid reaction rate in addition to the metal-free catalysis.[26] The reaction is also relatively mild, exhibiting compatible with a range of functionalities, including the relatively labile polyester bonds of PLA (Fig. 5). In this approach, difunctional PLA was prepared by ring

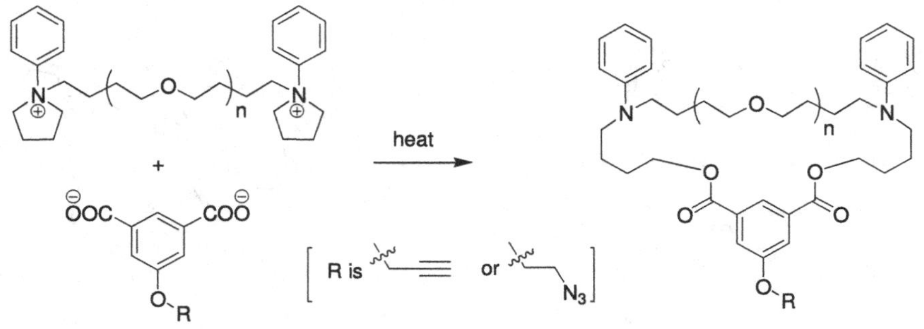

Fig. 4. The preparation of cyclic polymers via the ESA-CF approach, bearing pendant alkyne or azide functional groups.[22]

opening polymerization of lactide using 4-(2-hydroxyethyl)-10-oxa-4-azatricyclo[5.2.1.0]dec-8-ene-3,5-dione as initiator, followed by the modification of hydroxyl end group by using pentanedioyl chloride mono-[2-(3,5-dioxo-10-oxa-4-azatricyclo[5.2.1.0]dec-8-en-4-yl)ethyl]-ester. Pure cyclic PLA can be obtained due to the extremely rapid thiol-ene "click" reaction between di-ene polylactide and 1,2-ethanedithiol. Concurrent slow addition (~0.4 mL/h) of a precisely measured solution of the di-ene PLA (~7mM) and an equal molarity solution of 1,2-ethanedithiol yielded the desired cyclic product at ambient temperature. Saturated sodium metabisulfite was added to the reaction in order to suppress thiol oxidation (which might generate a disulfide byproduct). The purity of the product was confirmed by both GPC and matrix-assisted laser desorption/ionization time of flight (MALDI-TOF) mass spectra. One of the most significant advantages of this particular approach is that the versatile thiol-ene "click" reaction is sufficiently mild and functional group tolerant to enable the retention of stereochemistry of polyester backbone, as well as other biodegradable polymer backbones. Although its only use to date has been with polyesters, this approach is likely compatible with a wide range of polymer backbones.

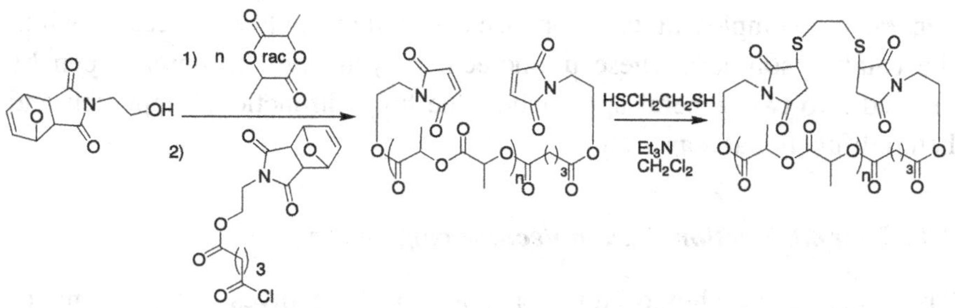

Fig. 5. The synthesis of cyclic poly(lactide) via bimolecular thiol-ene"click" reaction.[25]

The bimolecular ring closure approach was one of the first techniques established to generate cyclic polymers with reasonably narrow polydispersity and was the synthetic workhorse for many earliest studies on the physical properties of cyclic polymers. Its primary drawbacks are the slow kinetics of dilute second order reactions, in addition to the likely generation of linear impurities. In order to yield cyclic polymers of reasonably purity the synthetic design must incorporate specific measures to combat these inherent complications. Owing to these impurities, this technique has generally become disfavored relative to the other methods; however, two notable exceptions (the ESA-CF method and the thiol-ene"click" cyclization) have been reported and have been sufficiently optimized to yield well-defined, high purity cyclic polymers with a range of functionality.

3. Unimolecular Ring Closure Approach

Unlike bimolecular approach, the unimolecular approach involves the intramolecular cyclization of a difunctional polymer. Because this reaction entails a coupling reaction between functional groups on opposite ends of the same linear polymer, it does not require the measurement of exact stoichiometries characteristic of the bimolecular approaches. In addition, the unimolecular approach offers the unique advantage that intramolecular cyclization can be easily favored (versus intermolecular oligomerization) by simply carrying out the coupling reaction under highly dilute conditions. For couplings that are irreversible and require a catalyst, the quantity of solvent can be minimized and high purity of cyclic polymer generated by employing the dropwise addition of the linear precursors into the catalyst solution. These unimolecular cyclization techniques can be divided into two general approaches: the homodifunctional approach and heterodifunctional approach.

3.1. *Homodifunctional unimolecular ring closure*

The unimolecular homodifunctional approach involves the coupling of identical functionalities on opposite ends of the same polymer chain. The linear precursors can be most readily obtained by polymerization from a bis-functional initiator followed by simultaneous modification of both end groups to yield a polymer with identical functional groups at opposite chain ends. While this approach typically requires much less synthetic effort than the heterodifunctional approach, the limited number of highly efficient homocoupling reactions has resulted in substantially fewer uses of this technique. However, a number of noteworthy and promising examples have been published recently, and are described below.

Tezuka and Komiya first reported an efficient homodifunctional cyclization reaction by using the ring-closing metathesis of allyl terminated polymers at 2002.[27] PolyTHF was prepared via the living cationic polymerization of THF by using trifluoromethanesulfonic anhydride as an initiator, and then the both ends of the polymer were capped by reaction with sodium allyloxide to generated polyTHF with two allyl end-groups. Then by using the first generation Grubbs metathesis catalyst, ruthenium (II) dichloride phenylmethylenebis-(tricyclohexylphosphine), the allyl chain ends underwent olefin metathesis to form a cyclic polymer (Fig. 6). The reaction was carried out under sufficiently low concentrations of linear polymer precursor (0.2 g/L) to favor the formation of cyclic polymer while minimizing oligomerization. The irreversible loss of ethylene gas acted to increase the efficiency of the coupling reaction. MALDI-TOF mass spectrum verified a quantitative shift in the polymer mass distribution corresponding to a characteristic mass loss of -28 Da (ethylene) in the MALDI-TOF mass spectra upon cyclization, while GPC confirmed the expected decrease in hydrodynamic volume.

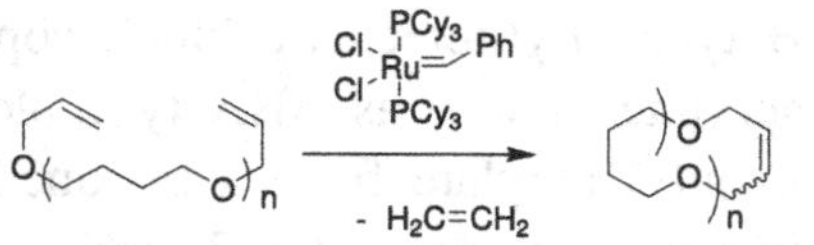

Fig. 6. The unimolecular cyclization of bis-allyl terminated polyTHF by using Grubbs metathesis catalyst.[27]

Since this initial work, Tezuka and coworkers also reported the combination of the olefin metathesis cyclization technique with atom transfer radical polymerization (ATRP) to prepare cyclic poly(methyl acrylate) in 2007.[28] ATRP is an attractive route to prepare the polymers with well-defined structure and a wide range of backbone functionalities, and if carried out at low conversion, offers high retention of the bromide end groups for conversion to other useful functionalities.[29] The linear polymer precursor is obtained by the ATRP of methyl acrylate from a dibromo initiator followed by a Keck allylation of both bromide end groups of the polymer. Using an analogous metathesis cyclization as described for the polyTHF, cyclic methacrylate polymers could be obtained (Fig. 7). Again MALDI-TOF mass spectra confirmed the successfully synthesis of cyclic poly(methyl acrylate) by the quantitative mass loss of ethylene across the polymer distribution, and the GPC retention time data exhibits a reduction in hydrodynamic volume of the product.

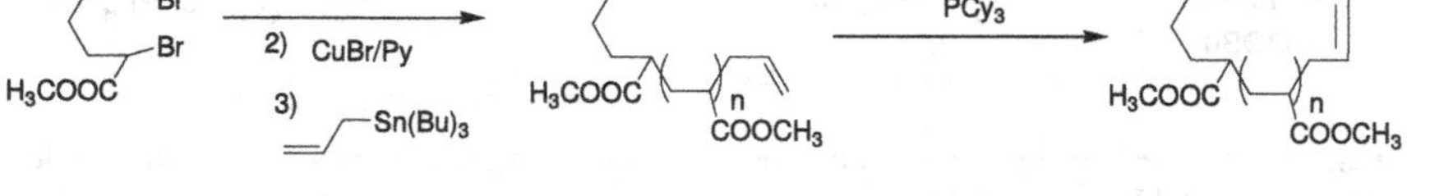

Fig. 7. The synthesis of cyclic poly(methyl acrylate) by the combination of ATRP and the Grubbs metathesis coupling reaction.[28]

An additional advantage of using ATRP combined with ring-closing metathesis techniques is the versatility of ATRP in preparing block copolymers. Adachi *et al.* demonstrated the synthesis of cyclic diblock copolymer at 2008.[30] By using ATRP, they demonstrated both the synthesis of the AB type allyl-telechelic block copolymers of methyl acrylate and butyl acrylate as well as ABA type block copolymers via telechelic growth of butyl acrylate from a bisfunctional poly(ethylene oxide) initiator. Both could be cyclized efficiently using the metathesis ring closing technique and the subsequent cyclic block copolymers exhibit unique properties relative to linear polymers of similar composition (Fig. 8). For example, recently Tezuka and coworkers, found that the thermal stability of a block copolymer self-assembled micelle was remarkably enhanced for the cyclic block copolymer, relative to their linear analogs.[31,32] Linear poly(butyl acrylate)-*block*-poly(ethylene oxide)-*block*-poly(butyl acrylate) and the ring closure product, cyclic poly(butyl acrylate)-*block*-poly(ethylene oxide), both can self-assembled to form flower-like micelles in water. Dynamic light scattering, atomic force microscopy, and transmission electron microscopy studies revealed that both micelles are spherical and approximately 20 nm in diameter. Although there is no significant difference between the chemical composition, size or shape of two micelles, the cloud point (T_c) was elevated by more than 40 °C for the cyclic topology relative to the linear polymer amphiphile. The improved thermal stability of the assemblies of the cyclic amphiphiles in addition to their biocompatibility and tunable structure offers promise for a range of biomaterials applications.

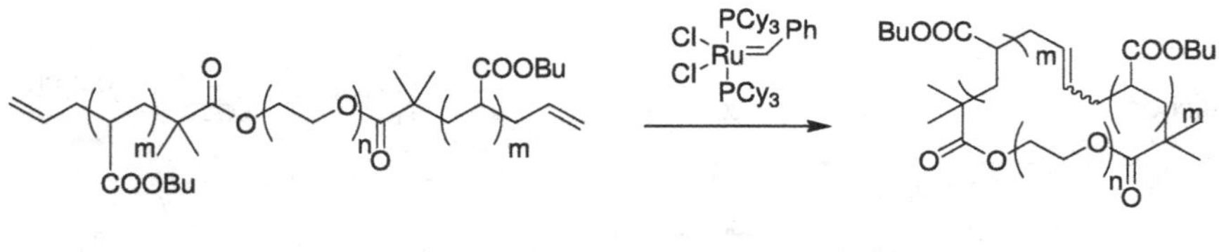

Fig. 8. The synthesis of cyclic ABA triblock polymer using metathesis ring closure yields more thermal stable micelles.[31]

Alternatively, the thiol group represents an appealing functionality for homodifunctional ring closures because a dithiol can easily undergo oxidative coupling to yield a disulfide linkage, yet the disulfide can be readily reduced to regenerate the dithiol. Whittaker *et al.* utilized reversible addition–fragmentation chain transfer (RAFT) polymerization to explore the utility of this homocoupling for preparation of cyclic polymers.[33] The dithiol PS was synthesized by polymerizing styrene from a difunctional RAFT agent 1,3-bis(2-(thiobenzylthio)prop-2-yl)benzene followed by conversion of the dithioester end groups to thiols via the addition of hexylamine. Oxidation (either aerial or with Fe(III)Cl$_3$) converted the thiols to disulfides, and could be used to generate high yields of the cyclic PS if dilute solutions of the dithiol PS were used. The concentration could also be increased to favor the generation of linear multiblocks if desired (Fig. 9). However, the slow dropwise addition of the precursor into a solution of FeCl$_3$ was a particularly useful technique to favor the generation of cyclic polymers. The most unique feature of this approach is its facile reversibility: both the ease with which the disulfide can be reduced (e.g. with Zn metal) to revert back to the linear dithiol precursor, as well as the simplicity of the oxidation of the dithiol to regenerate disulfide-linked cyclic polymer.

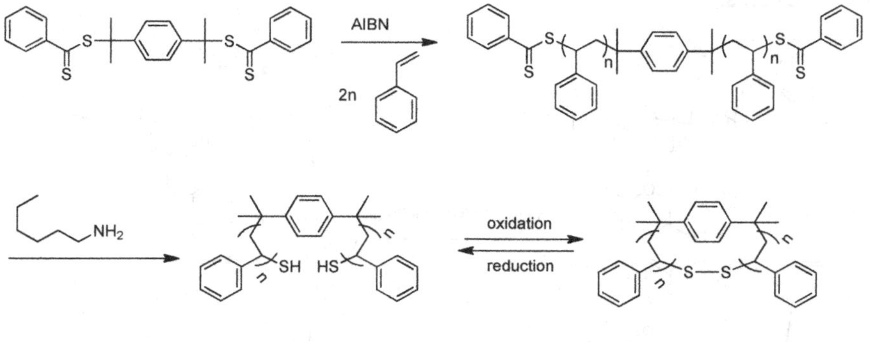

Fig. 9. Reversible cyclization and ring opening through oxidization and reduction of linear polymers with two thiol end groups.[33]

Because the homodifunctional approach typically requires high dilution to favor the cyclic products, such techniques are wasteful of solvent. In addition, long polymer chains can be difficult to cyclize because of entropic factors related to distance between polymer chain ends resulting a low probability of end-to-end coupling. Hu *et al.* reported a clever methodology to overcome these complications in 2009.[34] By using block copolymers with contrasting solubilities, similar to a technique first reported by the research group Shiyong Liu,[35] self-assembly of the poorly soluble chain ends could be used to template the end-to-end coupling (Fig. 10). In order to further increase the probability of coupling for larger polymers (>100,000 Da) blocks of random copolymers of 2-cinnamoyloxyethyl methacrylate and 2-trifluoroacetoxyethyl methacrylate) (the A block) were polymerized from the terminal ends to generate, in essence, an ABCBA pentablock copolymer with multiple reactive functionalities at each end. A micellar solution was first prepared in a solvent carefully selectively to be a poor solvent for the cinnamoyloxyethyl methacrylate/trifluoroacetoxyethyl methacrylate terminal "A" blocks. The solution of pre-formed micelles was then added dropwise to a reaction flask under constant stirring and irradiation. In the reactor, the micelles dissociate quickly into end-associated rings which rapidly undergo photocyclization. As with other "dropwise addition" approaches, as long as the rate of cyclization is faster than the rate of addition, the linear polymer concentration remains low throughout the coupling reaction, favoring conversion to macrocycles.

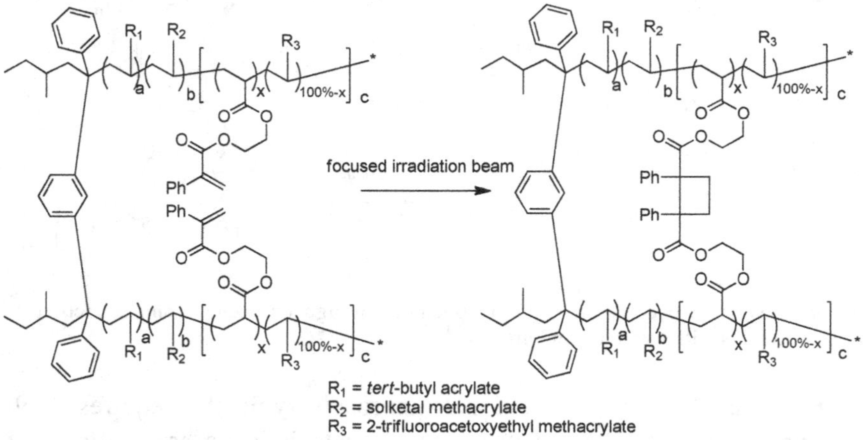

R_1 = *tert*-butyl acrylate
R_2 = solketal methacrylate
R_3 = 2-trifluoroacetoxyethyl methacrylate

Fig. 10. The synthesis of macrocyclic copolymer by using ABCBA block copolymer to template cyclization.[34]

The homocoupling of isocyanate groups also has been utilized for making cyclic polymers. Using 1,3-dimethyl-3-phospholene oxide as catalyst, two isocyanate groups can be coupled together to yield a carbodiimide linkage.[36] The precursor polymer was prepared by reacting the hydroxyl end groups of poly(propylene oxide) with an excess (2-4 equivalents) of terephthalate diisocyanate. Subsequent reaction with acetic acid enables generation of the cyclic polymer via the carbodiimide linkage (Fig. 11). Because of the relative lability of the carbodiimide linkage, heating above 120 °C leads to macrocyclic ring opening to a linear amide-isocyanate intermediate which readily undergoes self-condensation to generate urethane-linked macropolymers. This self-curing process is a particularly attractive way of preparing high molecular weight polyurethanes without the inclusion of volatile and toxic low molecular weight isocyantes.

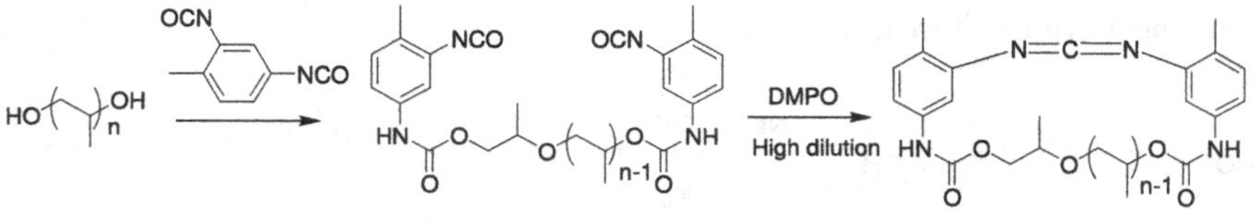

Fig. 11. The synthesis of cyclic PPG by using isocyanate-terminated prepolymer,(DMPO = 1,3-dimethyl-3-phospholene oxide).[36]

The Glaser coupling,[37] which enables the efficient homocoupling of two terminal alkynes to yield a 1,3-diyne, has also been successfully applied to prepare cyclic polymers. Recently Zhang and Huang reported that Glaser coupling reaction of alkynyl groups could be used to synthesize both cyclic poly(ethylene oxide) and cyclic PS (Fig. 12).[38] Hydroxy-terminated linear poly(ethylene oxide) was prepared by ring-opening polymerization (ROP) of ethylene oxide (EO) using 2,2-dimethyl-1,3-propanediol and diphenylmethylpotassium as co-initiators, while hydroxyl-terminated linear PS was prepared by anionic polymerization using lithium naphthalenide as initiator followed by termination with EO. The hydroxyl-telechelic polymers (HO-PEO-OH and HO-PS-OH) were then treated with sodium hydride and propargyl bromide, to yield the telechelic bis-propargyl precursors of PEO and PS. The intramolecular cyclization reactions were carried out by slow dropwise addition of linear precursor to catalyst solution of Cu(I)Br/PMDETA at room temperature under mild conditions with oxygen in the air as oxidant, leading to a near-quantitative conversion. Using this combination of polymerization and cyclization techniques, well-defined cyclic polymers could be prepared, (c-PEO: polydispersity index (PDI) = 1.20 and c-PS: PDI = 1.04). This Glaser cyclization has also been used to prepare amphiphilic cyclic poly(ethylene oxide)-*block*-polystyrene[c-(PEO-*b*-PS)] (PDI=1.08) from propargyl-telechelic poly(ethylene oxide)-*block*-polystyrene-*block*-poly(ethylene oxide) (PEO-*b*-PS-*b*-PEO).[39] In addition, the Glaser coupling is amenable to the preparation of more complex architectures including "figure-eight-shaped" multi-cyclic polymers.[40] This method is most attractive because of its simple and mild reaction conditions, and the ability to carry out the coupling in ambient reaction environments (e.g. room temperature and a non-inert atmosphere).

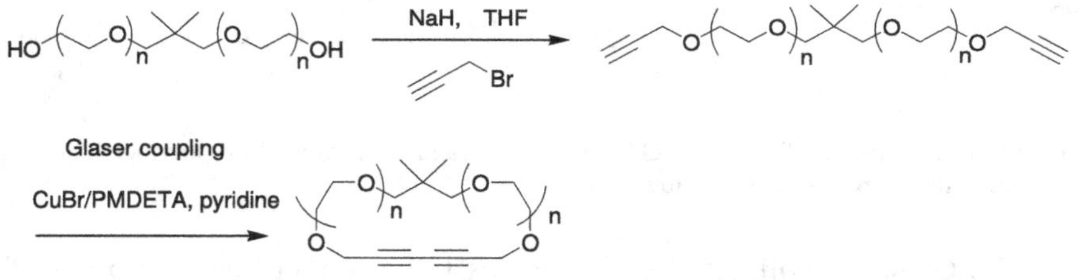

Fig. 12. The unimolecular cyclization of PEG by Glaser alkyne coupling.[38]

In addition to covalent end-to-end couplings, cyclic polymers have recently been prepared making use of metal coordination. Schappacher and Deffieux reported the end-to-end coupling of PS bearing two meso-tetraphenylporphyrin end groups and subsequently bridging them by forming the bis[iron(III) μ-oxoporphyrin] dimer.[41] The linear precursor was synthesized by living anionic polymerization of styrene from an acetal protected initiator followed by termination with an acetal functionalized alkyl chloride, to yield PS with two terminal acetals. It could then be functionalized with the porphyrin groups via acidic hydrolysis and transacetalization with diethyl acetal-functionalized tetraphenylporphyrin. Metalation could then be achieved by reaction with $FeCl_2$. The corresponding macrocyclic PS was obtained readily and selectively by intramolecular condensation of the bis-[chloroiron(III) *meso-*

tetraphenylporphyrin] polymer ends via slow addition of a small amount of sodiummethanolate/methanol solution as ring-closing catalyst to yield a diiron(III)-μ-oxobis-(porphyrin) dimer as the bridging unit. Addition of dilute HCl was shown to rapidly reconvert the diiron(III)-μ-oxobis(porphyrin) unit into the linear bis[chloroiron(III)porphyrin] PS precursor, demonstrating the selectivity and complete reversibility of the cyclization process. All NMR and GPC data suggest that this unimolecular chain-end coupling reaction is relatively independent of the dilution during the cyclization and is very efficient for the end-to-end cyclization of polymers. After this initial report, the reversible switching between linear and cyclic PEOs (Fig. 13) bearing iron tetraphenylporphyrin end groups was also explored using solvent, pH, and redox stimuli.[42]

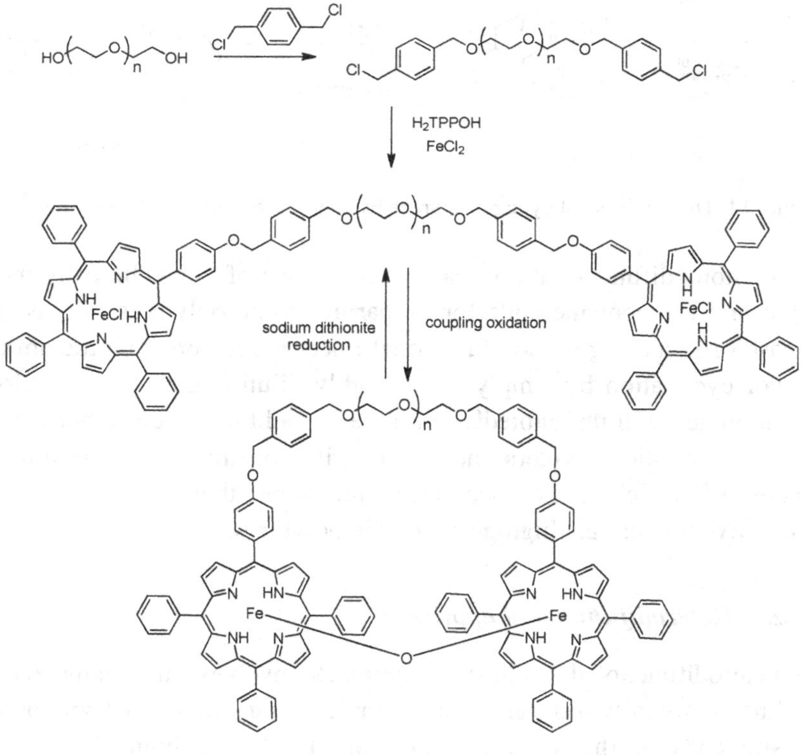

Fig. 13. The reversible cyclizaton by using porphyrin complexes to bridge the chain ends (PEG), (H₂TPPOH = 5-(4-Hydroxyphenyl)-10,15,20-triphenylporphyrin).[42]

In 2010 Voter and Tillman introduced a particular efficient and elegant approach to making cyclic polymers by combining ATRP with intramolecular atom transfer radical coupling (ATRC).[43] In this approach, the ATRP of styrene from a dibromointiator was quenched at low conversion to yield a well-defined telechelic polymer with two benzylic bromo end groups. The isolated telechelic polymer was then added dropwise to a similar Cu(I) catalytic system, employing the high dilution to favor an intramolecular cyclization. Using this technique, well-defined cyclic polymers could be prepared in just two steps from the commercially available initiator (Fig. 14).

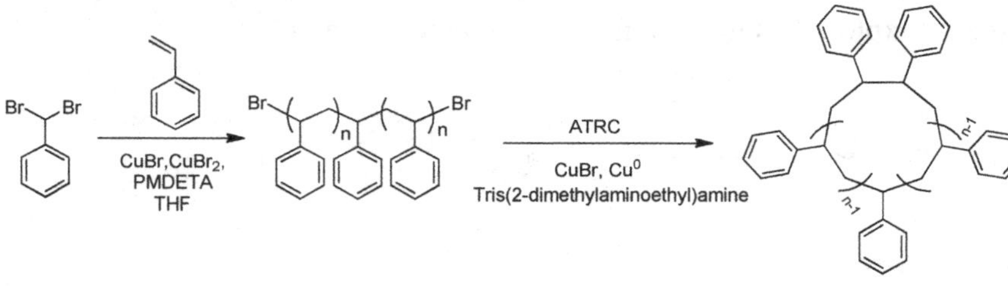

Fig. 14. The synthesis of cylic polystyrene by using ATRP and ATRC method.[43]

The homodifunctional approach remains one of the technically most simple and efficient methods for preparing cyclic polymers because of the ease of preparing homodifunctional linear precursors, and the ability to favor cyclization by simply using highly dilution during cyclization. This approach will undoubtedly continue to yield major contributions to the field of cyclic polymers and is primarily constrained by the limited number of efficient homocoupling reactions that are sufficiently quantitative to generate high purity cyclic polymers.

3.2. α,ω-Heterodifunctional approach

The heterodifunctional cyclization approach involves an intramolecular coupling between two different but complementary functional groups on opposite ends of the same polymer. Similar to the homodifunctional approach, the presence of both reactive functionalities on the same polymer chain circumvents the need to carefully balance reactant stoichiometries, one of the weaknesses of the bimolecular approaches. Also like the homodifunctional approach, high dilution can be used to suppress the intermolecular oligomerization reactions, and thereby favor the production of cyclic polymers. However, unlike the homodifunctional approach, the synthesis can sometimes be challenging, since it requires the nearly quantitative installation of different functional groups at opposite ends of each polymer. The heterodifunctional nature does, however, provide a slight advantage during coupling, as the likelihood of forming intermolecular dimers is further cut in half, because the end group of one polymer has only half the chance of being complementary to, and therefore coupling with, the end group of another polymer. Perhaps the most significant advantage of this approach is the wealth of highly efficient coupling reactions that involve two differing functionalities, and this has been verified by the diversity of heterodifunctional coupling reactions which have been used to prepare cyclic polymers.[44]

The seminal work that introduced the heterodifunctional cyclization technique was reported by Schappacher and Deffieux in 1991.[45] Extremely well-defined linear polymer precursors with low polydispersities were synthesized by using the living cationic polymerization of 2-chloroethyl vinyl ether (CEVE) (Fig. 15). The polymerization was initiated from a vinyl ether bearing a pendant styrenyl group. Addition of hydroiodic acid across the vinyl ether bond, afforded the initiator for the Lewis acid ($ZnCl_2$) catalyzed polymerization of CEVE to yielded linear poly(CEVE) with molecular weights ranging from 1.1 kDa to 2.9 kDa and PDI's ranging from 1.13 to 1.19. Reaction of the terminal iodo end-group with $SnCl_4$ yielded the carbocation which efficiently reacted with the styrenic end-group at the opposite end of the polymer to yield the more stable benzylic cation, as well as cyclize the polymer. Quenching with sodium methoxide in methanol yielded the stable cyclic polymer.

Crude GPC traces confirmed the generation of significant amounts of the desired cyclic poly(CEVE) as well as smaller amounts of linear byproducts (~20%), which ranged in molecular weights from 1 kDa to 3 kDa. Using fractionation, the intermolecular condensation product could be removed to enable isolation of high purity cyclic polymer. The cyclic nature of the product was also confirmed by the observation of an increased T_g, as predicted by theory. In subsequent studies, these authors also demonstrated that the same approach could be used to prepare more complicated cyclic topologies, such as figure-eight-shaped,[46] tadpole-shaped, and theta-shaped polymers.[47]

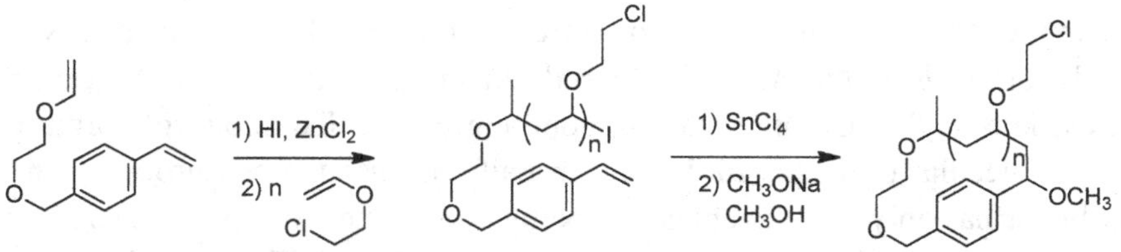

Fig. 15. The heterodifunctional cyclization of poly(chlorethylvinylether) reported by Schappacher and Deffieux.[45]

Rique-Lurbet *et al.* then extended the application of this cyclization chemistry to the living anionic polymerization of styrene of high molecular weight. The polymerization of styrene was initiated from the anion of a diethyl acetal.[48] After polymerization, first 1,1-diphenylethylene and then *p*-chloromethylstyrene were added to quench the polymerization, and place the complementary styryl end group opposite the acetal end group. Iodotrimethylsilane was used to activate the acetal with an iodo functionality, that could then couple with the styryl unit under Lewis acidic (SnCl₄) conditions to yield the cyclic polymer. This ring closure reaction was carried out by dropwise addition of the linear precursor into the Lewis acid catalyst, in order to maintain high dilution during the coupling to favor formation of the cyclic polymer (Fig. 16). This optimized technique afforded high purity cyclic PS in a wider range of molecular weights (2–12 kDa) yet with polydispersity indexes as low as 1.2. The cyclic polymer was obtained in yields greater than >95%, reducing the need for rigorous purification techniques such as fractionation or preparative GPC. When calibrated against linear PS standards, the ratio of the M_n's ($M_n$cyc/$M_n$lin) determined by GPC was 0.85 which is in agreement with measurements previously reported for cyclic PS prepared via the bimolecular coupling. This coupling technique is also sufficiently versatile to enable the preparation of block copolymers comprising of both PS and poly(CEVE).[49]

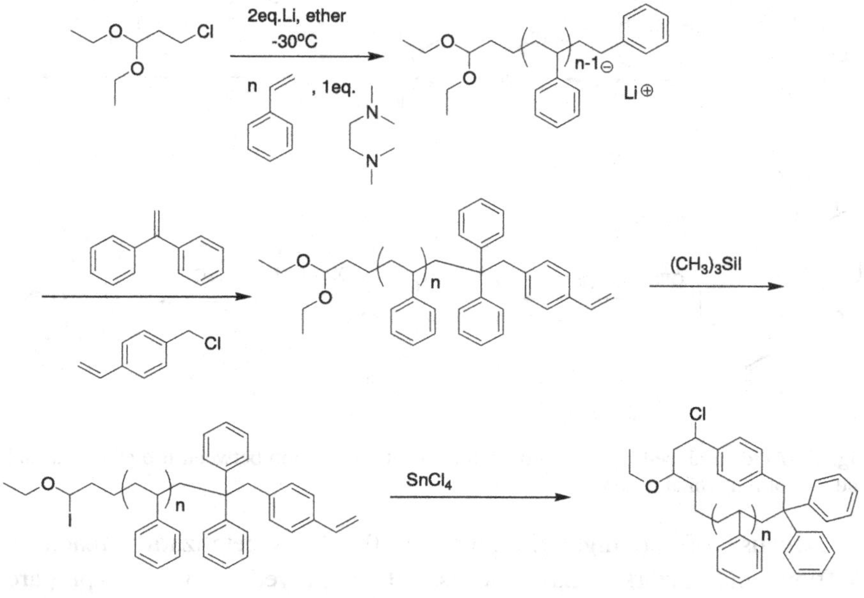

Fig. 16. The preparation of cyclic PS using cationic ring closure reaction.[48]

Schappacher and Deffieux have also reported using the acid-catalyzed transacetalization reaction as a means of generating cyclic polymers.[50] By installing a diethoxyacetal group on the initiator, and terminating with a diol, high purity cyclic polymer could be obtained under highly dilute conditions (Fig. 17). Linear PS was first prepared via the living anionic polymerization of styrene using 3-lithiopropionaldehyde diethyl acetal as the initiator. The polymerization reaction was terminated with isopropylidene-2,2-bis(hydroxymethyl)-1-(2-chloroethoxyethoxy)butane. Hydrochloric acid could be used to deprotect the diol, and provide the required functionalties for cyclization. In order to favor the cyclic polymer, transacetalization was carried out in dilute conditions, and GPC, T_g and MALDI-TOF mass spectrum data all confirmed the generation of cyclic polymer.

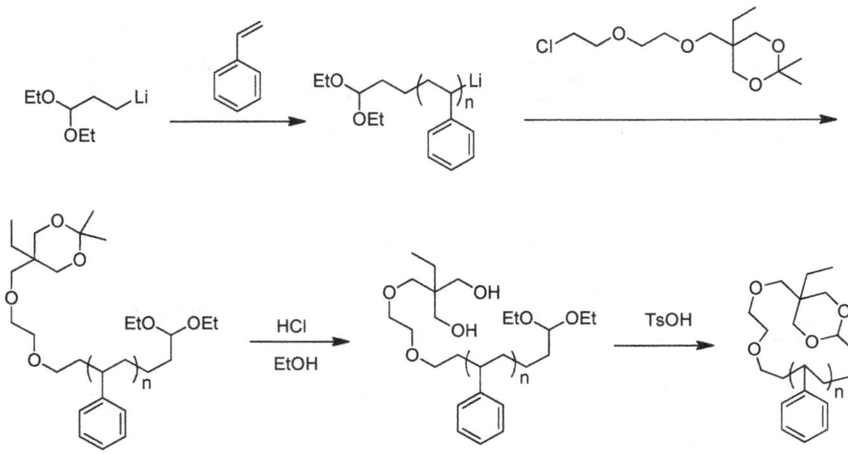

Fig. 17. Acid-catalyzed transacetalization cyclization reaction between α-diethoxy acetal and ω-diol functional groups.[50]

Because of the high efficiency of the transacetalization reaction, Deffieux and coworkers have successfully employed this route to prepare more complex cyclic topologies, including figure-eight-shaped, tadpole-shaped, and theta-shaped polymers.[51]

Most recently, Schappacher and Deffieux have modified this approach in order to prepare substantially larger ring structures using multiple complementary functional groups per polymer chain.[52,53] Initially, ABC triblock copolymers were prepared which exhibited a long central B block (CEVE) surrounded by two short A and C sequences that consisted of monomer units bearing complementary reactive functionalities (silyl protected alcohols and vinyl ethers). The silyl protect group could be removed using tetrabutylammonium fluoride and subsequently the ABC triblocks were reacted with acid catalyst under high dilution to form a macrocycle via multiple intramolecular couplings between the pendant functionalities of the A and C blocks (Fig. 18). The GPC and NMR data confirm that exceptionally large cyclic polymers (DP = 500 ~ 900) can be prepared with low polydispersities (PDI < 1.1), and shifts in the GPC retention time confirm the cyclic nature of the product. One particularly attractive feature of using CEVE as the B block is the ease with which still living anionic polymers can couple to this repeating unit. This was demonstrated by grafting linear PS onto the cyclic core to yield a cyclic-core comb polymer. AFM visualization of the product verified the presence of cyclic comb polymers, as well as minor amounts of linear, tadpole, and figure-eight-shaped comb polymers, presumably due to the likelihood of incomplete cyclizations, as well as couplings between separate chains.

Likewise, reaction the cyclic poly(CEVE) can be reacted simultaneously with both the living anions of PS and polyisoprene to yield comb polymers with a combination of these two polymer side chains. AFM images show that drying solutions of these brushes from heptane, (a good solvent for PI, but not PS) yielded tubular aggregates with diameters of about 100 nm and lengths of up to 700 nm. This structure is consistent with a collapse of the PS block within a corona of solubilizing polyisoprene, followed by subsequent self-assembly of these macromolecular discs into tubular assemblies.

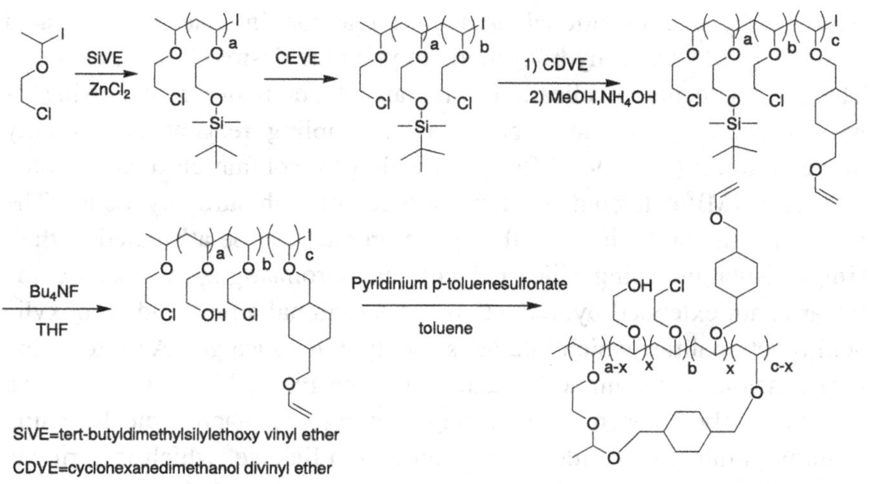

Fig. 18. The synthesis of cyclic triblock ABC polymer by using a coupling reaction between the functionalities of the A and C end blocks.[52]

These pioneering studies by Schappacher and Deffieux have demonstrated the utility of the hetero-coupling ring closure method for preparing cyclic polymers of high purity and their versatility in preparing more complex architectures. However, since their initial reports, a broad diversity of ring closure reactions have been investigated, the most significant of which will be detailed below.

Kubo first reported the use of activated ester chemistry to prepare cyclic PS with an amide linking functionality.[54] The polymerization of styrene was initiated with 3-lithiopropionaldehyde diethyl acetal in benzene with tetramethylethylenediamine as ligand at room temperature. The amino group was introduced by reacting the living polystyryllithium anion with 2,2,5,5-tetramethyl-1-(3-bromopropyl)-1-aza-2,5-disilacyclopentane. Pure amine-terminated PS could be isolated by column chromatography in a 92% yield. Before hydrolyzing the diethyl acetal function, the amino group was protected with benzyloxycarbonyl chloride. The hydrolysis of the diethyl acetal group was carried out in THF containing aqueous hydrochloric acid under reflux and the aldehyde group was oxidized to the carboxylic acid with *m*-chloroperoxybenzoic acid. Finally, the benzyloxycarbonyl group was removed by a conventional acidolysis reaction in glacial acetic acid containing 30 wt% of hydrobromide to yield the desired linear precursor. The intramolecular cyclization was carried out under reflux using 1-methyl-2-chloropyridinium iodide as a coupling reagent in a highly dilute solution (Fig. 19). After purification by column chromatography, GPC and NMR data confirmed the isolation of high purity cyclic PS. The reason for the high purity of the cyclic product can be attributed to their simple isolation using silica-gel column chromatography because the linear chain-extended byproducts possess terminal amine and carboxylic acid functionalities which interact strongly with silica gel. An interesting repercussion of the amide linkage within the PS backbone is that it can be selectively reduced without ring-opening the macrocycle by using lithium aluminum hydride to yield an amino linkage, which in turn can be readily functionalized. For example, reaction of the macrocyclic amine with a carboxyl-terminated linear polymer yielded a tadpole-shaped polymer whereas figure-eight-shaped polymers can be obtained by the one-step coupling reaction between two macrocyclic amines and a dicarboxylic acid such as glutaric acid.[55]

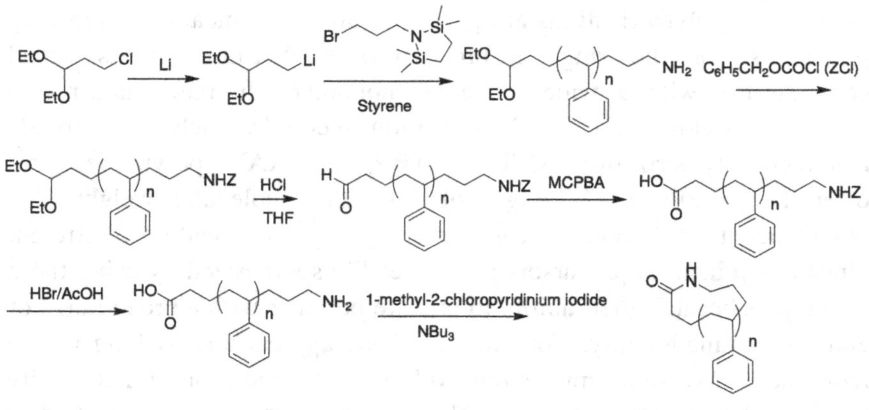

Fig. 19. The synthesis of cyclic PS using an amidation reaction for cyclization.[54]

Because trace amount of linear impurities are common using the unimolecular heterodifunctional coupling, the isolation of high purity cyclic polymer requires either extremely efficient coupling reactions, or the incorporation of an efficient means of removing linear impurities. One elegant concept for addressing the removal of linear impurities was developed by the Semlyen group by growing the linear precursors from a solid support.[56] In this report, cyclic polyester was prepared via the intramolecular coupling of a bromo-carboxylic acid using an anion-exchange resin as a solid support. The bromo-undecanoate monomer was bound to a tetraalkyl ammonium functionalized support via an electrostatic interaction. Upon heating, polymerization of the monomer could occur, but the product would remain electrostatically bound, until cyclization, at which point the neutral cyclic polymer was released. Although this synthetic design appears to be an ideal approach for reducing linear impurities and easing product isolation, it suffers from a few technical complications. Because the polymerization involves a step-growth mechanism from the solid phase, the reaction rate is slow and the polydispersities are high. For example, in this initial report, the final cyclic product shows a low degrees of polymerization (DP = 5–15). Tedious fractionation was still necessary to produce pure cyclic materials with narrow polydispersities.

Radical polymerizations are particular attractive means of preparing precursors for the ring closure approach, because of its broad compatibility with a wide range of monomers. In particular, recent developed living radical polymerization methods, such as nitroxide mediated polymerization (NMP),[57] ATRP,[58] and RAFT polymerization[59] offer the additional advantages of controlled molecular weight, low polydispersity and good control over end group functionalities. Different kinds of cyclization precursors can be readily synthesized by using these techniques because their ability to prepare polymer with a broad range of end chain functionality. For most of these approaches, as long as the monomer conversion remains relatively low, the end group functionality can be retained nearly quantitatively, and then subsequently modified to the desired functionalities for high yielding cyclization reactions.

The first application of living radical polymerizations to prepare cyclic polymer was demonstrated by Lepottevin *et al.* who used NMP to form a linear precursor containing complementary alcohol and carboxylic acid functionalities.[60] The controlled free radical polymerization of styrene was carried out using 4,4'-azobis(4-cyanovaleric acid) as the initiator and 4-hydroxy-TEMPO (2,2,6,6,-tetramethyl-1-piperidinyloxy) radical to mediate the polymerization. When the polymerization is terminated, PS was obtained with molecular weights between 1000 and 10,000 g mol^{-1} with a carboxylic acid on the initiating end, and an alcohol from the TEMPO reagent on the terminal end. Cyclization was carried out in high dilute dichloromethane solution using 2-chloro-1-methylpyridinium iodide to activate the carboxylic acid (Fig. 20). The yield of low molar mass (1,000 g mol^{-1}) macrocycles was close to 95%; however, for much higher molecular weight PS, linear contaminants complicated the isolation of pure cyclic polymer. These complication may results from the inherent thermal instability of the alkoxyamine linkage in the backbone of the cyclic polymer, suggesting that alternative living radical polymerizations may be more amenable to generating larger cyclic polymers.

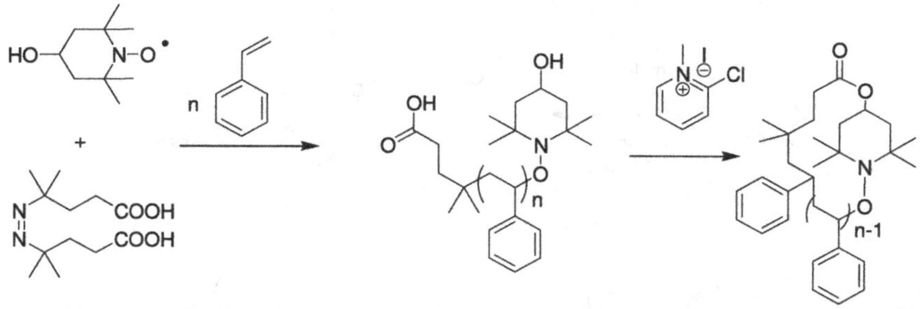

Fig. 20. Ester coupling of NMP generated liner precursors to afford cyclic polymers.[60]

The Huisgen copper catalyst azide-alkyne cycloaddition (CuAAC) "click" reaction[61,62] represents an important technological advance in the synthesis of polymers due to its exceptional coupling efficiency as well as its broad functional group compatibility.[63] Using the combination of CuAAC and living radical polymerization, high purity cyclic polymers with a diversity of backbone functionality can be easily prepared. Laurent and Grayson first reported the use of CuAAC to prepare cyclic polymers in 2006.[64] Linear PS precursors were prepared by using ATRP because the terminal benzylic bromide represents an ideal substrate for a nearly quantitative nucleophilic substitution with azide. By using an initiator bearing a pendant alkyne, the complementary functionalities can then be easily introduced onto opposite ends of the polymer chain for CuAAC cyclization. Using propargyl 2-bromoisobutyrate as initiator, PS could be obtained using a Cu(I)Br/pentamethyldiethylenetriamine (PMDETA) catalytic system. The polymerization was carried out in bulk to yield PS with a PDI less than 1.2. By terminating the polymerization at relatively low monomer conversion (typically less than 50%) near quantitative retention of the terminal bromine could be achieved, and efficiently converted to the azide by reaction with sodium azide. By adding this linear precursor dropwise via a syringe pump to the Cu(I) catalyst, high purity cyclic PS could be obtained through formation of the hardy triazole linkage via coupling of the alkyne and azide end groups (Fig. 21). The MALDI-TOF mass spectra verified that the molecular weight of the polymer samples remained unchanged after CuAAC "click" cyclization, however, because the cyclization leads to a more compact polymer structure, the macrocyclic polymers exhibit a longer size exclusion retention time than the parent linear polymer when analyzed by GPC.

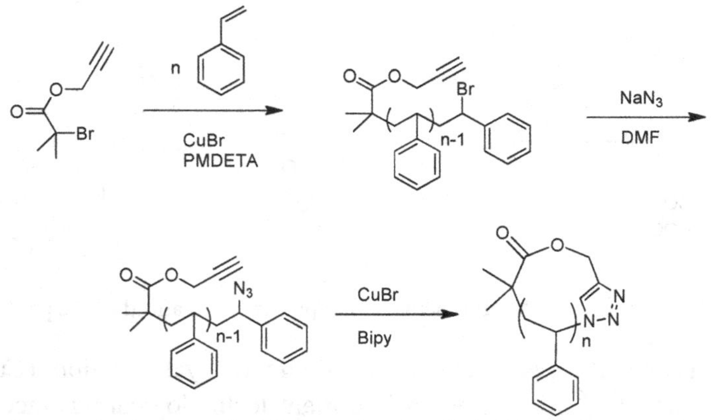

Fig. 21. The synthesis of cyclic PS by using the CuAAC coupling reaction between terminal azide and alkyne groups.[64]

The tremendous versatility of the CuAAC coupling combined with ATRP can be seen by the diversity of polymer backbones that have been cyclized using this technique. For example, this technique has also been used to prepare cyclic poly(*N*-isopropylacrylamide) (PNIPAM),[65] a backbone which is of interest due to its lower critical solution temperature (LCST) in water of about 30°C and its subsequent ramifications for biomedical materials. As expected, the cyclic topology exerts stringent restrictions on backbone conformation for cyclic-PNIPAM, leading to a lower LCST value, stronger concentration dependences of LCST and T_c values, and a smaller enthalpy value associated with thermal phase transitions, when compared to linear-PNIPAM. Cylic poly{6-[4-(4-methoxyphenylazo)phenoxy]hexyl methacrylate},[66] cyclic poly(4-vinylbenzyl)carbazole),[67] and cyclic poly(tert-butyl acrylate)[68] can also be prepared by using same method.

In addition to homopolymers, the combination of ATRP and CuAAC is particularly amenable to preparing block copolymers. Eugene and Grayson demonstrated this during the preparation of cyclic diblock copolymers of PS and poly(methyl acrylate).[69] Because the bromide end group is retained after polymerization of the first block, re-initiation with a second monomer enables generation of the linear diblock. After conversion of the bromide end group to an azide, the block copolymer can be cyclized using an analogous technique to those described above (Fig. 22).

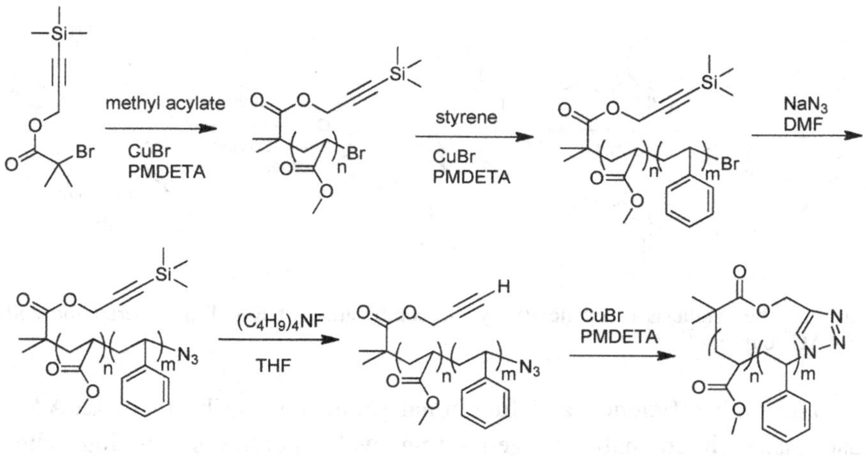

Fig. 22. The synthesis of cyclic diblock poly(methyl acrylate)-polystyrene by using the CuAAC cyclization.[69]

In addition to the CuAAC cyclization technique being highly compatible with ATRP, it has also been successfully applied with RAFT polymerizations. Winnik *et al.* prepared cyclic PNIPAM by initiating RAFT polymerization from an azide functionalized chain transfer agent. After terminating the polymerization, the trithiocarbonate chain transfer agent could be reduced via aminolysis to yield a terminal thiol functionality.[70] Then the bisfunctional polymer with complimentary chain ends was generated by the Michael addition of an α,β-unsaturated propargyl acrylate ester with terminated thiol. The CuAAC cyclization was carried out under extremely dilute conditions using Cu(II)SO$_4$ with sodium ascorbate as an *in situ* reductant to afford cyclic PNIPAM (Fig. 23) with molecular weights as high as 19 kDa and excellent control over polydispersity (PDI = 1.11). Goldmann *et al.* also used the combination of RAFT and CuAAC to produce cyclic PS from an azide functionalized RAFT agent.[71] The end-group was modified to the desired alkyne by reacting the living polymer chains with an excess of propargyl functionalized azo-bis(4-cyano valeric acid). The "click" cyclization was then performed using analogous procedures described above to yield the corresponding cyclic PS.

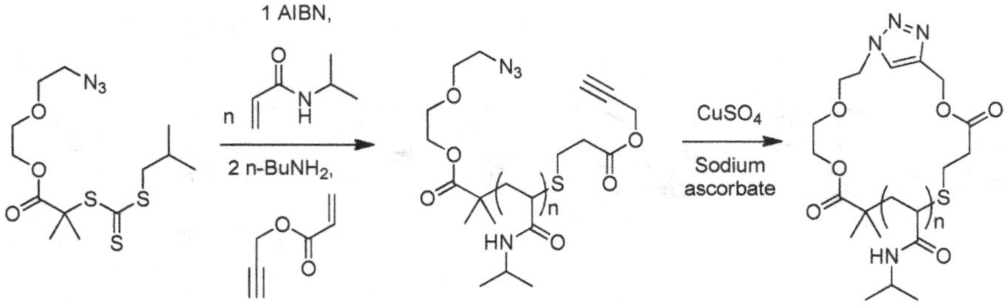

Fig. 23. The synthesis of cyclic PS by the combination of RAFT polymerization and CuAAC coupling.[71]

The high efficiency and functional group compatibility of CuAAC also makes it amenable to generating cyclic polymers utilizing other polymerization methods. For example, non-radical methods are required for polyesters, which are of interest for biological applications because of their biocompatibility and low toxicity. While much work has been done preparing cyclic polyesters using ring expansion methods,[72] a ring closure route that is compatible with living ring opening polymerizations would provide exceptional low polydispersity and access to exact linear and cyclic analogs. Hoskins and Grayson first reported the use of CuAAC to prepare cyclic poly(ϵ-caprolactone) (PCL) in 2009.[73] Using tin(II) ethylhexanoate (SnOct$_2$) as catalyst, CL was polymerized in bulk from an azido functionalized alcohol. Esterification of the terminal hydroxyl group with pentynoic anhydride yielded the desired linear precursor. MALDI-TOF mass spectra data could be used to verify near quantitative yields for the end group transformation, and the polymer could be cyclize using the same Cu(I)Br/PMDETA catalytic system previously reported (Fig. 24). The topological effects on cyclic polyester degradation were probed for the first time by monitoring the mass loss with both MALDI-TOF mass spectra and GPC during an acid catalyzed hydrolysis. As expected, the cyclic polymer exhibited a retarded loss of mass, as the first scission along the back bone converts the cyclic polymer into a linear polymer but does not results in any decrease in mass. Interestingly, the GPC data shows an initial increase in molecular size upon degradation. This results from the fact that the initial ring scission converts a cyclic polymer to a linear polymer, which actually exhibits an increased hydrodynamic volume. Only after the second ester hydrolysis on a given polymer chain does the size of the polymer actually exhibit a decrease in hydrodynamic volume. The retardation in molecular weight loss, as well as the unique initial increase in size upon degradation, gives these cyclic biodegradable polymers unique and complementary properties to those already available for biomedical applications. Other research groups have also examined the CuAAC ring closure approach with PCL and polyvalerolactone backbones.[74,75]

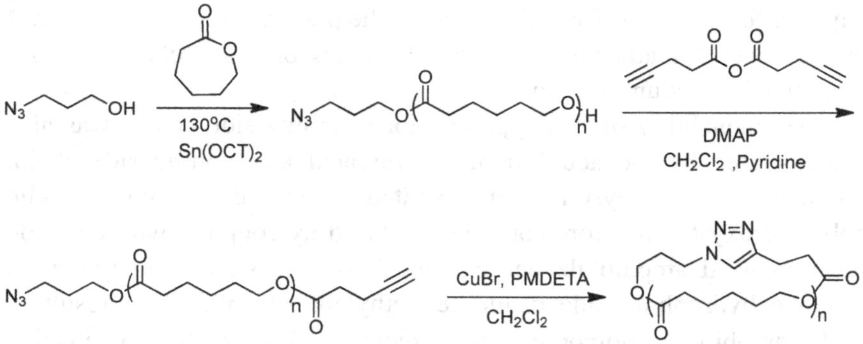

Fig. 24. The synthesis of cyclic polycaprolactone by the combination of lactone ring opening polymerization and CuAAC coupling.[73]

The CuAAC conjugation reaction can also be used to attach diverse side chains onto cyclic polymers via a tandem CuAAC cyclization, CuAAC functionalization methodology. This was first demonstrated by Laurent and Grayson during their preparation of cyclic dendronized polymers.[76] Poly-(4-acetoxystyrene) linear precursors were prepared using ATRP, followed by conversion of the terminal end group to an azide via nucleophilic displacement of the bromide. Cyclic polymers could be prepared by carrying out the CuAAC end-to-end intramolecular cyclization at high dilution, which also provided access to exactly comparable linear and cyclic analogues for control studies. The convergent "graft to" method was explored by initially functionalizing the cyclic polymer backbone "core" via hydrolysis of the acetyl protecting group, and esterification with pentynoic acid to yield an alkynyl group on each repeat unit. The pendant alkyne functionalities could then be coupling using the CuAAC reaction to any of a library of dendrons bearing a single azide group at the focal point (Fig. 25). This method allows for the efficient preparation of up to third generation dendronized cyclic materials showing approximately 90% coupling efficiency and retention of the low PDI. For comparison, a similar "graft from" route was explored using the same deprotected poly(hydroxystyrene) core, but using the iterative coupling, and deprotection steps to build the dendrons divergently in a layer by layer fashion. The increased number of post-cyclicization synthetic steps required for the "graft from" highlighted the practical advantage of "graft to" approach for attaching diverse side-chains onto a cyclic core via a single CuAAC coupling step.

The amenability of this approach for attaching side chains was also demonstrated by the addition of a bifurcated amphiphilic side chain. Laurent and Grayson demonstrated that the same cyclic poly(hydroxystyrene) core could be modified by coupling with an azide functionalized amphiphile bearing one lipophilic side chain (dodecyl) and one hydrophilic side chain (tetraethylene glycol).[77] The resulting cyclic amphiphilic homopolymers demonstrated the ability to selective encapsulate polar dyes in non-polar solvent, and exhibited a slow release of the guests into an aqueous phase when extracted with water.

Intramolecular cyclization reactions typically need to be conducted at high dilution (<0.1 g/L) to avoid interchain coupling, although the throughput has been improved substantially by the slow-addition technique. Ge et al. reported an elegant technique for making use of the selective solubilities in block copolymers to improve the cyclization rates in reduced solvent volumes.[35] Copolymers of poly(2-(2-methyoxyethoxy)ethyl methacrylate) and poly(oligo(ethylene glycol) methyl ether methacrylate) bearing complementary alkyne and azide end groups were assembled in aqueous media into micelles, separating the activated end groups. However, the small percentage that remains free from the micelle experienced high dilution, and could efficiently cyclized. Intramolecular CuAAC reactions occur exclusively for unimers, the concentration of which is well-known to be the critical micelle

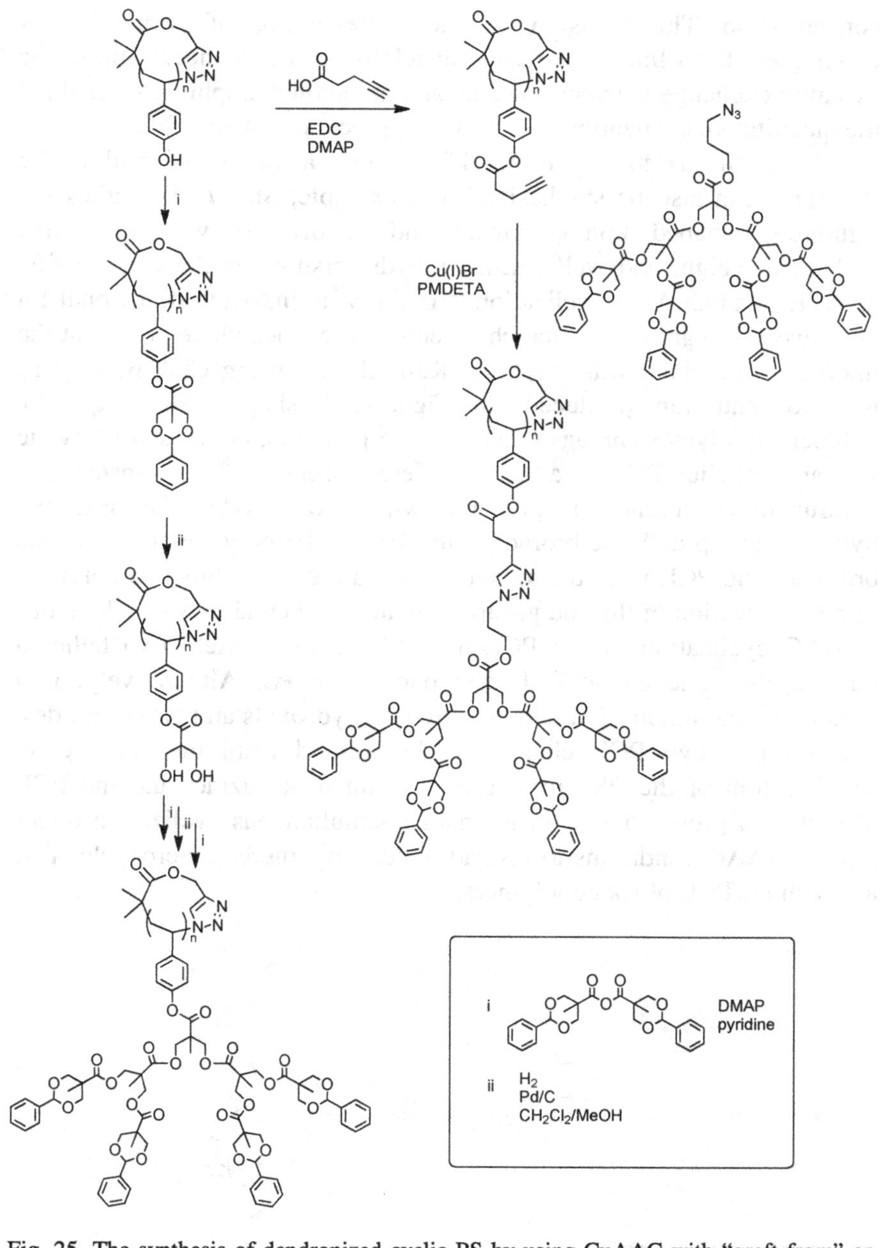

Fig. 25. The synthesis of dendronized cyclic PS by using CuAAC with "graft from" or "graft to" methods.[76]

concentration. This leads to the facile preparation of cyclic diblock copolymers from linear precursors at relatively high concentration. The dynamic exchange between the free and assembled amphiphiles enabled the quantitative cyclization of their linear precursors within 24 h.

Using this azido-alkyne "click" method, a variety of multicyclic structures can also be synthesized. For example, Shi *et al.* synthesized figure-eight-shaped homopolymers and copolymers with controlled molecular weight and well-defined polydispersities by the combination of ATRP and CuAAC cyclization.[78] Initially, a linear tetrafunctional PS with two azido groups at each chain end and two acetylene groups at the middle of the chain was prepared. Ring closure using CuAAC at very low concentration produced the figure-eight-shaped PS (Fig. 26). Diblock copolymer analogs could also be prepared, composed of cyclic PS and cyclic PCL via two different routes.[79,80] If instead, a tetrafunctional initiator is prepared with two acetylene groups, one hydroxyl group and one bromo group, PS can be polymerized from the bromide, and PCL from the alcohol to yield the linear block copolymer. Functionalization of the end groups with azides then allows simultaneous CuAAC cyclization of the PCL and PS blocks to yield two tethered macrocycles, one cyclic PCL and one cyclic PS. Alternatively, if a tetrafunctional intiator is used, bearing two hydroxyls and two bromides, two PS and two PCL chains can be grafted from the same core. Modification of the PS bromide end group to an azide, and the PCL alcohol end group to an alkyne enables simultaneous cyclization under dilute CuAAC conditions to instead yield two tethered macrocycles that are each PS/PCL block copolymers.

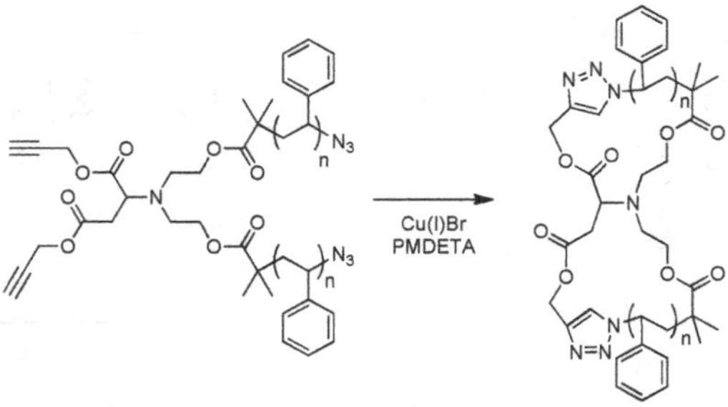

Fig. 26.The synthesis of figure-eight-shaped cyclic PS by using the CuAAC coupling of terminal azide and alkyne functionalities.[78]

Similarly, Lonsdale and Montiero demonstrated the synthesis a functional PS ring by using CuAAC cyclization that enabled the synthesis of a range of more complex cyclic architectures.[81,82] Using a trifunctional initiator bearing a hydroxyl group as well as the alkyne and the labile bromide, polymerization of styrene followed by cyclization yielded a cyclic PS with addressable alcohol functionality (Fig. 27). This alcohol could be functionalized to exhibit an alkyne or azide functionality, which in turn could be coupled with complementary functionalized linear and cyclic polymers to yield tad-pole or multicyclic polymers. Peng *et al.* also used a similar technique to prepare cross-linked cyclic polymer[83] while Wan *et al.* demonstrated the synthesis of well-defined amphiphilic and thermo-responsive tadpole-shaped linear-cyclic diblock copolymers.[84]

Although the versatility of the CuAAC cyclization route has been proven for a range of polymer backbones, some of the subtleties involving the kinetics of cyclization need to be better understood in order to increase the throughput and potential commercial viability of this route. The primary concern is that typical cyclization conditions have made use of dropwise addition into an excess of solvent using a high concentration of catalyst. Such techniques are wasteful of time, solvent, and catalyst. In order to maximize the yield of cyclic polymer while minimizing the time, solvent and catalyst, Monteiro and coworkers conducted a detailed kinetic study relying on both experimental and theoretical data to gain important insights into the optimization of the CuAAC cyclization reaction.

The Jacobson-Stockmayer theory[85] asserts that in a one-pot reaction the percentage of cyclic product formed should be independent of most reaction parameters, (e.g. catalyst concentration and temperature) but dependent instead only on polymer concentration. However, if the linear precursor is added dropwise, and at a rate below that of the rate of cyclization, a sufficiently low concentration of linear precursor can be maintained to assure high cyclic purity. To further elucidate the role of each reaction parameter, the effect of PS concentration, temperature, feed rate, Cu(I)Br concentration, and molecular weight were both modeled

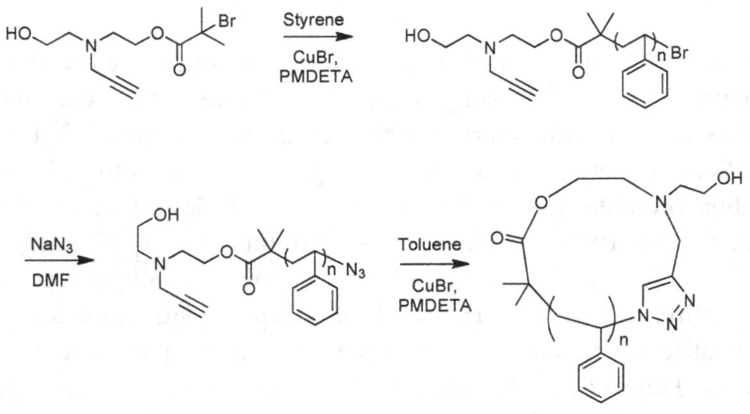

Fig. 27. The synthesis of cyclic PS with a single hydroxyl group for further functionalization.[81]

and investigated experimentally by Monteiro et al.[86,87] Because of the rapid nature of the CuAAC reaction, the use of dropwise addition vastly improved the throughput and purity of cyclic polymers, even at accelerated feed rates. For example, at 25 °C when adding a 1 mL solution of 20 mg of linear PS precursor (with 51 monomer units, $M_n = 5.1$ kDa), a feed rate of 0.124 mL min^{-1} into a 1 mL solution of Cu(I)Br (50 molar excess) yielded high purity (> 95%) cyclic polymer in less than 9 min. For higher molecular weight PS (with 104 or 136 repeat units, $M_n = 10.8$ kDa or 14.4 kDa), the feed rate had to be decreased to 0.012 mL min^{-1}, in order to maintain high levels of cyclic purity (~ 90%), and increasing the temperature to 80°C appeared to boost the rate of the CuAAC reaction, providing a slight improvement of cyclic purities. These studies confirmed that the ring closure route to cyclized polymers depended primarily on the interplay between feed rate and the rate of the coupling reaction. However, as long as the reaction rate is sufficiently rapid, as seen with the CuAAC reaction, the feed rate can be increased to enable the production of high purity macrocycle in reasonable quantities but without excessive solvent, offering promise to make the ring closure approach scalable for commercial applications.

Of the many reported heterodifunctional cyclization chemistries, the highly efficient and functional group tolerant CuAAC "click" coupling approach has been the one most frequently used because of its ability to produce cyclic polymers with high purity and narrow polydispersity. However, perhaps the most significant feature is that its functional group compatibility provides a route to prepare cyclic polymers with diverse backbone functionality and easily modified side chains, as well as facilitates the synthesis of more complex cyclic architectures.

However, this CuAAC cyclization approach, like all ring closure approaches, requires a quantitatively functionalizing both chain ends and suffers from increasing entropic challenges during the cyclization of higher molecular weight polymer. One clever technique for overcoming these types of entropic penalties at very high molecular weights was demonstrated by Schappacher and Deffieux by replacing a single pair of complementary functionalities at opposite chain ends with whole blocks of complementary functionality in an ABC triblock copolymer. Although this increases the likelihood of cyclization, it also increase the likelihood of side reactions, as cyclized chains still bear reactive functionalities, yielding mixtures of linear, cyclic, and multicyclic polymers. However, at present, the heterofunctional unimolecular coupling technique appears most useful of the methods yet reported for providing of well-defined, low molecular weight cyclic polymers with broad functional group compatibility.

4. Ring-Closure versus Ring-Expansion

The ring expansion technique offers a number of complementary features relative to the ring closure approach. Ring-expansion polymerizations involve the repeated insertion of cyclic monomer into a cyclic catalyst or initiator to yield a polymer macrocycle.[88] Typically thermodynamic factors, such as ring strain in the monomer, are the predominant driving force of polymerization to form large polymer macrocycles. Because the growing polymer maintains a cyclic conformation throughout the polymerization, it is not subject to the high entropic penalties associated with the ring closure methods. As an additional consequence, high dilution is not required, circumventing the need for excess solvent. In addition, this approach is less susceptible to the formation of linear by-products, as long as the monomer and initiator are rigorously purified to remove any linear contaminants. As a net result, this technique is particularly useful for preparing extremely high molecular weight polymer macrocycles as seen during ring expansion metathesis polymerizations[89] and N-heterocyclic carbene catalyzed lactone cyclopolymerizations,[90] yielding cyclic polymers with molecular weights as high as M_n = 6,000 kDa.[91] The primary disadvantages for the ring-expansion approach are the limited compatibility monomers (usually strained cyclic olefins or lactones with limited side chain functionality) and the lack of control over chain termination, which leads to cyclic polymers of broad polydispersity (>1.3). For much larger polymers, the ring expansion approach remains the most appealing synthetic approach, whereas for smaller, well-defined, and functional polymers, the ring closure approach is generally preferred.

5. Conclusions

The synthesis, modification, and physical characterization of polymer macrocycles still remains an area of interest due to the unique properties of these structures and the synthetic challenges associated with their preparation. While the ring expansion approach has come to the fore as the preferred approach for preparing extremely large cyclic polymers, the technical ease and synthetic precision exhibited by the ring closure approach have assured its popularity as the most frequently used technique at present. The most critical feature for a successful ring closure approach is the use of a highly efficient coupling reaction that exhibits negligible side reactions. While the homodifunctional ring closure represents the most technically simple route to preparing precursors, the lack of highly efficient, functional group tolerant homo-coupling reactions has limited its more broad usage. The heterodifunctional ring closure exhibits the most versatility, because of the breadth of the coupling reactions which can be used, and foremost among those investigate to date is the CuAAC ring closure. The ease of installing azide and alkyne functional groups, the high efficiency of the reaction, and the broad functional group compatibility have led to its application for the preparation of a wide variety of cyclic polymers and more complex cyclic architectures. It is expected that the synthetic versatility exhibited by the CuAAC reaction and other highly efficient "click" ring closure techniques, will lead to many future innovations in this field of cyclic polymers and complex cyclic topologies.

Acknowledgments

This work was supported by the National Science Foundation (NSF-CAREER 0844662), the Louisiana Alliance for Simulation-Guided Materials Applications (NSF EPSCOR). The authors would like to thanks Profs. Yasuyuki Tezuka and Michael Monteiro for valuable feedback during the writing of this chapter.

References

1. P. Ruggli, *Liebigs. Ann. Chem.*, **392**, 92 (1912).
2. K. Ziegler, H. Eberle, and H. Ohlinger, *Liebigs. Ann. Chem.*, **504**, 94 (1933).
3. L. Ruzicka, *Chem. Ind. (London)*, **54**, 2 (1935).
4. A. Stern, W. A. Gibbons, and L. C. Craig, *P. Natl. Acad. Sci. Usa.*, **61**, 734 (1968).
5. F, J.; EL, W. *Sym. Soc. Exp. Biol.*, **12**, 75 (1958).
6. S. D. Ross, E. R. Coburn, W. A. Leach, and W. B. Robinson, *J. Polym. Sci.*, **13**, 406 (1954).
7. H. R. Kricheldorf, *Macromolecules*, **36**, 2302 (2003).
8. K. Dodgson, and J. A. Semlyen, *Polymer*, **18**, 1265 (1977).
9. K .Dodgson, D. Sympson, and J. A. Semlyen, *Polymer*, **19**, 1285 (1978).
10. H. C. Lee, H, Lee, W, Lee, T, Chang, and J. Roovers, *Macromolecules*, **33**, 8119 (2000).
11. S. Singla, T. Zhao, and H. W. Beckham, *Macromolecules*, **36**, 6945 (2003).
12. B. A. Laurent, and S. M. Grayson, *Chem. Soc. Rev.*, **38**, 2202 (2009).
13. H. Oike, T. Mouri, and Y. Tezuka, *Macromolecules*, **34**, 6592 (2001).
14. D. Geiser, and H. Höcker, *Macromolecules*, **13**, 653 (1980).
15. G. Hild, A. Kohler, and P. Rempp, *Eur. Polym. J.*, **16**, 525 (1980).
16. B. Vollmert, and J. Huang, *Makromol. Chem-Rapid.*, **1**, 333 (1980).
17. G. Hild, C. Strazielle, and P. Rempp, *Eur. Polym. J.* **19** , 721 (1983).
18. H. Oike, H. Imaizumi, T. Mouri, Y. Yoshioka, A. Uchibori, and Y. Tezuka, *J. Am. Chem. Soc.*, **122**, 9592(2000).
19. Y. Tezuka, K. Mori, and H. Oike, *Macromolecules*, **35**, 5707 (2002).
20. H. Oike, M. Hamada, S. Eguchi, Y. Danda, and Y. Tezuka, *Macromolecules*, **34**, 2776 (2001).
21. Y. Tezuka, *J. Polym. Sci. Polym. Chem.*, **41**, 2905 (2003).
22. N. Sugai, H. Heguri, K. Ohta, Q. Meng, T. Yamamoto, and Y. Tezuka. *J. Am. Chem. Soc.*, **132**, 14790 (2010),.
23. C. E. Hoyle, and C. N. Bowman, *Angew. Chem.-Int. Edit.*, **49**, 1540 (2010).
24. C. Hoyle, T. Lee, and T. Roper, *J. Polym. Sci. Polym. Chem.*, **42**, 5301 (2004).
25. M. J. Stanford, R. L. Pflughaupt, and A. P. Dove, *Macromolecules*, **43**, 6538 (2010).
26. A. Dondoni, *Angew. Chem.-Int. Edit.*, **47**, 8995 (2008).

27. Y. Tezuka, and R. Komiya, *Macromolecules*, **35**, 8667 (2002).
28. S. Hayashi, K. Adachi, and Y. Tezuka, *Chem. Lett.*, **36**, 982 (2007).
29. V. Coessens, and K. Matyjaszewski, *J. Macromol. Sci.-Pure Appl. Chem.*, **A36**, 667 (1999).
30. K. Adachi, S. Honda, S. Hayashi, and Y. Tezuka, *Macromolecules*, **41**, 7898 (2008).
31. S. Honda, T. Yamamoto, and Y. Tezuka, *J. Am. Chem. Soc.*, **132**, 10251 (2010).
32. T. Yamamoto, and Y. Tezuka, *Eur. Polym. J.*, **47**, 535 (2011).
33. M. R. Whittaker, Y.-K. Goh, H. Gemici, T. M. Legge, S.Perrier, and M. J. Monteiro, *Macromolecules*, **39**, 9028 (2006).
34. J. Hu, R. .Zheng, J. Wang, L. Hong, and G. Liu, *Macromolecules*, **42**, 4638 (2009).
35. Z. Ge,; Y. Zhou, J. Xu, H. Liu, D. Chen, and S. Liu, *J. Am. Chem. Soc.*, **131**, 1628 (2009).
36. C.W. Chen, C.C Cheng, and S. A. Dai, *Macromolecules*, **40**, 8139 (2007).
37. P. Siemsen, R. Livingston, and F. Diederich, *Angew. Chem.-Int. Edit.*, **39**, 2633 (2000).
38. Y. Zhang, G. Wang, and J. Huang, *Macromolecules*, **43**, 10343 (2010).
39. Y. Zang, G.Wang, and J. Huang, *J. Polym. Sci. Polym. Chem.*, **49**, 4766 (2011).
40. G. Wang, X. Fan, B. Hu, Y. Zhang, and J. Huang, *Macromol. Rapid Commun.*, **32**, 1658 (2011).
41. M. Schappacher, and A. Deffieux, *J. Am. Chem. Soc.*, **133**, 1630 (2011).
42. M. Schappacher, and A. Deffieux, *Macromolecules*, **44**, 4503 (2011).
43. A. F. Voter, E. S. Tillman, *Macromolecules*, **43**, 10304 (2010).
44. Z. Jia, and M. J. Monteiro, *J. Polym. Sci. Pol. Chem.*, **48**, 4496 (2010).
45. M. Schappacher, and A. Deffieux. *Makromol. Chem-Rapid*, **12**, 447 (1991).
46. M. Schappacher, and A. Deffieux. *Macromolecules*, **28**, 2629 (1995).
47. M. Schappacher, and A. Deffieux. *Angew. Chem.-Int. Edit.*, **48**, 5930 (2009).
48. L. Rique-Lurbet, M. Schappacher, and A. Deffieux. *Macromolecules*, **27**, 6318 (1994).
49. A..Deffieux, M. Schappacher, and L. Rique-Lurbet, *Macromol. Symp.*, **95**, 103 (1995).
50. M. Schappacher, and A. Deffieux. *Macromolecules*, **34**, 5827 (2001).
51. M. Schappacher, and A. Deffieux. *Macromolecules*, **28**, 2629 (1995).
52. M. Schappacher, and A. Deffieux. *Science*, **319**, 1512 (2008).
53. M. Schappacher, and A. Deffieux. *J. Am. Chem. Soc.*, **130**, 14684 (2008).
54. M. Kubo, T. Hayashi, H. Kobayashi, K. Tsuboi, and T. Itoh, *Macromolecules*, **30**, 2805 (1997).
55. M. Kubo, T. Hayashi, H. Kobayashi, and T. Itoh, *Macromolecules*, **31**, 1053 (1998).
56. B. Wood, P. Hodge, and J. Semlyen, *Polymer*, **34**, 3052 (1993).
57. C. Hawker, G. Barclay, A. Orellana, J. Dao, and W. Devonport, *Macromolecules*, **29**, 5245 (1996).
58. J. Wang, and K. Matyjaszewski, *J. Am. Chem. Soc.*, **117**, 5614 (1995).

59. J. Chiefari, Y. Chong, F. Ercole, J. Krstina, J. Jeffery, T. Le, R. Mayadunne, G. Meijs, C. Moad, G. Moad, E. Rizzardo, and S. Thang, *Macromolecules*, **31**, 5559 (1998).
60. B. Lepoittevin, X. Perrot, M. Masure, and P. Hemery *Macromolecules*, **34**, 425 (2001).
61. H. Kolb, M. Finn, and K. Sharpless, *Angew. Chem.-Int. Edit.* **40**, 2004 (2001)
62. C. Tornoe, C. Christensen, and M. Meldal, *J. Org. Chem.* **67**, 3057 (2002)
63. J.F. Lutz, *Angew. Chem.-Int. Edit.*, **46**, 1018 (2007).
64. B. Laurent, and S. Grayson, *J. Am. Chem. Soc.*, **128**, 4238 (2006).
65. J. Xu, J. Ye, and S. Liu, *Macromolecules*, **40**, 9103 (2007).
66. D. Han, X. Tong, Y. Zhao, T. Galstian, and Y. Zhao, *Macromolecules*, **43**, 3664 (2010).
67. X. Zhu, N. Zhou, Z. Zhang, B. Sun, Y. Yang, J. Zhu, and X. Zhu, *Angew. Chem.-Int. Edit.*, **50**, 6615 (2011).
68. F. Chen, G. Liu, and G. Zhang, *J. Polym. Sci. Polym. Chem.*, **50**, 831 (2012).
69. D. M. Eugene, and S. M. Grayson, *Macromolecules*, **41**, 5082 (2008).
70. X.P Qiu, F, Tanaka, F. M. Winnik, *Macromolecules*, **40**, 7069 (2007).
71. A. S. Goldmann, D. Quemener, P. E. Millard, T. P. Davis, M. H. Stenzel, C. Barner-Kowollik, and A. H. E.Wuller, *Polymer*, **49**, 2274 (2008).
72. H. Kricheldorf, and G. Schwarz, *Macromol. Rapid Commun.*, **24**, 359(2003).
73. J. N. Hoskins, and S. M. Grayson, *Macromolecules*, **42**, 6406 (2009).
74. H. Misaka, R. Kakuchi, C. Zhang, R.Sakai, T. Satoh, and T. Kakuchi, *Macromolecules*, **42**, 5091 (2009).
75. M. Xie, J. Shi, L. Ding, J. Li, H. Han, and Y. Zhang, *J. Polym. Sci. Polym. Chem.*, **47**, 3022 (2009).
76. B. A. Laurent, and S. M. Grayson, *J. Am. Chem. Soc.*, **133**, 13421 (2011).
77. B. A. Laurent, and S. M. Grayson, *Polym. Chem.*, (2012) in press 10.1039/c1py00378j.
78. C. Y. Pan, and G. Y. Shi, *Macromol. Rapid Commun.*, **29**, 1672 (2008).
79. G. Y. Shi, L. P. Yang, and C. Y. Pan, *J. Polym. Sci. Polym. Chem.*, **46**, 6496 (2008).
80. G. Y. Shi, and C. Y.Pan, *J. Polym. Sci. Polym. Chem.*, **47**, 2620 (2009).
81. D. E. Lonsdale, and M. J. Monteiro, *Chem. Commun.*, **46**, 7945 (2010).
82. D. E. Lonsdale, and M. J. Monteiro, *J. Polym. Sci. Polym. Chem.*, **49**, 4603 (2011).
83. Y. Peng, H. Liu, X. Zhang, S. Liu, and Y. Li, *Macromolecules*, **42**, 6457 (2009).
84. X. Wan, T. Liu, and S. Liu, *Biomacromolecules*, **12**, 1146 (2011).
85. H. Jacobson, and W.H. Stockmayer, *J.Chem. Phys.* **18**, 1600 (1950).
86. D. E. Lonsdale, C. A. Bell, and M. J. Monteiro, *Macromolecules*, **43**, 3331 (2010).
87. D. E. Lonsdale, and M. J. Monteiro, *J. Polym. Sci. Pol. Chem.*, **48**, 4496 (2010).
88. H. Kricheldorf, and S.Lee, *Macromolecules*, **28**, 6718 (1995).
89. C. Bielawski, D. Benitez, R. Grubbs, *Science*, **297**, 2041 (2002).
90. D. A. Culkin, W. Jeong, S. Csihony, E. D. Gomez, N. R. Balsara, J. L. Hedrick, R. M. Waymouth, *Angew. Chem. Int. Edit*, **46**, 2627 (2007).
91. Y. Xia, A. J. Boydston, and R. H. Grubbs, *Angew. Chem.-Int. Edit.*, **50**, 5882 (2011).

CHAPTER 11

CYCLIC MACROMONOMERS: SYNTHESIS AND PROPERTIES

Masataka Kubo

Graduate School of Regional Innovation Studies, Mie University
1577 Kurima-machiya, Tsu 514-8507, Japan
E-mail: kubo@chem.mie-u.ac.jp

Polymerization of a monomer in the presence of cyclic macromonomer (a cyclic polymer with polymerizable group) in high concentration leads to mechanically linked three-dimensional network polymer. The network structure is formed by the threading of the cyclic moiety by a segment of another polymer chain during the copolymerization process. This class of materials has attracted a great deal of interest as new functional materials with unusual chemical, physical and mechanical properties due to the high degrees of freedom in segmental movement since cross-linking points can move in these three-dimensional structures.

1. Introduction

Cross-linking is to link one polymer chain to another to lead three-dimensional polymer network. Two most important methods in cross-linking process are chemical and physical cross-linking. Chemical cross-linking is a process that introduces covalent bonding between polymer chains using two or more functionalized compounds as chemical cross-linking agents (cross-linkers). Chemical covalent cross-links are stable mechanically and thermally, so once formed are difficult to break. Physical cross-linking utilizes weaker physical interactions such as ionic interaction, hydrogen bonding, and crystallization to connect polymer chains each other. Another recently approach for constructing polymer network structures has been developed utilizing mechanical cross-linking through chain threading of ring segment. Typical examples of such network structures are shown in Fig. 1. The three-dimensional structures are quite different from those of conventional network polymers formed by chemical or physical cross-linking. Mechanical cross-linking can introduce movable cross-linking points into the network structures. These network polymers are expected to exhibit good swelling properties and high impact strength due to high degrees of freedom in the segmental movement since cross-linking points can move.

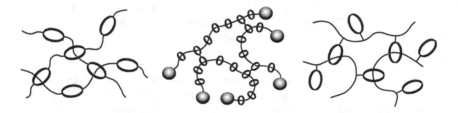

Fig. 1. Examples of network polymers formed by mechanical cross-linking.

2. Preparation of Mechanically Cross-Linked Polymer

There are several synthetic methods to introduce mechanical cross-linking into polymer network depending on the type of three-dimensional structure. Delaviz and Gibson carried out the direct polycondensation reactions between diamines and dicarboxylic acid carrying 32-crown-10 moiety to obtain insoluble poly(amide crown ether)s.[1] The insolubility is attributed to in situ threading of linear segments of one polymer chain through the macrocyclic cavity of another during polymerization (Fig. 1, left). Okumura and Ito successfully prepared mechanically cross-linked polymer by introducing covalent bonding between ring segments of polyrotaxane consisting of α-cyclodextrin (α-CD) and polyethylene glycol (PEG) (Fig. 1, center).[2] Three-dimensional structure composed of mechanical cross-linking can be obtained by a copolymerization of a monomer with a cyclic macromonomer which is a cyclic polymer carrying polymerizable functional group (Fig. 1, right). Figure 2 shows the mechanism for this type of cross-linked polymer. The threading of the cyclic moiety by a segment of another polymer chain during the copolymerization leads the cross-linked structure. In this case a cyclic macromonomer can be regarded as a non-covalent cross-linking agent to introduce movable cross-linking points into polymer network.

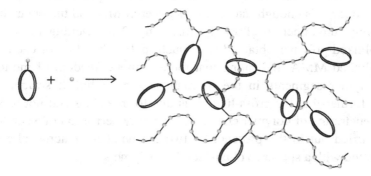

Fig. 2. Preparation of mechanically cross-linked polymer using cyclic macromonomer.

3. Cyclic Macromonomer Based on Polyether

The first example of cyclic macromonomer as a noncovalent cross-linking agent was reported by Zada *et al.*[3] They prepared cyclic octaethylene glycol fumarate and carried out its radical copolymerization with styrene (St) or methyl methacrylate (MMA) to obtain insoluble cross-linked products. Since the homopolymer of the cyclic octaethylene glycol fumarate is soluble, the cross-linking could have originated from the threading of the vinyl polymer chain in the ring of the fumaryl ester in the copolymer.

Another cyclic macromonomer based on a cyclic octaethylene glycol is a cyclic octaethylene glycol 5-methacrylamide-isophthalate.[4,5] While the copolymerization of the cyclic methacrylamide monomer with styrene lead a soluble polymer, its copolymerization with MMA gave insoluble cross-linked polymer. Since MMA is polar monomer, it is expected there is enough interaction between MMA and the polar ring of the cyclic monomer to allow the threading. The swelling ratio of the cross-linked polymer obtained was much higher than that of chemically cross-linked MMA, indicating that there is more freedom of motion of the polymer segments in the mechanically cross-linked structure. To prove further that the cross-linking is due to threading and not to some interpenetrating of polymer chains, radical polymerization of St or MMA was carried out in the presence of two 30-memberd macrocyclic rings connected with a spacer to obtain soluble polymers.[5]

Oike *et al.*, designed a methacrylate-functionalized cyclic poly (tetrahydrofuran) (PTHF) for an efficient noncovalent cross-linking agent.[6] Its copolymerization with MMA in benzene using 2,2'-azobisisobutyronitrile (AIBN) as a radical initiator gave copolymer product with cyclic PTHF branches initially. Gelation took place at a later stage of the reaction. In contrast, no gelation was observed either in the copolymerization of MMA with a related methacrylate-functionalized, open-chain PTHF or in the homopolymerization of MMA in the presence of cyclic PTHF without methacrylate functionality. These experimental results indicate that the gel formation was not due to the side reactions but to the threading of the cyclic PTHF by the propagating polymer backbone to lead to three-dimensional structure.

4. Cyclic Macromonomer Based on Polystyrene

Cyclic macromonomers based on a cyclic polystyrene (PSt) were developed. Since PSt chain is harder than aliphatic polyether chains, it is expected to form a relatively larger diameter for easy threading. The key compound for cyclic PSt-based macromonomers is a cyclic PSt carrying amine functionality which was synthesized from a well-defined α-amino, ω-carboxyl heterodifunctional PSt.[7]

To introduce polymerizable group into the macrocyclic PSt amine, the condensation reaction with vinylbenzyl chloride was carried out to obtain a cyclic macromonomer, St-*c*-pSt.[8,9]

St-*c*-pSt

Its radical copolymerization with *tert*-butyl acrylate in benzene gave an insoluble cross-linked polymer with a good swelling property. The degradation of the mechanically linked polymer network into soluble polymers can be possible if the pendant cyclic polystyrene segments are detached from the backbone chain. This will be achieved by C–N bond cleavage. Actually, the insoluble polymer became soluble in organic solvent after treatment with 1-chloroethyl chloroformate as a debenzylation agent of tertiary amines. Further, macrocyclic polystyrene amine was recovered from the degradation products. This reveals that the chain threading of the cyclic polystyrene by another chain is responsible for the gelation to form a mechanically cross-linked polymer.

Cross-linked PSt gels are widely used in ion exchange resins, solid supports in combinatorial chemistry, various paints, and fillers. However, it was found that AIBN-initiated radical copolymerization of St-*c*-PSt with St gave soluble polymers instead of gel. To give a network structure, every chain must have at least one threaded cyclic segment. One approach is to increase macromonomer concentration in order to introduce enough amounts of cyclic segments into a polymer chain. However, increasing St-*c*-PSt concentration resulted in the decrease of the molecular weight of the resulting polymer due to the chain transfer reaction. Another approach is to lengthen the polymer chain utilizing an emulsion polymerization technique. Table 1 summarizes the emulsion polymerization of St in the presence of St-*c*-PSt.

Table 1. Emulsion copolymerizations[a] of St-*c*-PSt with St.

[St-*c*-PSt]/[St]	THF-insoluble part Yield (%)	THF-soluble part Yield (%)	$M_n \times 10^{-5}$ [b]
	M_n of St-*c*-PSt = 2500 (DP = 22)		
1/960	30	62	6.8
1/480	37	56	6.6
1/320	33	59	6.0
	M_n of St-*c*-PSt = 3600 (DP = 32)		
1/960	56	38	6.2
1/480	47	51	6.0
1/320	68	31	5.9
	M_n of St-*c*-PSt = 4900 (DP = 45)		
1/960	61	34	4.0
1/480	64	30	3.8
1/320	63	31	3.0

[a] Conditions: St = 1.0 g; $K_2S_2O_8$ = 5 mg; $CH_3(CH_2)_{16}O_2Na$ = 50 mg; 1% PVAL = 5 mL; Temperature = 80 °C; Time = 10 h. [b] Determined by GPC.

The emulsion copolymerization successfully gave THF-insoluble cross-linked polymers. Judging from the molecular weight of the THF-soluble fraction, sufficient amounts of cyclic macromonomer were seemed to be incorporated into the polymer chain. It is apparent that cyclic macromonomer with larger ring size is more effective to form cross-linked polymer probably due to the easier threading of larger ring. Varying the cyclic macromonomer concentration had a negligible effect on the yield of gel fraction. One possible explanation may be an insufficient supply of the cyclic macromonomer into the micelle where the polymerization actually occurs.

It was found that efficient cross-linking is possible utilizing a thermal self-initiated copolymerization of St-*c*-PSt with St. In this case the gel yield was greatly dependent on the cyclic macromonomer concentration. The polymerization results are summarized in Table 2.

Table 2. Thermal self-initiated copolyerizations[a] of St-*c*-PSt with St.

| [St-*c*-PSt]/[St] | THF-insoluble part | THF-soluble part | |
	Yield (%)	Yield (%)	M_n x 10^{-5}[b]
0	0	97	5.8
1/1900	82	16	5.0
1/960	84	13	4.2
1/480	97	2	1.6

[a] Conditions: St = 1.0 g; Temperature = 80 °C; Time = 48 h; M_n of St-*c*-PSt = 4900.
[b] Determined by GPC.

In order to introduce movable cross-linking points into unstabilized vinyl polymers such as poly(vinyl acetate) (PVAc), resonance-unstabilized cyclic macromonomer, VE-*c*-PSt, was prepared.[10] VE-*c*-PSt is based on a cyclic PSt carrying vinyl ether moiety as a polymerizable group.

VE-*c*-pSt

It is known that radical polymerization of vinyl acetate (VAc) in solution gives relatively short and branched polymers due to chain transfer reaction. Actually, radical copolymerization of VE-*c*-PSt with VAc gave cross-linked poly(vinyl acetate) (PVAc) with relatively low yield (below 22%). Especially the gel yield was very low when the copolymerization was carried out at high temperature at 80 °C. To improve the gel yield, emulsion polymerization of VAc was carried out in the presence of VE-*c*-PSt. The polymerization results are summarized in Table 3. Although emulsion polymerization technique improved the chance of chain threading for network formation, the gel yield was modest (ca. 50%). In the radical copolymerization between 2-chloroethyl vinyl ether (M_1) and VAc (M_2), r_1 and r_2 are reported to be 0.16 and 2.36, respectively,[11] suggesting that VE-*c*-PSt may be reluctant to participate in the copolymerization.

Table 3. Emulsion copolymerizations^a of VE-*c*-PSt with VAc.

[VE-*c*-PSt]/[VAc]	MeOH-insoluble part	MeOH-soluble part	
	Yield (%)	Yield (%)	$M_n \times 10^{-5 \, b}$
0	0	99	1.5
1/1800	16	76	1.0
1/1200	49	51	1.1

^{a} Conditions: VAc = 1.0 g; $K_2S_2O_8$ = 3 mg; $C_{12}H_{25}OSO_3Na$ = 100 mg; 1% PVAL = 5 mL; Temperature = 70 °C; Time = 12 h; M_n of VE-*c*-PSt = 2800. ^{b} Determined by GPC.

The mechanically cross-linked PVAc can be converted to poly(vinyl alcohol) (PVAL) by the base-catalyzed hydrolysis reaction.

5. Cyclic Macromonomer with Pentamethylcyclotrisiloxane

Poly(dimethylsiloxane) (PDMS) and other silicone polymers have been widely used in various fields due to their elastic behavior, good thermal stability, low surface energy, and bio-compatibility.[12,13] Silicone elastmers are high-molecular-weight linear polymers and can be cross-linked by free-radical catalysts or by curing reactions including the reaction of Si-H with Si-vinyl polymers to give covalently bonded three-dimensional structures. Introduction of movable cross-linking points into PDMS network is expected to improve flexibility of the resulting silicone rubber due to the increased segmental movement of the PDMS chains. Such materials will find potential applications including vibration dampers, shock absorbers, and rubbers. Garrido *et al.* reported on an original approach for the synthesis of PDMS network having no chemical cross-linking (Olympic Gel) by carrying out end-linking reaction of difunctional linear PDMS in the presence of cyclic PDMS.[14]

Two cyclic macromonomers for introducing movable cross-linking points into PDMS were designed.[15] D₃-*c*-PSt and D₃-*c*-PDMS are based on a cyclic PSt and cyclic PDMS, respectively. These cyclic macromonomers carry a cyclic trisiloxane moiety as a polymerizable group.

D₃-*c*-PSt   D₃-*c*-PDMS

Anionic ring-opening polymerization of octamethylcyclotetra-siloxane (D₄) was carried out in the presence of D₃-*c*-PSt using potassium hydroxide as an initiator. The polymerization results are summarized in Table 4. Gel formation was observed when nitrobenzene was employed as the solvent. The obtained insoluble polymer was elastic solid with good elongation. It was strong and flexible enough to be stretched to over ten times its initial length and completely relax back to its original shape probably due to the nature of movable cross-linking points. After recovering the formed insoluble gel, the remaining soluble products were analyzed by gel permeation chromatography (GPC) and matrix-assisted laser desorption/ionization time-of-flight mass spectroscopy (MALDI TOF MS). The results suggested that the moderate gel yield (below 60%) was due to the waste of D_4 monomer by ring-chain equilibration and depolymerization reactions. The reluctant threading between the incompatible components (PSt and PDMS) may be another reason.

Table 4. Copolymerizations[a] of D_3-c-PSt with D_4.

[D_3-c-PSt]/[D_4]	Solvent	Temperature, °C	Gel, %	% swelling[b]
1/500	THF	60	0	
1/500	o-dichlorobenzene	110	0	
1/500	nitrobenzene	140	59	2,400
1/250	nitrobenzene	140	54	2,100
1/100	nitrobenzene	140	47	730

[a] Conditions: D_4 = 2.0 g; KOH = 1.6 mg; Solvent = 2 mL; Time = 12 h; M_n of D_4-c-PSt = 2500. [b] In hexane.

The major drawback of the copolymerization between D_4 and D_3-c-PSt is lack of processability. The resulting cross-linked polymers precipitated from the reaction system making it difficult to process these materials into shaped article. The cyclic macromonomer, D_3-c-PDMS, is based on a cyclic PDMS to provide compatibility between linear and chain segments. D_3-c-PDMS is a viscous oily material and is miscible with PDMS. Bulk ring-opening polymerization of D_4 was carried out in the presence of D_3-c-PDMS using potassium methoxide as an initiator. The reaction mixture was changed from a liquid to transparent elastic solid. The cross-linked polymers were obtained in good yields. The polymerization results are summarized in Table 5. The swelling property of the gel decreased with cyclic macromonomer content, indicating that higher amounts of D_3-c-PDMS gave PDMS gel with higher cross-linking density.

Table 5. Copolymerizations[a] of D_3-c-PDMS with D_4.

[D_3-c-PDMS]/[D_4]	Gel, %	% swelling[b]
1/500	88	1,200
1/400	93	1,100
1/300	92	920
1/200	86	790

[a] Conditions: D_4 = 2.0 g; CH_3OK = 1.0 mg; Temperature = 140 °C; Time = 12 h; M_n of D_4-c-PSt = 2900. [b] In hexane.

6. Water-Soluble Cyclic Macromonomer

The preparation of hydrogels has received significant interest in recent years with promising application in biosensors, membranes, molecular imprinting, and drug delivery devices.[16–18] Environmentally responsive hydrogels show drastic changes in their swelling ratio in response to a variety of environmental stimuli such as external pH, temperature, ionic strength, and solvent composition.[19–21] Most such hydrogels have been prepared by either chemical or physical cross-linking of water-soluble polymers. Introduction of mechanical cross-linking into hydrogel can enhance the swelling properties due to the high degrees of freedom in the segmental movement. Polymerizations of water-soluble cyclic macromonomer with a water-soluble vinyl monomer will give a hydrogel with movable cross-linking points.

A water-soluble cyclic macromonomer, St-c-PKA, based on a cyclic potassium polyacrylate (PKA) was prepared.[22] The cyclic PKA chain is expected to provide a stiff and extended macrocyclic moiety because of the intramolecular electrostatic repulsions between charged segments (carboxylate anions) located along the polymer backbone in an aqueous solution. This should increase the chance of threading leading a mechanically cross-linked structure.

St-*c*-PKA

The cyclic macromonomer, St-*c*-PKA, was copolymerized with acrylamide (AAm) in water using 2,2'-azobis[2-methyl-*N*-(2-hydroxyethyl)propionamide] (VA-086; Wako) Table 6 summarizes the results of the polymerization. Gel formation was observed when more than 0.1 mol% of cyclic macromonomer was present in the polymerization system.

Table 6. Copolymerizations[a] of St-*c*-PKA with AAm.

[St-*c*-PKA]/[AAm]	Gel, %	% swelling[b]
0	0	-
1/2000	Trace	-
1/1000	88	270
1/500	93	150

[a] Conditions: AAm = 0.86 g; H_2O = 1.0 mL; VA-086 = 14 mg; Temperature = 85 °C; Time = 25 min; M_n of St-*c*-PKA = 1770. [b] In water.

The obtained polyacrylamide gel was subject to an acid treatment. A mixture of the swollen gel, methanol, water, and hydrochloric acid was heated under reflux. The gel gradually dissolved and finally gave a clear, homogeneous solution. Since the gel remained unchanged after stirring in boiling water, the degradation of the three-dimensional structure was due to the acid-catalyzed hydrolysis of the amide linkage in the cyclic PKA.

Ricka and Tanaka reported the swelling behavior of acrylic acid-acrylamide copolymer gel in water solution of Cu(II) salt.[23] The observed volume shrinking is induced by complexation between Cu^{2+} and carboxyl groups. To investigate the effect of mechanical cross-linking, a mechanically cross-linked polyelectrolyte gel which was composed of polyacrylamide and sodium polyacrylate with 8:2 composition was prepared. The obtained polyelectrolyte gel exhibited a slow but large volume shrinkage in copper(II) chloride solution.

The threading ability of the ring plays an important role in the forming of mechanical cross-linking. A cyclic macromonomer, St-*c*-

PAA, based on a cyclic polyacrylic acid (PAA) was prepared and copolymerized with *N*-isopropylacrylamide (NIPAAm) in various solvents.[24] The results are summarized in Table 7. The gel yield and swelling property were greatly dependent on the solvent used. The most effective solvent for cross-linking was methanol which is a good solvent for PAA. In methanol, the conformation of cyclic segment of St-*c*-PAA should adopt a more expanded conformation, which explains effective threading.

St-*c*-PAA

Table 7. Copolymerizations[a] of St-*c*-PAA with NIPAAm.

[St-*c*-PAA]/[NIPAAm]	Solvent	Gel, %	% swelling[b]
0	DMSO	0	
1/300	DMSO	0	
1/100	DMSO	87	5,600
1/100	MeOH	93	3,100
1/100	DMF	68	7,500
1/100	DMAc	56	8,200

[a] Conditions: NIPAAm = 0.23 g; Solvent = 150 μL; BPO = 0.4 mg; *N, N*-Dimethylaniline = 0.4 μL; Temperature = rt; Time = 12 h; M_n of St-*c*-PAA = 2500. [b] In methanol.

In the case of mechanically cross-linked polymer, the swelling property is greatly dependent on the mobility of the cross-linking points. Figure 2 shows the relationship between the swelling ratio and pH for the PNIPAAm gel prepared by the copolymerization of St-*c*-PAA with

NIPAAm. The equilibrium swelling ratio of the gel increased with increasing pH. The hydrogel exhibited much higher swellability than that in pure water. This means that cross-linking points can move easily in basic solution. Such an easy movement of the cross-linking points can be explained by considering the shape of the cyclic PAA segment. An charged cyclic polymer chain in basic solution will adopt more expanded conformation due to the repulsion between carboxylate anions on the polymer chain.

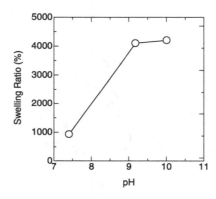

Fig. 2. Relationship between swelling ratio and pH at 15 °C.

Among environmentally responsive hydrogels, PNIPAAm hydrogel is the best known temperature sensitive material.[25-27] As a temperature sensitive hydrogel, the response rate upon external temperature changes is important for special application including artificial organs,[28] actuators,[29,30] and on-off switches.[31] In order to improve the response dynamics, various strategies were reported including phase separation technique,[32] introduction of macroporous structure,[33,34] and grafting of freely mobile hydrophilic chain into network.[35,36] Recently, a novel type of temperature sensitive gel based on PNIPAAm was fabricated using a hydrophilic polyrotaxane-based movable cross-linker.[37]

The cyclic macromonomer, AAm-*c*-PEG, was designed to introduce movable cross-linking points into three-dimensional PNIPAAm network structure.[38,39] AAm-*c*-PEG is based on a water-soluble cyclic PEG and carries an acrylamino group as a polymerizable component.

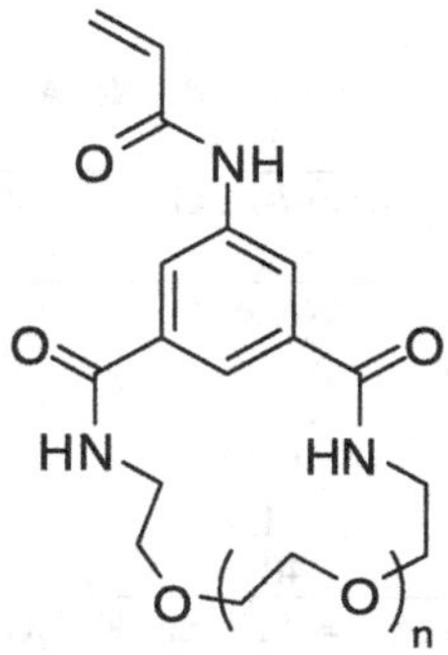

AAm-*c*-PEG

The PNIPAAm hydrogel was prepared by the copolymerization of AAm-*c*-PEG with NIPAAm. Since introduction of PEG chains into PNIPAAm network is reported to enhance gel deswelling rates,[40] chemically cross-linked PEG-grafted PNIPAAm gel with the same PEG content was prepared as a control hydrogel to compensate the effect of PEG segment. Radical copolymerizations of NIPAAm with AAm-*l*-PEG were carried out in the presence of methylene bis(acrylamide) (MBAAm) as a cross-linking agent to obtain chemically cross-linked PEG-grafted PNIPAAm gels.

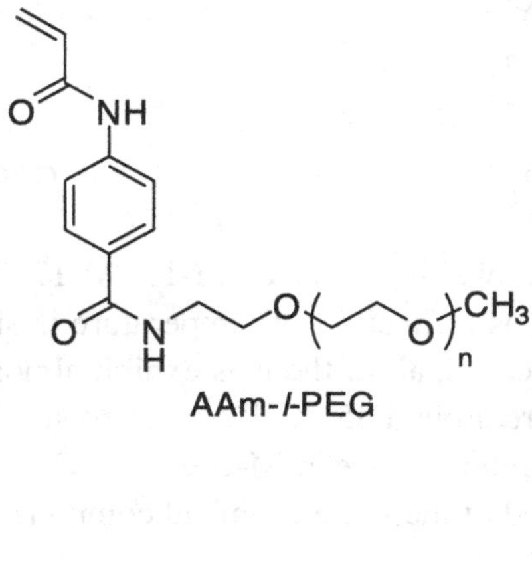

AAm-*l*-PEG

The composition of the gels are summarized in Table 8. The mechanically cross-linked PNIPAAm gel and chemically cross-linked PEG-grafted PNIPAAm gel are designated M and C, respectively.

Table 8. Feed composition of PNIPAAm hydrogels.[a]

Sample	AAm-c-PEG, mg	AAm-l-PEG, mg	MBAAm, mg	PEG content, wt%
M-13	17	None	None	13
M-17	24	None	None	17
M-25	40	None	None	25
C-13	None	17	1.5	13
C-17	None	24	1.5	17
C-25	None	40	1.5	25

[a] Polymerization conditions: NIPAAm = 110 mg; $K_2S_2O_8$ = 6 mg; TMEDA = 3 µL; H_2O = 400 µL; Temperature = 10 °C; Time = 24 h; M_n of AAm-c-PEG = 1200.

M-Gel   C-Gel

The equilibrium swelling ratio of M-13, M-17, M-25, C-13, C-17, and C-25 hydrogels as a function of temperature is shown in Fig. 3. As the temperature increased, all of the gels exhibit almost the same volume transition temperature around 35 °C. There is no significant deference in the transition temperature between M-Gels and C-Gels. This is because the gels are composed of the same chemical components.

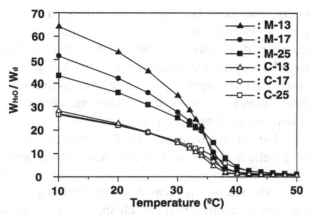

Fig. 3. Equilibrium swelling ratio of M-13, M-17, M-25, C-13, C-17, and C-25 hydrogels.

The deswelling kinetics of hydrogels upon temperature jumping from 10 to 50 °C was studied and the data are shown in Fig. 4. It is obvious that M-Gels exhibit faster shrinking rate than C-Gels. For example, the water retention reduces from 100% to about 10% within 10 min in the case of M-13. However, C-13 gel exhibits slow deswelling rate, reducing from 100% to about 40% within 30 min.

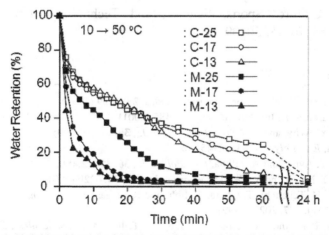

Fig. 4. Shrinking kinetics of M-13, M-17, M-25, C-13, C-17, and C-25 hydrogels.

7. Conclusions and Outlook

Numerous investigations have been reported on linear macromonomers for constructing of various macromolecular architectures including highly branched or graft polymers. However, only a limited examples have been reported on cyclic macromonomers. This article has focused almost exclusively on the application of cyclic macromonomer as noncovalent cross-linking agents for introducing movable cross-linking points into three-dimensional network polymers.

Copolymerization of a monomer in the presence of a cyclic macromonomer (cyclic polymer with a polymerizable group) in high concentration leads mechanically linked three-dimensional network polymer. The network structure is formed by the threading of the cyclic moiety by a segment of another polymer chain. This class of materials has attracted a great deal of interest as new functional materials with unusual chemical, physical and mechanical properties due to the high degrees of freedom in segmental movement since cross-linking points can move in these three-dimensional structures.

Acknowledgments

This work was financially supported by a Grant-in Aid from Ministry of Education, Culture, Sports, Science, and Technology, Japan (No. 11650908 and No. 18550109).

References

1. Y. Delaviz and H. W. Gibson, *Macromolecules*, **25**, 4859 (1992).
2. Y. Okumura and K. Ito, *Adv. Mater.* , **13**, 485 (2001).
3. A. Zada, Y. Avny, and A. Zilkha, *Eur. Polym. J.* , **35**, 1159 (1999).
4. A. Zilkha, *Eur. Polym. J.* , **37**, 2145 (2001).
5. A. Zada, Y. Avny, and A. Zilkha, *Eur. Polym. J.* , **36**, 351 (2000).
6. H. Oike, T. Mouri and Y. Tezuka, *Macromolecules*, **34**, 6229 (2001).
7. M. Kubo, H. Takeuchi, T. Ohara, T. Itoh, and R. Nagahata, *J. Polym. Sci: Part A: Polym. Chem.* , **37**, 2027 (1999).
8. M. Kubo, T. Hibino, M. Tamura, T. Uno, and T. Itoh, *Macromolecules* , **35**, 5816 (2002).
9. M. Kubo, N. Kato, T. Uno, and T. Itoh, *Macromolecules* , **37**, 2762 (2004).

10. M. Kubo, N. Hayakawa, Y. Minami, M. Tamura, T. Uno, and T. Itoh, *Polym. Bull.* , **52**, 201 (2004).
11. A. I. Kurbanov, A. B. Kucharov, and A. B. Alovitdinov, *Tr. Tashk. Politekh. Inst.* , **90**, 133 (1972).
12. J. E. Mark, *Adv. Polym. Sci.* , 44, 1 (1982).
13. J. E. Mark, Elastmers and Rubber Flasticity, J. E. Mark and J. Lal Eds.; American Chemical Society: Washington, DC (1982).
14. L. Garrido, J. E. Mark, S. J. Clarson, and J. A. Semlyen, *Polym. Commun.* , **26**, 53 (1985).
15. K. Miki, Y. Inamoto, S. Inoue, T. Uno, T. Itoh, and M. Kubo, *J. Polym. Sci: Part A: Polym. Chem.* , **47**, 5882 (2009).
16. C. L. Bell and N. A. Peppas, *Adv. Polym. Sci.* , **122**, 125 (1995).
17. A. M. Lowman and N. A. Peppas, In *Encyclopedia of Controlled Drug Delivery*, Ed. E. Mathiowitz (Wiley: New York, 1999), p. 397.
18. T. Miyata, In *Supramolecular Design for Biological Applications*, Ed. N. Yui (CRC Press: Boca Raton, FL, 2002), p. 95.
19. T. Tanaka, *Phys. Rev. Lett.* , **40**, 820 (1778).
20. O. Hirasa, S., Itoh, A. Yamauchi, S. Fujishige, and H. Ichijo, Polymer gels, fundamentals and biomedivcal application (Plenum Press, New York, 1991), p. 247.
21. N. A. Peppas, P. Bures, W. Leobandung, and W. Ichikawa, *Eur. J. Pharm. Biopharm.*, **50**, 27 (2000).
22. M. Kubo, T. Matsuura, H. Morimoto, T. Uno, and T. Itoh, *J. Polym. Sci: Part A: Polym. Chem.* , **43**, 5032 (2005).
23. J. Ricka and T. Tanaka, *Macromolecules* , **18**, 83 (1985).
24. K. Miki, K. Ishida, T. Uno, T. Itoh, and M. Kubo, *Kobunshi Ronbunshu* , **68**, 685 (2011).
25. Y. Hirokawa and T. Tanaka, *J. Chem. Phys.* , **81**, 6379 (1984).
26. C. J. Wang, Y. Li, and Z. B. Hu, *Macromolecules* , **30**, 4727 (1997).
27. H. Hirose and M. Shibayama, *Macromolecules* , **31**, 5336 (1998).
28. Y. Osada, H. Okuzaki, and H. Hori, *Nature* , **355**, 242 (1992).
29. D. J. Beebe, J. S. Moore, J. M. Bauer, Q. Yu, R. Liu, C. Devadoss, *Nature* , **404**, 588 (2000).
30. J. Hoffmann, M. Plotner, D. Kuckling, and W. J. Fischer, J. *Sens. Actuat. A* , **77**, 139 (1999)
31. Y. H. Bae, T. Okano, R. Hsu, and S. W. Kim, *Makromol. Rapid Commun.* , **8**, 481 (1987).
32. T. Okajima, I. Harada, K. Nishio, and S. Hirotsu, *J. Chem. Phys.* , **116**, 9068 (2002).
33. S. X. Cheng, J. T. Zhang, and R. X. Zhuo, *J. Biomed. Mater. Res., Part A* , **67**, 96 (2003).
34. T. Kaneko, T. Asoh, and M. Akashi, *Macromol. Chem. Phys.*, **206**, 566 (2005).
35. R. Yoshida, K. Uchida, Y. Kaneko, K. Sakai, A. Kikuchi, Y. Sakurai, and T. Okano, *Nature* , **374**, 240 (1995).

36. H. K. Ju, S. Y. Kim, and Y. M.Lee, *Polymer* , **42**, 6851 (2001).
37. A. B. Imran, T. Seki, K. Ito, and Y. Takeoka, *Macromolecules* , **43**, 1975 (2010).
38. M. Kubo, K. Ishida, T. Uno, and T. Itoh, *Polym. Prep. Jpn.* , **59**, 929 (2010).
39. M. Kubo, K. Ishida, T. Uno, and T. Itoh, *Polym. Prep. Jpn.* , **60**, 5525 (2011).
40. Y. Kaneko, S. Nakamura, K. Sakai, T. Aoyagi, A. Kikuchi, Y. Sakurai, and T. Okano, *Macromolecules* , **31**, 6099 (1998).

CHAPTER 12

TOPOLOGICAL EFFECTS ON THE STATISTICAL AND DYNAMICAL PROPERTIES OF RING POLYMERS IN SOLUTION

Tetsuo Deguchi and Kyoichi Tsurusaki[†]

Department of Physics, Graduate School of Humanities and Sciences
Ochanomizu University
Bunkyo-ku, Tokyo 112-8610, Japan
deguchi@phys.ocha.ac.jp

[†] *Chemical Engineering Division, Kanagawa Industrial Technology Center*
705-1, Shimoimaizumi, Ebina, Kanagawa 243-0435, Japan

We review recent theoretical studies on the statistical and dynamical properties of ring polymers in solution, in particular, those associated with topological effects. We first define the knot probability by the probability that a given random polygon of N nodes (i.e. vertices) has a fixed knot type. We show a formula for expressing it as a function of N. Through a physical argument using the knot probability we show that topological properties are significant for ring polymers in θ solution or in a good solvent close to the θ temperature, i.e. for ring polymers with small excluded volume. Furthermore, we consider the following topics: topological effects on the mean square radius of gyration of a ring polymer in solution, two-point correlation functions and scattering functions of knotted ring polymers, diffusion constants of knotted ring polymers in good solution, and the probability of random linking. We evaluate fundamental physical quantities of ring polymers in solution through numerical simulation using knot invariants explicitly, and derive theoretical results on them by applying physical interpretation such as the renormalization group arguments.

1. Introduction

Novel knotted structures of polymers have been found recently in various fields of researches such as DNA, proteins and synthetic polymers.[1-7] For a ring polymer in solution its three-dimensional conformation is described by

[†]Affiliation footnote.

a closed curve with no ends. The topology of a closed curve is represented by a knot. For an illustration, knot diagrams are shown in Fig. 1 for some knots with their symbols.[8]

Among various systems of knotted polymers, dilute solutions of ring polymers should be one of the most fundamental ones in statistical mechanics of polymers.[9,10] Ring polymers in solution have topological constraints: their topological types do not change under thermal fluctuations. Polymer chains do not break during time evolution, since the chemical bond energy is much larger than the thermal energy. Statistical and dynamical properties of a dilute solution of ring polymers may depend on the knot type of the ring polymers. It is the case, if all the ring polymers have the same topology such as the trivial knot and the solution is near the θ temperature.

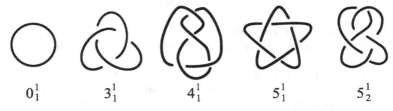

$$0_1^1 \qquad 3_1^1 \qquad 4_1^1 \qquad 5_1^1 \qquad 5_2^1$$

Fig. 1. Knot diagrams of the trivial knot (0_1^1), the trefoil knot (3_1^1), the figure-eight knot (4_1^1), 5_1^1 and 5_2^1. There are two knots 5_1^1 and 5_2^1 whose minimal crossing number is given by five. Here symbol 5_2^1 means that the knot diagram corresponds to the second knot type among all the knots for which the minimal number of crossing points is given by five and the number of component (i.e. spatial curve) of the graph is given by one.[8]

The problem of topological effects on ring polymers in association with knots was first addressed in the 1960s.[11,12] Pioneering theoretical researches on topological effects of ring polymers[13,14] were performed in the 1970s.[15,16] The topological interaction between a pair of ring polymers was studied in order to explain the anomalous osmotic pressure observed for the solution of ring polymers at the θ temperature of corresponding linear polymers.[17–19] However, knotted ring polymers were explicitly derived first in DNA experiments for more than a quarter-century ago.[20,21] Circular DNAs of various knot types are separated into knotted species by gel electrophoresis.[20–23] Moreover, recent progress in experiments of ring polymers is quite remarkable. Diffusion constants of linear, relaxed circular and supercoiled DNAs in solution have been measured quite accurately.[24,25] Here the DNA double helices are unknotted. Ring polymers of large molecular weights are synthesized not only quite effectively[26] but also with small dispersions and of extreme high purity.[27-29] Recently, synthetic ring polymers with nontrivial knots have been synthesized.[27,29] Furthermore, polymers with various different topologies have been systematically synthesized.[30] Knotted structures are also studied for proteins.[5,31,32]

Fundamental properties of ring polymers in dilute solution were studied theoretically several decades ago.[33-36] However, no topological constraint on ring polymers was considered explicitly at that time. It is recent that the effects of topological constraints on statistical properties of ring polymers are explicitly studied in theoretical studies.[13,37] For instance, it was proven mathematically that every infinitely long self-avoiding walk is knotted;[38,39] the probability of random knotting was measured in DNA experiments.[40,41] It was shown that the concept of tight knots, i.e. *ideal knots*, plays a central role in the dynamics of ring polymers.[42-44] In this review we shall show several numerical results on the probability of random knotting,[45-63] and the mean square radius of gyration of knotted ring polymers in solution.[16,64-75] We discuss for knotted ring polymers in solution the two-point correlation functions, the scattering functions,[35,76-79] and the diffusion constants.[69,80-84] We derive the probability of random linking,[14,52,85-87] which is related to the anomalous osmotic pressure in θ solution.[88-90] Furthermore, there are many interesting topics related to topological effects of ring polymers: the probability of forming a local knot in a random polygon,[91,92] flat knots (i.e. knotted ring polymers in two dimensions),[93,94] the mean square radius of gyration of ring polymers in melt,[95-100] the minimal step number of a lattice knot,[101,102] relaxation dynamics of a knotted ring polymer in solution,[103] and etc.[104,105]

In order to investigate statistical properties of ring polymers we consider a model of random polygons (or self-avoiding polygons).[106] The topological constraint that polygons should have a fixed knot type restricts the number of possible configurations of polygons (i.e., conformations of ring polymers) so that the available volume in the phase space of such polygons can be much smaller than that of no topological constraint. The rate of reduction of the volume gives the probability of random knotting. We also call it the knot probability, briefly.

The average size of random polygons under a topological constraint can become much larger than that of no topological constraint, if the number of nodes N is large enough. The enhancement should be a consequence of "repulsive entropic forces" acting among the segments of each polygon under the topological constraint. The entropic forces are also suggested by the fact that no bond crossing occurs in the time evolution of ring polymers in solution under topological constraints.

Thus, the effect of entropic reduction results not only in the probability of random knotting but also in the effective repulsions among polygonal segments so that the average size of random polygons of a knot enhances.

The contents of the present review consist of the following. In §2 we introduce the knot probability, which corresponds to the reduced entropy of a ring polymer under a given topological constraint. In §3 we review theoretical studies on the average size of a knotted ring polymer in solution. We show that the mean square radius of gyration of a ring polymer under a topological constraint can be larger than that of no topological constraint. In §4 we show first the numerical plots of the scattering functions of knotted ring polymers in θ solution. We then introduce a fitting formula for the distribution function of the distance between two nodes of a ring polymer with fixed knot in θ solution. From the fitting formula we derive analytic expressions of the scattering functions. In §5 we evaluate diffusion constants of knotted ring polymers in good solution through the Brownian dynamics with hydrodynamic interaction. Here, no bond crossing is allowed in time evolution. We show that the ratios of the diffusion constants of knotted ring polymers are universal and independent of the number of segments. They are determined by the average crossing number of ideal knots. In §6 we define the linking probability and give an intuitive derivation of the fitting formula as a function of the distance between the centers of mass for a pair of ring polymers with fixed knot type.

Hereafter we abbreviate the symbols of knots such as 0_1 and 3_1 for the trivial and trefoil knots, respectively, by suppressing the superscript showing the number of component.

2. The Probability of Random Knotting

Let us consider a polymer chain of molecular weight M_{WT} consisting of N statistically independent segments with molecular weight M_0 in solution. Here we have $M_{WT} = NM_0$. The spatial conformations of the polymer chain are described in terms of random walks (RW) or self-avoiding walks (SAW).

We consider a random model which gives an ensemble of polygons in random configurations as a theoretical model of ring polymers in solution (Fig. 2). If each segment of the polygons has excluded volume, we call them self-avoiding polygons (SAP), while if each segment has no excluded

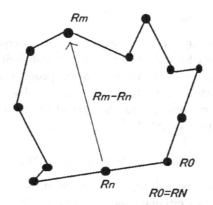

Fig. 2. Random polygon of N nodes corresponds to a conformation of a ring polymer in solution consisting of N statistical segments with molecular weight M_0.

volume, we call them random polygons (RP). For an illustration we explain the crank-shaft algorithm for generating off-lattice SAP of N nodes in the three-dimensional space in Appendix A.

2.1. *Definition of the knot probability*

We now define the probability of random knotting for a model of random polygons (or SAPs) of N nodes (i.e. vertices).[49-51] We assume that the model has an algorithm by which we can generate an ensemble of polygons of N nodes. Let us consider the probability that a polygon of N nodes generated by the algorithm has a given knot type K. We call it the random knotting probability of knot K for the model of random polygons, and denote it by $P_K(N)$. We also call it the knot probability, in short.

For lattice polygons the total number of polygons of N nodes is finite. We denote it by $Z(N)$. Then, we denote by $Z_K(N)$ the total number of polygons of N nodes with a given knot type K. It is clear that $Z_K(N) < Z(N)$. We evaluate $P_K(N)$, i.e. the probability of a random polygon with N nodes having the knot type K, by

$$P_K(N) = Z_K(N)/Z(N). \tag{1}$$

The knot probability $P_K(N)$ depends on the model of random polygons or SAPs. However, the N-dependence of the knot probability has universal properties, which do not change for different models of random polygons.

2.1.1. *Historical remarks*

The question on the topological constraints on macromolecules such as circular DNA was first formulated by Delbrück,[11] Frisch and Wasserman[12] in the 1960s. It was conjectured that if we produce or synthesize long ring polymers in dilute solution they will be knotted with high probability. The conjecture was proven as a mathematical theorem later by Sumners, Whittington and Pippenger.[38,39]

2.1.2. *Experiments of measuring the knot probability*

The random knotting probability can be measured in experiments. For the model of worm-like circular chains the theoretical values of the knot probability of the trivial and the trefoil knots were measured and compared with the experiments of randomly closing nicked DNA chains, where linear DNA chains are closed by controlling the temperature of the solution.[40,41] However, for knots other than the trivial and the trefoil knots, it is an interesting open problem how one can compare the chromatographic data with theoretical values of the knot probability for some proper model of random polygons.

Furthermore, quite recently, some nontrivial knots have been synthesized through chemical reactions.[29] The theoretical values of the knot probability are useful for estimating the fractions of ring polymers of some species of knots.

2.2. **Numerical method for evaluating the knot probability**

We evaluate the knot probability numerically as follows. Suppose that an ensemble of random polygons are constructed by the algorithm of a random model. We denote by W_{all} the number of all generated polygons. We calculate knot invariants such as the Vassiliev invariants for each of the polygons, and select such polygons that will have a given knot type K.[48,50,52] If a polygon has the same set of values of some knot invariants as the knot K, then we assume that the topology of the polygon is given

Table 1. Values of the two knot invariants for some knots: the determinant of a knot (the Alexander polynomial evaluated at $t = -1$, $\Delta_K(t = -1)$,) and the Vassiliev invariant of the second degree, $v_2(K)$.

Knot type	0_1	3_1	4_1	5_1	5_2	$3_1 \# 3_1$	$3_1 \# 4_1$
Determinant of knot	1	3	5	5	7	9	15
2nd order Vassiliev invariant	0	1	−1	3	2	2	0

by the knot K. Here, the number of selected polygons, W_K, can be much smaller than that of all polygons, W_{all}. By the ratio W_K/W_{all} we evaluate the knot probability for knot K, $P_K(N)$.

In particular, we calculate the Vassiliev invariant of the second degree and the determinant of a knot for each polygon (for the definition of the knot invariants, see the textbook[107]). With the values of the two knot invariants we detect the knot type of the polygon (see Table 1).[50,52] The Vassiliev invariants can be calculated in polynomial time by the method of the R-matrix[48,50] or by the method of the Gauss diagrams.[108]

We remark that the present method for evaluating the knot probability gives practically accurate values for the knot probability. There can be a certain complex knot that would have the same set of values of the knot invariants employed in the method. However, the number of wrong identifications of knot types should be small if the number of nodes N is not very large or knot types are not very complex.

2.3. *Formula for the N-dependence of the knot probability*

Through numerical simulation we find that the N-dependence of the probability of a knot K is well approximated by the following formula:[51]

$$P_K(N) = C_K \left(\frac{N - N_{ini}(K)}{N_K} \right)^{m(K)} \exp\left(-\frac{N - N_{ini}(K)}{N_K} \right). \quad (2)$$

Here there are four fitting parameters: $N_K, C_K, m(K)$, and $N_{ini}(K)$. We call $m(K)$ the exponent of the knot probability for knot K.

The random knotting probabilities of various knots were evaluated for the Gaussian random polygon and the rod-bead model with several different values of bead radius, and it was shown that the four-parameter formula (2) gives good fitting curves to the simulation data as functions of N with respect to the χ^2 values.[51,52]

In Figs. 3 and 4 formula (2) gives good fitting curves to the knot probabilities of some prime knots and composite knots for equilateral random polygons generated by the crankshaft method with bond-interchange move (see Appendix A for the algorithm).[62] We observe that for any given non-trivial knot the fitting curve has a maximum. It is derived from eq. (2) if exponent $m(K)$ is positive. Moreover, each curve decays exponentially for large N. It is consistent with the asymptotic expansion of the partition function $Z_K(N)$ of knotted polygons, as we shall see in eq. (3).

We call N_0 the *characteristic length of random knotting*. The estimates of the parameters N_K are given by almost the same value with respect to

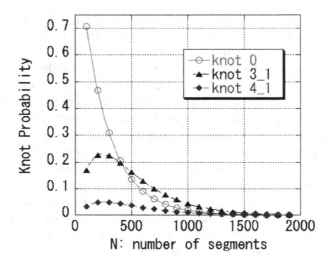

Fig. 3. The random knotting probabilities of the trivial (0_1), the trefoil (3_1) and the figure-eight (4_1) knots for equilateral random polygons generated by the crankshaft method. The number of nodes, N, is given by $N = 100, 200, \ldots, 1900$. The data of the knot probability were calculated by Akita for 10^5 SAPs.[62] Formula (2) gives good fitting curves, where the χ^2-values with 19 points are given by 31.6, 22.5 and 20.0 for 0_1, 3_1 and 4_1, respectively.

statistical errors for several different knots.[51] Here the suffix 0 of N_0 denotes the trivial knot, 0_1. The estimate of characteristic length N_0 depends on the model of random polygons or self-avoiding polygons. It is approximately given by $N_0 = 300$ for random polygons such as Gaussian polygons and equilateral random polygons.[51] On the other hand, the value of N_0 becomes quite large such as the order of 10^5 for self-avoiding polygons on lattices.[46,57]

If we neglect the cut-off parameter $N_{ini}(K)$, formula (2) corresponds to the asymptotic expansion of the partition function $Z_K(N)$ with respect to N as follows.

$$\log\left(Z_K(N)/Z(N)\right) = -\frac{N}{N_0} + m(K)\log N/N_0 + \log C_K + O(1/N). \quad (3)$$

It is natural to assume that the exponent $m(K)$ should be given by a universal value for each knot K, by applying the renormalization group arguments.[10] Moreover, for SAP on the cubic lattice, it is suggested that the exponents for the asymptotic expansion $m(K)$ should be given by integers at least approximately.[53,54] In fact, it is consistent with the additivity of exponents $m(K)$'s, as shown in eq. (5). However, for random polygons of a finite number of nodes N are concerned, it seems that the best estimates of $m(K)$ are different from integers, particularly, if we introduce the cut-off parameters $N_{ini}(K)$.[51] Thus, formula (2) for the knot probability does not necessarily give the asymptotic behavior of the partition function of the knotted ring polymer.

Formula (2) thus generalizes the exponentially decaying behavior of the trivial knot to the case of nontrivial knots, by considering exponents $m(K)$ and cut-offs $N_{ini}(K)$. In previous numerical studies[45–47] it was found that the knot probability of the trivial knot 0_1 decays exponentially, i.e. $m(0_1) = 0$. The exponential decay was shown for the rod-bead model with different values of bead radius.[47] It was then shown in the simulation[49] that the knot probability of a nontrivial knot has a maximum as a function of N.

Let us consider the worm-like chain model for ring polymers. The model produces an ensemble of self-avoiding polygons consisting of cylinder-shape segments of unit length with cylindrical radius a, and is considered suitable for studying statistical properties of circular DNA. For the model the characteristic length N_0 as a function of cylindrical radius a is evaluated as[55]

$$N_0(a) = N_0(0) \exp(\alpha a) \,. \tag{4}$$

Here the parameters are estimated as $N_0(0) = 292 \pm 5$ and $\alpha = 43.5 \pm 0.6$.

The parameters $N_{ini}(K)$ give cut-offs in the N value of the knot probability $P_K(N)$ for small N. It is shown rigorously that any SAP of N nodes on the cubic lattice has only the trivial knot if $N < 24$.[101] Therefore, the knot probability of the trefoil knot (3_1) for SAP on the cubic lattice vanishes if $N < 24$. The theorem was extended also for other knots.[102]

We remark that the ratios between knot probabilities of two different knots are studied for lattice polygons for asymptotically large N.[63]

2.4. *Random knotting probabilities of composite knots*

Let us denote by $K_1 \# K_2$ the composite knot for a given pair of knots K_1 and K_2 (e.g. $3_1 \# 3_1$ in Fig. 5). Numerically it has been observed that the exponent $m(K)$ of knot K satisfies the additivity for a composite knot[50]

$$m(K_1 \# K_2) = m(K_1) + m(K_2) \,. \tag{5}$$

Let us now introduce an empirical property of C_Ks. From the list of the best estimates of C_K s[51,62] we observe that the estimates of C_K for composite knots satisfy the factorization property:

$$C_{K_1 \# K_2} = C_{K_1} C_{K_2} \,. \tag{6}$$

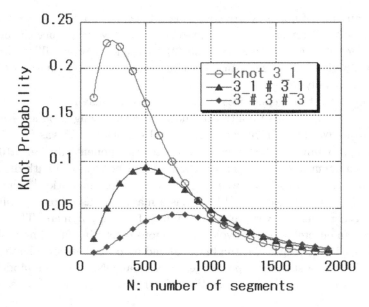

Fig. 4. Fitting curves to the knot probabilities of the trefoil knot (3_1) and its composite knots $3_1\#3_1$ and $3_1\#3_1\#3_1$. The data of the knot probabilities were obtained by Akita.[62]

We denote by $K^{\#n}$ the n-fold composite knot of a knot K such as $K^{\#n} = K(\#K)^{n-1} = K\#K\#\cdots\#K$. For an n-fold composite knot $K^{\#n}$ we have the inverse of the n-factorial, i.e. $n!$:

$$C_{K^{\#n}} = (C_K)^n/n!. \tag{7}$$

The factorization property for the knot probability of a composite knot has been found for SAP on the cubic lattice through simulation.[61]

It is an important consequence of the factorization property (6) and (7) that we can estimate the knot probability of a composite knot $K_1\#K_2$ by

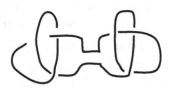

Fig. 5. Knot diagram of a composite knot consisting of two trefoil knots ($3_1\#3_1$).

making use of the additivity of exponents (5) together with eqs. (6) and (7) from the knot probabilities of constituent knots K_1 and K_2, for models of random polygons or self-avoiding polygons. It is particularly useful in experiments for estimating the fractions of composite knots.

3. Mean Square Radius of Gyration for Ring Polymers in Solution

3.1. *Definition of the mean square radius of gyration*

Let us consider ring polymers consisting of N statistical segments in dilute solution. In order to investigate the statistical properties we assume a model of random polygons of N nodes such as the equilateral random polygon generated by the crank-shaft method. We denote by r_j the position vector of the jth segment of a random polygon. We express the position vector of the center of mass as r_C; i.e. $r_C = \sum_{j=1}^{N} r_j/N$. We define the mean square radius of gyration, $R_{g,K}^2$, of the ring polymer with fixed knot type K by

$$R_{g,K}^2 = \frac{1}{N} \sum_{j=1}^{N} \langle (r_j - r_C)^2 \rangle_K \,. \tag{8}$$

Here the symbol $\langle A \rangle_K$ denotes the ensemble average of a given physical quantity A over all possible configurations of polygons with fixed knot K.

We calculate numerically $\langle A \rangle_K$ as follows. We first construct a large number of random polygons or self-avoiding polygons of N nodes. To each of them we calculate the two knot invariants, the Vassiliev invariant of the second degree and the determinant of knots, and with the values we effectively determine the knot type. We select such polygons that have the same values of the two knot invariants as knot K, and calculate the average value of quantity A over the selected polygons.

We denote by $R_{g,all}^2$ the mean square radius of gyration for random polygons of N nodes under no topological constraint.[9]

$$R_{g,all}^2 = \frac{1}{N} \sum_{j=1}^{N} \langle (r_j - r_C)^2 \rangle_{all} \,. \tag{9}$$

Here *all* denotes that all possible polygons of N nodes under no topological constraint.

The mean square radius of gyration plays a fundamental role in the scaling behavior of ring polymers in solution from the viewpoint of statistical physics. For an illustration, let us consider random polygons with unit bond length. For large N we have[34]

$$R_{g,all}^2 = N/12 \qquad \text{for} \quad N \gg 1. \tag{10}$$

Let us consider ring polymers with a fixed knot type K in solution. For asymptotically large N the mean square radius of gyration is given by

$$R_{g,K}^2 = A_K N^{2\nu_K} \left(1 + O(1/N)\right). \tag{11}$$

In a good solvent we expect that the exponent ν_K is given by that of SAW: $\nu_K = \nu_{SAW} = 0.588.$[10]

3.2. Topological swelling of ring polymers in a good solvent close to the θ temperature

The mean square radius of gyration of random polygons (or SAPs) under a topological constraint, denoted by $R_{g,K}^2$, can become significantly larger than that of no topological constraint $R_{g,all}^2$. We call the phenomenon *topological swelling*.

In Fig. 6 it is shown that the mean square radius of gyration with fixed knot K, denoted by $R_{g,K}^2$, is larger than that of no topological constraint, denoted by $R_{g,all}^2$, for the trivial knot ($K = 0_1$) and the trefoil knot ($K = 3_1$). The ratio $R_{g,K}^2/R_{g,all}^2$ is greater than 1.0 for the two knots if $N > N_0$. The $R_{g,K}^2$ are evaluated for 10^5 equilateral random polygons generated by the crank-shaft and bond-interchange moves.[62] In the plots, the number of nodes is less than or equal to $N = 1000$.

In the simulation[67,68] we observe that topological swelling occurs for ring polymers in a good solvent if it is close to the θ temperature; If it is not close to the θ temperature, topological swelling does not occur. Let us consider cylindrical self-avoiding-polygons (SAP) consisting of cylindrical segments. Here each of the segments is given by a cylinder of unit length whose radius is given by r. The mean square radius of gyration was evaluated for cylindrical SAPs in the simulations.[67,68] It was then shown that topological swelling occurs if the value of r is very small such as 0.01.

In §3.3 we shall argue that topological swelling vanishes under the excluded volume effect by making use of the characteristic length N_0.

Quite recently, the topological swelling has been observed in experiments of synthetic ring polymers.[75]

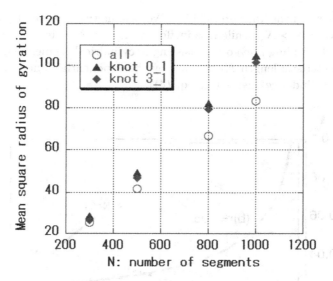

Fig. 6.   The mean square radius of gyration for equilateral random polygons with fixed knot K, $R^2_{g,K}$, and that of no topological constraint, $R^2_{g,all}$, plotted against the polygonal number N. The equilateral random polygons were generated by the crankshaft method with bond-interchange move. Open circles correspond to $R^2_{g,all}$; filled upper triangles and filled diamonds $R^2_{g,K}$ for $K = 0_1$ and 3_1, respectively. The numerical data were obtained by Akita.[62]

3.3. *Criterion of topological swelling in a good solvent*

Making phenomenological assumptions we shall argue that topological swelling appears only when the solution is close to the θ temperature.[67,68] In §3.3 we denote the characteristic length N_0 by N_c. We express the radius a of the cylindrical radius as r.

We first note that the volume of a cylindrical segment with radius r and of unit length is given by πr^2. We define N_{ex} by $N_{ex} = 1/\pi r^2$. Here we assume that if the number of segments N is large enough so that the total excluded volume is larger than 1, then the excluded-volume effect should be effective, while if it is less than 1, the excluded-volume effect is not effective. Therefore, the excluded-volume effect appears when $N > N_{ex}$. Here we also assume that the excluded-volume effect makes the topological effect diminish. If the polymer chain is thick enough, it will not be knotted easily.

Secondly, we recall that the characteristic length N_c describes the exponential decay of the knot probability. We assume that the topological effect is effective if $N > N_c$, while it is ineffective if $N < N_c$. Here we recall that for the self-avoiding polygons consisting of cylindrical segments of radius r, the characteristic length N_c is given by the exponential function of r, the radius of cylinder segments (see eq. (4)).

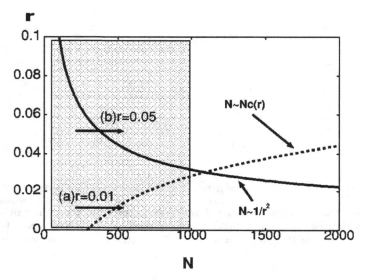

Fig. 7. The region of topological swelling in the diagram of the cylindrical radius r and the number of nodes N ($N - r$ diagram), as presented in Ref. 68.

It follows from the assumptions on N_{ex} and N_c that the topological swelling occurs if $N > N_c$ and $N < N_{\text{ex}}$. Thus, the topological swelling should appear for the self-avoiding polygons with small excluded volume.

It explains the reason why we observe knotted ring polymers in DNA. DNA chains are quite stiff, and hence are considered very thin polymers, i.e. they have small excluded volume. Thus, they are likely to be knotted. Here we assume very long circular DNA chains, and regard them as polygons of unit bond length corresponding to the persistence length (or Kuhn length).

Here we remark that the two assumptions on N_{ex} and N_c are phenomenological. However, they are consistent with various numerical observations. They should thus be useful at least as working hypotheses.

For an illustration, let us consider two graphs in the $N - r$ plane:[68] $N = N_{ex}(r)$ and $N = N_c(r)$. In Fig. 7, the vertical line expresses the

r-axis and the horizontal one the N-axis. The graph $N_c(r) = N$ reaches the N-axis at $N = N_c(0) \approx 300$. Here we recall that the function $N_c(r)$ is given by $N_c(r) = N_c(0) \exp(\alpha r)$. There is a crossing point for the two curves. The coordinates of the crossing point are approximately given by $N^* = 1300$ and $r^* = 0.03$. For a simulation of the ratio R_K^2/R^2 with a fixed radius r, we have a series of data points located on a straight line parallel to the N-axis. Let us first consider the case of small r such as $r = 0.003$ and $r = 0.01$. From the simulation it is shown that the effective expansion due to the topological constraint is large.[68] This is consistent with the following interpretation of the $N - r$ diagram: if we start from the region near the r-axis and move in the direction of the N-axis, then we cross the curve $N = N_c(r)$ before reaching another one $\sqrt{N}r = 1$; thus, we expect that the excluded volume remains.

3.4. *Anomalous "scaling behavior" of ring polymers in the θ solution of corresponding linear polymers*

It was suggested by des Cloizeaux[16] that topological constraints on ring polymers in θ solution may play a similar role as the excluded volume. The interesting suggestion has been studied rather recently by several authors through numerical simulations.[65,70–73]

Let us consider the average size of ring polymers in a θ solvent. Here we should remark that the θ temperature is given by that of linear polymers corresponding to the ring polymers, where the excluded-volume interaction among segments is effectively screened. From the graphs of "3_1/ave" ($\langle R_g^2 \rangle_{3_1}/\langle R_g^2 \rangle_{\text{all}}$) and "$4_1$/ave" ($\langle R_g^2 \rangle_{4_1}/\langle R_g^2 \rangle_{\text{all}}$) of Fig. 8 we have the following observation: For a nontrivial knot, the average size, R_K, is smaller than that of no topological constraint, R_{all}, if N is small such as $N < N_0$. However, it becomes larger than R_{all} if N is large enough such as $N \gg N_0$.

The mean square radius of gyration $R_{g,K}^2$ of ring polymers with fixed knot K has been numerically evaluated in some models of random polygons for a few knots, and the scaling exponent ν_K has been numerically estimated.[65,70–73] In Refs. 65 and 71–73 the numerical estimates of the scaling exponent ν_K are definitely larger than $\nu_{RP} = 0.5$, i.e. the scaling exponent of random polygons under no topological constraint. In fact, in Refs. 71–73 the estimates are given by the scaling exponent of self-avoiding walks, $\nu_{SAP} = 0.588$, while in Ref. 70 the estimate is compatible with the scaling exponent of random walks, $\nu_{RP} = 0.5$.

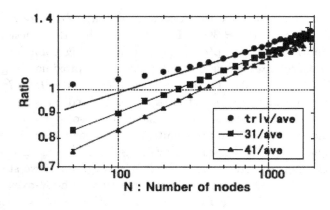

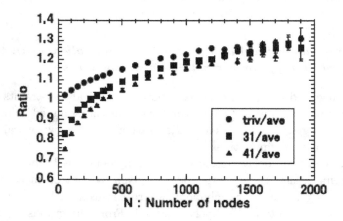

Fig. 8.   Topological swelling of Gaussian random polygons: The ratio of the mean square radius of gyration with fixed knot K to that of no topological constraint, $\langle R_g^2 \rangle_K / \langle R_g^2 \rangle_{\text{all}}$ versus the number N of polygonal nodes: (i) Double logarithmic plots (the upper panel); (ii) Linear plots (the upper panel). They are given in Ref. 70. The symbols "triv", "31", "41" and "ave" denote the trivial knot 0_1, the trefoil knot 3_1, the figure-eight knot 4_1 and all knots all.

It is not trivial to determine the value of the scaling exponent ν_K through numerical simulation. The exponent ν_K should characterize the asymptotic behavior of random polygons with topological constraint K where the number of nodes N is very large . However, in the simulations the number of polygonal nodes N is not very large: we have $N \leq 3000$ for all the simulations performed so far.[65,70–73] The range of N ($N \leq 3000$) is not large enough to determine the value of the scaling exponent ν_K. Furthermore, we introduce three or four fitting parameters in order to have good fitting curves by considering finite-size effects as follows

$$R_{g,K}^2 = A_K N^{2\nu_K} \left(1 + B_K N^{-\Delta_K} + O(1/N)\right). \tag{12}$$

Here we have four fitting parameters A_K, ν_K, B_K and Δ_K. Since we have many fitting parameters and N is not large enough, the same data points can be compatible with two different values of the scaling exponent ν as far as the χ^2 values are concerned:[72] one is close to 0.5 and another close to 0.6 . Thus, it is not certain that any of the estimates of ν_K investigated so far is valid for the asymptotic behavior.

We shall show in §4.3 that a formula close to the Gaussian case gives good fitting curves to the data of the distribution function of the distance between two segments of a ring polymer in solution.[62,76,78,79] Thus, considering the two-point functions of knotted ring polymers, we suggest that the exponent ν should be close to the Gaussian value but not the same, such as $\nu = 0.52$ or 0.53.

4. Scattering and Correlation Functions of Random Knots

4.1. *Radial distribution functions and scattering functions*

Let us consider a random walk or polygon of N nodes. We denote by $\boldsymbol{R}_m$ the position vector of the mth node for $m = 1, 2, \ldots, N$. We first introduce

$$g_n(\boldsymbol{r}) = \sum_{m=1}^{N} \langle \delta(\boldsymbol{r} - (\boldsymbol{R}_m - \boldsymbol{R}_n)) \rangle. \tag{13}$$

Here the average $\langle \cdot \rangle$ is taken over all possible configurations of the polygon of N nodes. The quantity: $g_n(\boldsymbol{r}) \, d^3 r / N$ describes the probability of finding another node in the volume element $d^3 r = dx\,dy\,dz$ located at position vector $\boldsymbol{r}$ relative to the position of the nth node, $\boldsymbol{R}_n$. We define the pair correlation function for a random walk or polygon of N nodes, taking the average of $g_n(\boldsymbol{r})$ with respect to n, as follows:[9]

$$g(\boldsymbol{r}) = \frac{1}{N} \sum_{n=1}^{N} g_n(\boldsymbol{r}) = \frac{1}{N} \sum_{m,n=1}^{N} \langle \delta(\boldsymbol{r} - (\boldsymbol{R}_m - \boldsymbol{R}_n)) \rangle.$$

For random polygons $g_n(\boldsymbol{r})$ does not depend on the number n due to the cyclic symmetry with respect to n.

Due to the rotational symmetry, the pair correlation function $g(\boldsymbol{r})$ depends only on the distance $r = |\boldsymbol{r}|$. We denote it by $g(r)$, and call it the *radial distribution function*.

We now introduce the Kratky plot for a linear (or ring) polymer consisting of N segments. We define a form factor $\widetilde{P}(\boldsymbol{q})$ as follows

$$\widetilde{P}(\boldsymbol{q}) = \frac{g(\boldsymbol{q})}{g(0)}. \tag{14}$$

Due to the rotational symmetry, the form factor is given by a function of $q = |\boldsymbol{q}|$ as follows

$$\widetilde{P}(\boldsymbol{q}) = f(qR_G). \tag{15}$$

Here R_G denotes the square root of the mean square radius of gyration of the polymer: $R_G = \sqrt{\langle R_G^2 \rangle}$. For models of polymers with no excluded volume, the form factor has the large-q asymptotic behavior: $f(qR_G) \propto (qR_G)^{-2}$. Thus, it is useful to make the plot of $(qR_G)^2 f(qR_G)$ versus qR_G. We call such a plot the *Kratky plot* of the form factor of the linear (or ring) polymer.

4.2. *Scattering functions of knotted ring polymers*

Let us now consider random polygons under a topological constraint. For a random polygon of N-nodes with fixed knot K we define the pair correlation function by

$$g_K(\boldsymbol{r}) = \frac{1}{N} \sum_{m,n=1}^{N} \langle \delta(\boldsymbol{r} - (\boldsymbol{R}_m - \boldsymbol{R}_n)) \rangle_K. \tag{16}$$

Here the average $\langle \cdot \rangle_K$ is taken over all possible configurations of the random polygon with knot K. We denote a physical quantity A under no topological constraint by suffix $K = all$ such as A_{all}.

We define the (single-chain) static structure factor $g_K(\boldsymbol{q})$ of the ring polymer under a topological constraint K by

$$g_K(\boldsymbol{q}) = \int d\boldsymbol{r} e^{i\boldsymbol{q}\cdot\boldsymbol{r}} g_K(\boldsymbol{r}) = \frac{1}{N} \sum_{m,n=1}^{N} \langle \exp(i\boldsymbol{q} \cdot (\boldsymbol{R}_m - \boldsymbol{R}_n)) \rangle_K. \tag{17}$$

We also call it the scattering function. In terms of the spherical coordinate systems (r, ϕ, θ) with $d\boldsymbol{r} = r^2 dr \, d\phi \sin\theta d\theta$, we calculate $\int e^{i\boldsymbol{q}\cdot\boldsymbol{r}} d\boldsymbol{r}$ as follows:

$$\int e^{i\boldsymbol{q}\cdot\boldsymbol{r}} g_K(r) d\boldsymbol{r} = \int_0^\infty g_K(r) \frac{\sin qr}{qr} dr \tag{18}$$

Here $q = |\boldsymbol{q}|$ and we take the z-axis so that we have $\boldsymbol{q} = q\boldsymbol{e}_z = (0, 0, q)$ and $\boldsymbol{q} \cdot \boldsymbol{r} = qr \cos \theta$.

4.2.1. *Double logarithmic plot of scattering functions*

We define the form factor $\widetilde{P_K}(q)$ for the Gaussian random polygon under a topological condition K as follows

$$\widetilde{P_K}(q) = \frac{g_K(q)}{g_K(0)}. \tag{19}$$

We have $\widetilde{P_K}(q) = g_K(q)/N$ from the definition (17). Let us define variable u by $u = qR_{g,K}$. Here we recall that $R_{g,K}$ denotes the square root of the mean square radius of gyration for the Gaussian random polygon under a given topological constraint K. Similarly as eq. (15), we define f_K, a function of variable u, by $\widetilde{P_K}(q) = f_K(qR_{g,K})$.

The double logarithmic plot of the form factor $\widetilde{P_K}(q)$ versus $u = qR_{g,K}$ (i.e. $f_K(u)$ versus u) is shown in Fig. 9. Here we recall that the form factor $\widetilde{P_{all}}(q)$ was evaluated analytically in terms of the Dawson integral.[35] In Fig. 9 we have evaluated the form factor $\widetilde{P_K}(q)$ for the five topological conditions through simulation.

In the region from $u = 0$ up to $u = 2$ or 3, the form factors $\widetilde{P_K}(q)$ for the five topological conditions overlap each other in the double-logarithmic scales. We note that the low-q part of the form factor $\widetilde{P_K}(q)$ is related to the large-r part of the radial distribution function $g_K(r)$ through the Fourier transformation. We shall show the correspondence precisely through the Kratky plot in Fig. 10.

For $u > 5$, the graphs of the different topological conditions make approximately straight lines in the main panel. In the inset of Fig. 9, the graphs except for that of the other knots (*other*) make distinct lines parallel to each other. The gradient is almost given by -2, which is consistent with the Gaussian asymptotic behavior.

It follows from the four parallel and distinct lines in the inset of Fig. 9 that for large q the form factor $\widetilde{P_K}(q) = f_K(qR_{g,K})$ is approximated by

$$f_K(u) \approx a_K/u^2. \tag{20}$$

Here constants a_K are distinct for different topological conditions.

The asymptotic behavior with exponent -2 is also consistent with the large-u behavior of the Kratky plot, as we shall see in Fig. 10.

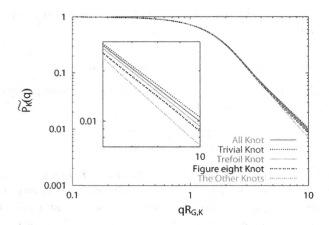

Fig. 9. Double logarithmic plot of the form factor $\widetilde{P_K}(q)$ of a Gaussian polygon of N nodes under a topological condition K versus the variable $u = qR_{g,K}$, as presented in Ref. 77. The curves correspond to the cases of no topological condition (all), the trivial knot (0), the trefoil knot (3_1), the figure-eight knot (4_1) and the other knots ($other$), respectively. In the inset, the main panel is enlarged from $qR_{g,K} = 6$ to 10; the graphs of 0, all, 3_1, 4_1 and $other$ are located from higher to lower positions. In the axis-title $qR_{g,K}$ is denoted by $qR_{G,K}$.

4.2.2. Kratky plots of knotted ring polymers in a θ solvent

Let us present the Kratky plots of the form factors $\widetilde{P_K}(q)$ of the Gaussian random polygon under a topological condition K for some different topological conditions. The plots of $(qR_{g,K})^2\widetilde{P_K}(q)$ versus the variable $u = qR_{g,K}$ are shown in Fig. 10. In other words, they are the plots of $u^2 f_K(u)$ versus u for the different topological conditions, K.

With the graphs shown in Fig. 10 we observe that the mean square radius of gyration, $\langle R_{g,K}^2 \rangle$, plays an important role in the scattering function of the Gaussian polygon under a topological condition, K. In fact, plotting the form factor in terms of variable u makes the graphs quite simple. In particular, in the small u region such as $u < 2$, the Kratky plots for the different topological conditions overlap completely. For $u > 2$ the graphs become separate gradually.

The peak positions of the Kratky plots are given by almost the same value of u for all the five topological conditions with respect to errors. The peak heights of the Kratky plots depend on the topological conditions, as shown in the inset of Fig. 10. The Kratky plot for the trivial knot has the smallest peak height. The peak height for the trefoil knot is a little larger

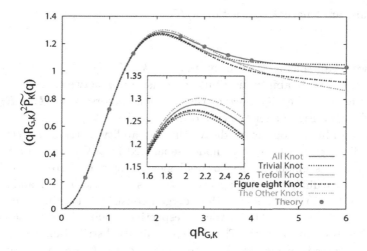

Fig. 10. Plot of $(qR_{g,K})^2 \widetilde{P_K}(q)$ versus $u = qR_{g,K}$: the Kratky plots of the trivial, trefoil and figure-eight knots are represented by the single-dotted, fine-dotted and double-dotted curves, respectively, as presented in Ref. 77. The solid curve corresponds to the case of no topological constraint. In the inset, the graphs near peak positions are enlarged. The filled circles are calculated by using the analytical expression of the scattering function of ring polymers given by Casassa.[35] In the axis-title $R_{g,K}$ is denoted by $R_{G,K}$.

than that of the trivial knot. However, for the Kratky plots of the trefoil and figure-eight knots, the peak heights are given by almost the same value.

We remark that the Kratky plots of all polygons and knotted polygons shown in Ref. 37 approximately correspond to the plots of no topological constraint and the trefoil knot, respectively, which are shown in Fig. 10.

4.3. *Analytic approaches to the scattering functions of knots*

4.3.1. *Distribution function of the distance between two nodes*

In subsection 4.3 we shall present an analytic approach to the scattering functions and two-point correlation functions of knotted ring polymers in a θ solvent. We give an analytic expression of the two-point correlation functions, from which we derive analytic expressions of the scattering functions.

Let us formulate the distribution function of the distance between two chosen nodes of a random polygon with fixed knot. We select a pair of nodes, say, j and k, out of the N nodes. If the two arcs of the polygon between them have n and $N - n$ segments, respectively, where $n \leq N/2$, we define arc length λ by fraction n/N:

$$\lambda = n/N. \tag{21}$$

Here we recall that n is given by $\min(|j - k|, N - |j - k|)$. We denote by $f(r; \lambda, N)$ the probability distribution of distance r between the two nodes. Here r denotes the difference of the position vectors: $r = R_j - R_k$, and $r = |r|$. (See Fig. 11 for the case of $j = 0$ and $k = n$.)

For an illustration, let us derive the probability distribution $f(r; \lambda, N)$ of the distance r between two nodes separated by λN bonds in a random polygon of N segments under no topological constraint. Here we recall that r denotes the difference of the position vectors of the two nodes, and $r = |r|$. With arc length parameter λ, the polygon consists of two linear chains of $n = N\lambda$ and $N - n = N(1 - \lambda)$ bonds, respectively, which have the same end-to-end vector r in common. Thus, the probability distribution $\rho(r; n, N - n)$ of the vector r is proportional to the product of the Gaussian distributions such as $\rho(r; n, N - n) \propto p(r; n)p(r; N - n)$. Here $p(r; N)$ denotes the probability distribution of the end-to-end vector r of a random walk of N steps. For very large N, it is given by the Gaussian distribution: $p(r, N) \propto \exp(-3r^2/2Nb^2)$. Considering the rotational symmetry, we therefore have

$$f(r; \lambda, N) \propto 4\pi r^2 \exp\left(-\frac{3r^2}{2\lambda(1 - \lambda)Nb^2}\right). \tag{22}$$

Hereafter we write the distribution function under no topological constraint, $f(r; \lambda, N)$, as $f_{all}(r; \lambda, N)$, and we set $b = 1$.

If the random polygon has a fixed knot type K, we denote by $f_K(r; \lambda, N)$ the distribution function of distance between two nodes with arc length $\lambda = n/N$. In a recent simulation it was shown that the distribution function $f_K(r; \lambda, N)$ is roughly approximated by the Gaussian one, although there exists some deviation from it.[79]

4.3.2. Correlation functions expressed in terms of the distance distributions

We now show that the pair correlation function of a knotted random polygon is related to the distribution function of distance between two nodes.

The quantity $4\pi r^2 g(r)\Delta r/N$ gives the probability of finding another segment in a spherical shell from radius r to $r + \Delta r$ centered at a given segment. In short, it gives the probability distribution of distance r between two nodes for all pairs of the nodes of the polygon. The distribution

$f(r; \lambda, N)$ is therefore related to the radial distribution function $g(r)$ by

$$4\pi r^2 g(r) = 2N \int_0^{1/2} f(r; \lambda, N) \, d\lambda. \tag{23}$$

Similarly, the distribution $f_K(r; \lambda, N)$ is related to the radial distribution function $g_K(r)$ by

$$4\pi r^2 g_K(r) = 2N \int_0^{1/2} f_K(r; \lambda, N) \, d\lambda. \tag{24}$$

In terms of the distribution $f_K(r; \lambda, N)$, the scattering function of a ring polymer with knot K is expressed as follows:

$$g_K(\boldsymbol{q}) = \int d\boldsymbol{r} e^{i\boldsymbol{q} \cdot \boldsymbol{r}} g_K(\boldsymbol{r}) = 2N \int_0^\infty \int_0^{1/2} f_K(r; \lambda, N) d\lambda \frac{\sin qr}{qr} dr. \tag{25}$$

4.3.3. *Formula of the distribution function of the distance between two nodes*

Generalizing distribution (22) we propose the following formula for distance distribution $f_K(r; \lambda, N)$ under a topological constraint K:

$$f_K(r; \lambda, N) = C_K(\lambda, N) \, r^{2+\theta_K(\lambda)} \exp\left[\frac{-3r^2}{2N\sigma_K(\lambda)^2} \right] \tag{26}$$

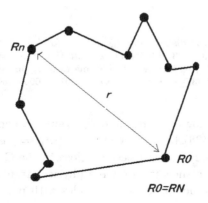

Fig. 11. Distance r between two nodes separated by arc length $n = \lambda N$. Here λ denotes the fraction of the arc length between them with respect to the total length. Here $r = \boldsymbol{R}_n - \boldsymbol{R}_0$.

where the normalization $C_K(\lambda, N)$ is given by

$$C_K(\lambda, N) = \left(\frac{3}{2N\sigma_K{}^2}\right)^{\frac{3+\theta_K}{2}} \frac{2}{\Gamma\left(\frac{3+\theta_K}{2}\right)}.$$

The constants θ_K and σ_K are functions of variable $z = \lambda(1-\lambda)$ as

$$\sigma_K(z; N) = z^{\frac{1}{2}} \exp(\alpha_K z), \qquad \theta_K(z; N) = b_K z^{\beta_K}. \qquad (27)$$

The parameters α_K, β_K and b_K depend on the knot K and the number of nodes, N. Fitting curves for $\lambda = 1/2$ and $N = 300$ are shown in Fig. 12.

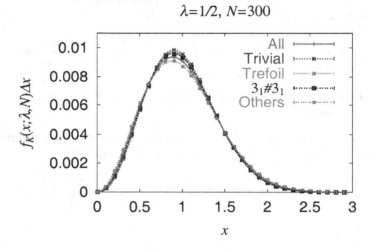

Fig. 12. The probability distribution $\tilde{f}_K(x; \lambda, N)$ for $\lambda = 1/2$ and $N = 300$. For topological conditions, 0, 3_1, $3_1\#3_1$, *others* and *all*, the χ^2 per datum are given by 3.19, 1.30, 0.31, 2.85 and 0.17, respectively; the estimates of θ_K are given by 0.300 ± 0.004, 0.225 ± 0.003, 0.169 ± 0.003, -0.164 ± 0.003 and 0.0007 ± 0.0005, respectively, as presented in Ref. 78.

Formulas (27) show that σ_K and θ_K vanish when $\lambda = 0$. Here we note that function (26) becomes Gaussian if $\sigma_K = \theta_K = 0$. In fact, the distribution $f_K(r; \lambda, N)$ should become close to the Gaussian one when λ is very small. We can show it through a physical argument.[78]

The average distance between two nodes with parameter λ, $r_K(\lambda, N)$, is given by

$$r_K(\lambda, N) = \int_0^\infty r \, f(r; \lambda, N) dr = \sqrt{\frac{2}{3}} N^{\frac{1}{2}} D(\theta_K) \sigma_K. \qquad (28)$$

Here $D(\theta_K)$ is given by $D(\theta_K) = \Gamma\left((4 + \theta_K)/2\right)/\Gamma\left(\frac{3+\theta_K}{2}\right)$. In terms of the ratio of distance r to the average distance, $x = r/r_K(\lambda, N)$, we express the distribution function $f(r; \lambda, N)$ as

$$\tilde{f}(x; \lambda, N) = \frac{2D^{3+\theta_K}(\theta_K)x^{2+\theta_K}}{\Gamma\left(\frac{3+\theta_K}{2}\right)} \exp\left[-D^2(\theta_K)x^2\right] . \tag{29}$$

Here we note that $\tilde{f}(x; \lambda, N)dx = f(r; \lambda, N)dr$. We also remark that $\tilde{f}(x; \lambda, N)$ has only one fitting parameter θ_K.

We find from the simulation data that formula (26) together with eq. (27) gives good fitting curves not only as a function of r but also as a function of λ.[78] We find that both of them give good fitting curves except for the case of other knots.

4.3.4. *Analytic expression of the scattering functions of knotted ring polymers*

The scattering function of a random polygon of N nodes with fixed knot K is given by[78]

$$g_K(q) = 2N \int_0^{1/2} d\lambda \frac{2}{a(\lambda)\Gamma((3 + \theta_K(\lambda))/2)}$$
$$\times \int_0^\infty y^{1+\theta_K(\lambda)} \exp(-y^2) \sin(a(\lambda)y)dy . \tag{30}$$

where $a(\lambda)$ is given by $a(\lambda) = r_K(\lambda, N)q/D(\theta_K)$. Here we recall that the scattering function is given by the Fourier transform of the correlation function (See, eqs. (17) and (25)).

In the small-q limit, we show from eq. (30) that the scattering function does not depend on the knot type, if it is expressed as a function of $qR_{g,K}$, as follows:[78]

$$g_K(q)/N = 1 - \frac{q^2}{3}R_{g,K}^2 + \cdots . \qquad (q \ll 1) . \tag{31}$$

The result is consistent with the previous numerical simulation.[77]

In the large-q case, however, the behavior might be nontrivial. Through the saddle point approximation, we have

$$g_K(q)/N \approx 2 \int_0^{1/2} \frac{\sqrt{2}\sin(\pi\theta_K(\lambda) + \pi/4)}{\Gamma((3 + \theta_K(\lambda))/2)/(\sqrt{\pi}/2)} \left(\frac{Nq^2}{6}\sigma_K^2(\lambda)\right)^{\theta_K(\lambda)/2}$$
$$\times \exp(-\frac{Nq^2}{6}\sigma_K^2(\lambda)) d\lambda . \tag{32}$$

Furthermore, with respect to a large parameter $A = Nq^2/12 = R_{g,all}^2 q^2$, we can show

$$g_K(q)/N \approx \frac{1}{A}\left(1 + O(A^{-\beta_K})\right) \qquad (q \gg 1). \tag{33}$$

For the trivial, trefoil and composite knots, we have $\beta_K \approx 0.5$ and we may have a very slow asymptotic expansion with respect to $A = Nq^2/12$.

4.3.5. *An improved fitting formula*

In §4.3.3 we have shown that formula (26) close to the Gaussian one gives good fitting curves to the numerical data of the distance distribution for knotted ring polymers in θ solution. Furthermore, it has lead to the analytic expression for the scattering functions of knotted ring polymers in θ solution in §4.3.4. However, a more rigorous numerical analysis suggests the following fitting formula:[79]

$$f_K(r; \lambda, N) = C_K(\lambda, N)\, r^{2+\theta_K} \exp\left[\frac{-3r^\delta}{2N\sigma_K{}^\delta}\right] \tag{34}$$

where the normalization factor $C_K(\lambda, N)$ is given by

$$C_K(\lambda, N) = \left(\frac{3}{2N\sigma_K{}^\delta}\right)^{(3+\theta_K)/\delta} \frac{\delta}{\Gamma\left((3+\theta_K)/\delta\right)}. \tag{35}$$

Here σ_K, θ_K and δ are fitting parameters. They are expressed by eqs. (27) as functions of z. According to the renormalization-group argument[10] if we assume that formula (34) holds also for large r, the exponent δ is given by the scaling exponent ν as $\delta = 1/(1 - \nu)$. In the data analysis[62] we have the estimate $\delta = 2.1$. It is close to the Gaussian value $\delta = 2$ but is slightly different. A precise data analysis should be discussed later.[79]

5. Dynamics of Ring Polymers in Solution

We now evaluate the diffusion constant of a ring polymer with fixed knot type in solution numerically via the Brownian dynamics with hydrodynamic interaction. In the dynamics we set the finite extensible non-linear elongational (FENE) potential so that bond crossing is effectively prohibited during time evolution.[82]

Let us denote by D_K the diffusion constant of a ring polymer with fixed topology K in good solution. Here we consider various knot types for K. We denote by D_L the diffusion constant of a corresponding linear polymer in solution such that it has the same molecular weight as the ring polymer.

We evaluate the ratio D_K/D_L or D_K/D_0. Here D_0 denotes the diffusion constant of a ring polymer with the trivial knot type. The ratios should correspond to a universal amplitude ratio of critical phenomena, and play a significant role in the dynamics of knotted ring polymers.

The ratio D_K/D_L should be universal if the number of monomers, N, is large enough, according to the renormalization group arguments.[109–111] The ratio D_K/D_L generalizes the C value in previous studies.[82,112,113] Here it is defined by the ratio $C = D_R/D_L$, where D_R denotes the diffusion constant of a ring polymer under no topological constraint and D_L that of the corresponding linear polymer.

Through simulation we find that ratio D_K/D_L or D_K/D_0 is approximately constant with respect to N for various knots. Thus, if we evaluate ratio D_K/D_L at some value of N, it is practically valid for other values of N. We can therefore predict the diffusion constant D_K of a polymer model at some value of N, multiplying the ratio D_K/D_L by the estimate of D_L for N. Here the value of D_L may depend on the number N and on details of the model.[80,81]

Furthermore, we show numerically that ratio D_K/D_L is a linear function of the average crossing number (N_{AC}) of the ideal knot of K. Here the ideal knot is given by an ideal configuration of knotted curve K, which will be defined shortly. Since the ratio D_K/D_L is almost independent of N, it follows that the linear fitting formula should be valid practically in a wide range of finite values of N. Thus, the ideal knot of a knotted curve K should play a fundamental role in the dynamics of finite-size knotted ring polymers in solution.

The ratio D_K/D_L may have some experimental applications. Ring polymers of different knot types can be separated experimentally with respect to their topologies by making use of the difference among the sedimentation coefficients, which can be calculated from the diffusion constants.[36]

5.1. *Ideal knots and the average crossing number N_{AC}*

Let us introduce the *ideal knot*, briefly. For a given knot K it is given by the trajectory that allows maximal radial expansion of a virtual tube of uniform diameter centered around the axial trajectory of the knot K.[42,43]

The pictures of linear and knotted ring polymers consisting of thick segments like cylinders are shown in Fig. 13. The ideal knot trajectory of a given knot is obtained by increasing the thickness of the segments gradually step-by-step, while keeping the total arc length of the trajectory fixed.

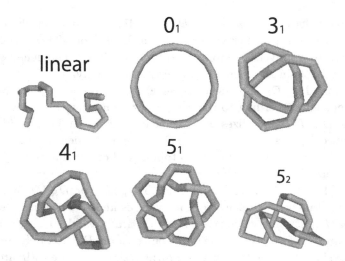

Fig. 13. Figures of a linear polymer and knotted ring polymers with the symbols of knots given in Rolfsen's textbook. They are part of the figures shown in Ref. 83.

We next introduce the average crossing number N_{AC}. We define the N_{AC} of a knotted curve as follows: We take its projection onto a plane, and enumerate the number of crossings in the knot diagram on the plane. Then, we consider a large number of projections onto planes whose normal vectors are uniformly distributed on the sphere of unit radius, and take the average of the crossing number (N_{AC}) over all the normal directions.

We shall show in §5.4 that the average crossing number N_{AC} of an ideal knot plays a central role in the dynamics of knotted ring polymers in solution. The importance of the N_{AC} of an ideal knot was first shown in the case of gel electrophoresis of knotted DNA.[23] The electrophoretic drift velocities of knotted DNA are linear functions of the average crossing numbers N_{AC} of ideal knots in gel electrophoresis experiments.

For a lattice model, electrophoretic drift velocities of knotted ring polymers in gel are approximated by a function of the N_{AC} of ideal knots.[114,115]

5.2. *Brownian dynamics with no bond crossing*

The ring polymer molecule is modeled as a cyclic bead-and-spring chain with N beads connected by N FENE springs with the following force law:

$$\boldsymbol{F}(\boldsymbol{r}) = -H\boldsymbol{r}/(1 - r^2/r_{max}^2),\qquad(36)$$

where $r = |\boldsymbol{r}|$. Let us denote by b the unit of distance. Here we assume that the average distance between neighboring monomers is approximately given by b. We set constants H and r_{max} as follows: $H = 30k_BT/b^2$ and $r_{max} = 1.3b$. We call the setting dynamics A.[82] There is no bond crossing allowed during time evolution of dynamics A. In the standard case we set $H = 3k_BT/b^2$ and $r_{max} = 10b$, which we call dynamics B.[82]

We assume the Lennard-Jones (LJ) potential acting among monomers as follows.[116]

$$V(r) = 4\epsilon_{\text{LJ}} \left[\left(\frac{\sigma_{\text{LJ}}}{r_{ij}} \right)^{12} - \left(\frac{\sigma_{\text{LJ}}}{r_{ij}} \right)^6 \right]. \tag{37}$$

Here r_{ij} is the distance of beads i and j, and ϵ_{LJ} and σ_{LJ} denote the minimum energy and the zero energy distance, respectively.[117] We set the Lennard-Jones parameters as $\sigma_{LJ} = 0.8b$ and $\epsilon_{LJ} = 0.1k_BT$ so that they give good solvent conditions.[118] Here k_B denotes the Boltzmann constant.

We employ the predictor-corrector version[119] of the Ermak-McCammon algorithm for generating the time evolution of a ring polymer in solution. The hydrodynamic interaction is taken into account through the Ronte-Prager-Yamakawa tensor[120,121] where the bead friction is given by $\zeta = 6\pi\eta_s a$ with the bead radius $a = 0.257b$ and a dimensionless hydrodynamic interaction parameter $h^* = (\zeta/6\pi\eta_s)\sqrt{H/\pi k_BT} = 0.25$.

5.3. *Diffusion constants of linear and ring polymers*

We define the diffusion constant of a polymer by the following:

$$D = \lim_{t\to\infty} \frac{1}{6t} \langle (\vec{r}_G(t) - \vec{r}_G(0))^2 \rangle. \tag{38}$$

Here $\vec{r}_G(t)$ denote the position vector of the center of mass of the polymer. Making use of eq. (38) we have evaluated diffusion constants for ring and linear polymers.

According to the Einstein relation, the diffusion constant of a polymer should be given by $D = k_BT/\zeta$ where ζ is given by $\zeta = 6\pi\eta R_H$ with viscosity η and the hydrodynamic radius R_H. Let us assume that the hydrodynamic radius R_H has the same asymptotic scaling behavior with the square root of the mean square radius of gyration: $\sqrt{\langle R_G^2 \rangle} \propto N^\nu$. Thus, in a dilute solution, we have the following large-N behavior:

$$D = \frac{k_BT}{6\pi\eta R_H} \propto N^{-\nu}. \tag{39}$$

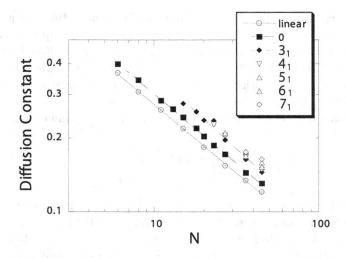

Fig. 14.   Diffusion constants of linear and knotted ring chains with knots 0, 3_1, 4_1, 5_1, 6_1 and 7_1, versus the number of segments N. Fitted by $D = aN^{-\nu}(1 + bN^{-\Delta})$ with the following best estimates: For a linear chain, $a = 0.90 \pm 0.23$, $\nu = 0.53 \pm 0.06$, $b = 0.51 \pm 0.93$, $\Delta = 1.14 \pm 2.39$, $\chi^2 = 17$; for the trivial knot (0), $\nu = 0.55 \pm 0.18$, $b = 0.14 \pm 0.78$, $\Delta = 0.60 \pm 6.09$, $\chi^2 = 28$; for the trefoil knot (3_1) $a = 1.00 \pm 3.87$, $\nu = 0.52 \pm 0.67$, $b = 1.18 \pm 1.12$, $\Delta = 0.77 \pm 6.09$, $\chi^2 = 27$.[83]

The diffusion constants D_L and D_K for $N = 45$ have been evaluated through the Brownian dynamics with no bond crossing (dynamics A). The estimates are plotted against N in Fig. 14. Here the errors of the diffusion constants are as small as 10^{-4}. The fitting curves to them are given by $D = aN^{-\nu}(1 + bN^{-\Delta})$. The estimates of the exponent ν are close to that of the self-avoiding walks.

Here we remark that D_R does not mean D_K of a knot K. By D_R we mean that when we evaluate the diffusion constant, the knot type of the ring polymer is not necessarily fixed in the time evolution.[82,112,113]

5.4.   Universal ratios for knotted ring polymers in solution

In simulation (dynamics A) ratio D_{K_1}/D_{K_2} of two different knots K_1 and K_2 is almost constant with respect to N, at least in the range investigated.[83] For instance, the graph of ratio D_{4_1}/D_0 versus N is almost flat, as shown in Fig. 15. Here 0, 3_1 and 4_1 denote the trivial, the trefoil and the figure-eight knot, respectively, as shown in Fig. 1. The numerical values of D_{3_1}/D_0 are given from 1.14 to 1.17. The values of D_{4_1}/D_0 are given from 1.14 to 1.21,

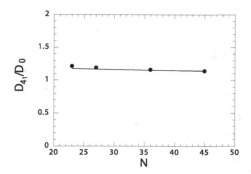

Fig. 15. Ratio D_{4_1}/D_0 of diffusion constants for the figure-eight knot (4_1) and the trivial knot (0) versus the number of segments N.[83] Fitting curve is given by $D_{4_1}/D_0 = a(1 + bN^{-c})$ where $a = 1.02 \pm 0.56$, $b = 1.76 \pm 8.26$, and $c = 0.70 \pm 2.58$ with $\chi^2 = 0.03$.

as shown in Fig. 15. Thus, the estimate of D_K/D_0 evaluated at a value of N, say $N = 45$, for some knot K should also be valid at other finite values of N, since it is almost independent of N.

It is shown[82] that ratio D_0/D_L is given by about 1.1 for the present polymer model and almost independent of N within the range investigated.

From the numerical observations and the RG arguments, we have two conjectures: (A) D_0/D_L should be given by 1.1 for some wide range of finite values of N and also in the large N limit; (B) ratio D_K/D_0 for a nontrivial knot K should remain almost the same value in a wide range of finite values of N, i.e. the N-dependence should be very small.

Quite interestingly we find that ratio D_K/D_L can be approximated by a linear function of the average crossing number (N_{AC}) of ideal knots, i.e. the ideal representations of the corresponding knots. In Fig. 16 the simulation data of D_K/D_L are plotted against N_{AC} of ideal knots. We find that the data points are fitted well by the following empirical formula:

$$D_K/D_L = a + b\,N_{AC}. \qquad (40)$$

Here, the estimates of a and b are given in the caption of Fig. 16. Thus, the diffusion constant D_K of a knot K can be estimated in terms of the N_{AC} of the ideal knot of K.

Let us discuss the χ^2 values. We have $\chi^2 = 2$ for the fitting line of Fig. 16, which is for the data of $N = 45$. For the data of $N = 36$ we have a good fitting line with $\chi^2 = 3$. The estimates of a and b for $N = 36$ are similar to those for $N = 45$. Thus, we may conclude that the graph of D_K/D_L

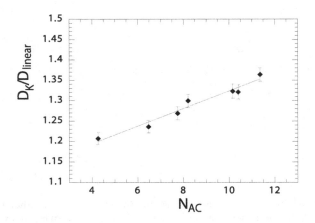

Fig. 16. D_K/D_L versus the average crossing number (N_{AC}) of ideal knot K for $N = 45$: The data are approximated by $D_K/D_L = a + bN_{AC}$ where $a = 1.11 \pm 0.02$ and $b = 0.0215 \pm 0.0003$ with $\chi^2 = 2$, as shown in Ref. 83.

versus N_{AC} is fitted by a linear line.

For a finite value of N, we can estimate the diffusion constant D_K of a knot K through formula (40) by the N_{AC} of the ideal knot of K. Here we have assumed that coefficients a and b of eq. (40) are independent of N since the graphs of D_K/D_0 and D_0/D_L are almost flat with respect to N. In fact, there is almost no numerical support for suggesting a possible N-dependence of a and b, directly.

We thus summarize the simulation results so far as follows: Ratio D_K/D_0 for a knot K should be almost constant with respect to N in a wide range of N and can be expressed by the linear function of N_{AC} of ideal knots. Eq. (40) should be useful in separating synthetic ring polymers into various knotted species by making use of the difference among sedimentation coefficients.

Ideal knots should play a fundamental role in the dynamics of knotted ring polymers in solution. In fact, we have shown it for the diffusion constants. In experiments of gel electrophoresis drift velocities of different knots formed on the same DNA molecules were shown to be simply related to the N_{AC} of ideal knots.[23] The two independent results suggest the importance of the N_{AC} of ideal knots in the dynamics of knotted ring polymers, although the physical situations are different.

Drift velocities in a dilute solvent have been evaluated for a lattice model.[114,115] However, the ratios of the drift velocity of the trefoil knot 3_1 to that of knots 4_1, 5_1 and 7_1, respectively, are about 3 percentage larger than the ratios of the diffusion constant of knot 3_1 to those of the same species of knots, respectively. The inconsistency should be related to the renormalization of the Monte-Carlo steps for each knot. Here we recall that there is no parameter to be adjusted in our first-principle simulation.

5.5. *Dynamics of open and closed chains of linked ring polymers in solution via Brownian dynamics*

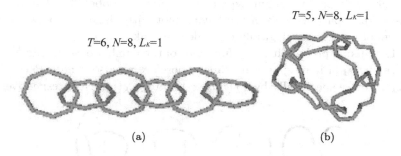

$T{=}5$, $N{=}8$, $L_K{=}1$

$T{=}6$, $N{=}8$, $L_K{=}1$

(a)   (b)

Fig. 17.   An open chain (left) and a closed chain (right) of T linked ring polymers, where each ring polymer consists of N monomers. Here, $T = 6$ and $N = 8$ for the open chain, $T = 5$ and $N = 8$ for the closed chain. The linking number between neighboring ring polymers is equal to 1, as shown in Ref. 84.

We can study the diffusion constant of a polymer with complex topology and characterize it by applying the results in §5. For instance, we consider a polymer chain composed of T ring polymers where each of the ring polymers consists of N monomers and the T ring polymers are linked to each other with linking number one (see Fig. 17).[84] We call the polymer chain a *linked ring polymers*. We evaluated the diffusion constants of open and closed chains of linked ring polymers in solution via Brownian dynamics, which we denote by D_L and D_R, respectively. The ratio D_R/D_L is almost independent of the number T and is given by 1.15.

6. Linking Probability

6.1. *Definition of linking probabilities*

Let us systematically define the linking probabilities. We define the *linking probability* of two SAPs of the trivial knot, $P_{link}(r)$, by the probability that a given self-avoiding pair of SAPs of the trivial knot where the centers of mass of the SAPs are separated by a distance r is topologically equivalent to a non-trivial link.

If a given self-avoiding pair of SAPs is topologically equivalent to the trivial link, we call it a *trivial link* of SAPs, otherwise we call it a *non-trivial link* of SAP. In a nontrivial link of SAP the two polygons are entangled each other (see Fig. 18). Here we have assumed that each of the two SAPs has the trivial knot type.

Let L denote such a link that consists of two trivial knots. We define the linking probability of link type L, $P_L(r)$, by the probability that a given self-avoiding pair of SAPs of the trivial knot with distance r between the centers of mass is topologically equivalent to link-type L.

The linking probabilities are functions of the number of segments N and the radius r_d of hard spherical beads. Therefore we also denote $P_L(r)$ and $P_{link}(r)$ more specifically by $P_L(r, N, r_d)$ and $P_{link}(r, N, r_d)$, respectively.

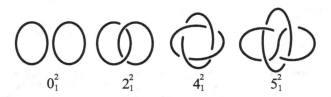

$$0_1^2 \qquad 2_1^2 \qquad 4_1^2 \qquad 5_1^2$$

Fig. 18.   Diagrams of some simple two-component links such as the trivial link 0_1^2 and the Hopf link 2_1^2. The superscript denotes the number of components.[8]

Through simulation we numerically evaluate linking probabilities of SAP of the trivial knot. We assume that all configurations of self-avoiding links of SAPs with the trivial knot have statistically the same probability of appearance. We thus calculate the linking probabilities by

$$P_i(r, N, r_d) = \frac{W_i}{W}, \qquad i = L \text{ or } link, \qquad (41)$$

where L is a link consisting of two trivial knots, and W_L and W_{link} is the number of self-avoiding links with link type L and that of nontrivial links, respectively. Here W is the total number of self-avoiding links of SAP of the trivial knot generated in the simulation. We remark that $W = 100,000$ in our simulation.

Calculating link invariants we can practically determine the link type of a given configuration of self-avoiding link.[52] We use two link invariants called the *linking number* and the *Alexander polynomial* evaluated at $t = -1$. For some simple links we can distinguish them by the values of the two link invariants (see Table 2). Thus, if the fraction of rather complicated links is small, we can evaluate the linking probabilities quite accurately.

Table 2. Values of the two link invariants, the linking number and the Alexander polynomial evaluated at $t = -1$ for simple two-component links.

Link type	0_1^2	2_1^2	4_1^2	5_1^2	6_1^2	6_2^2	6_3^2
Linking number	0	1	2	0	3	2	3
Alexander polynomial	0	1	2	4	3	6	5

Hereafter in this section, we abbreviate the superscript 2 in the symbols of links. For instance, we denote the Hopf link 2_1^2 simply by 2_1.

6.2. Good fitting curves with respect to χ^2 values

Let us introduce a fitting formula for the linking probability as a function of the distance r between the centers of mass for a given pair of SAP.[87]

$$P_i(r) = \exp(-\alpha r^{\nu_1}) - C \exp(-\beta r^{\nu_2}), \qquad i = L \text{ or } link, \qquad (42)$$

where C, α, β, ν_1 and ν_2 are fitting parameters. An intuitive derivation of the fitting formula (42) will be given later. We note that a similar formula was introduced for the linking probability of the Hopf link where ν_1 and ν_2 are fixed as $\nu_1 = \nu_2 = 3$.[52]

We have applied formula (42) to the data-points of $P_{link}(r)$ and $P_L(r)$ obtained through simulation.

The fitting curves are depicted together with the data points in Figure 19. Each panel contains five fitting curves for P_{link}, P_{2_1}, P_{4_1}, P_{5_1}, and P_{others}, which correspond to the linking probabilities of the following link types: all the nontrivial links, 2_1, 4_1, 5_1, and all the nontrivial links other than 2_1, 4_1 and 5_1, respectively.

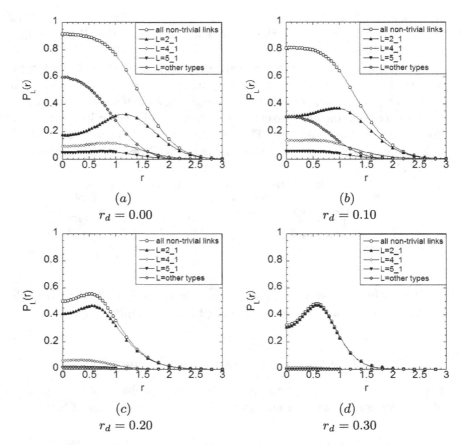

Fig. 19.   Linking probability $P_L(r, N, r_d)$ with link type L versus distance r for SAP of the number of nodes $N = 256$ with excluded volume parameter $r_d = 0.0, 0.1, 0.2, 0.3$, as represented in Ref. 87.   Open circles correspond to the probabilities of nontrivial links, P_{link}; closed triangles, open diamonds, and closed inverted triangles correspond to the probabilities P_{2_1}, P_{4_1}, and P_{5_1}, respectively.   Open crosses correspond to the probabilities P_{others}, which is the probability having more complicated link types.   Solid lines show the fitting curves given by eq. (42).

6.3.   Physical derivation of the linking probability formula

Let us discuss an intuitive derivation of the linking probability of $P_{link}(r)$ for $r_d = 0$, i.e. the probability of a given pair of random polygons separated by distance r between the centers of mass being equivalent to a nontrivial link.[52]   The derivation is not rigorous but may give some hints to the reason why formula (42) gives good fitting curves.

Let us now discuss the large r-dependence of $P_{link}(r)$ for $r_d = 0$. We assume an ensemble of random polygons of N nodes.

(1) Choose a pair of polygons of N nodes, randomly, from the ensemble. Place the two polygons in a way such that the distance between their centers of mass is given by r.

(2) If they are unlinked, they should be unlinked when they are put so that the distance between the centers of mass is changed to $r + dr$. If they are linked, then they may become unlinked when they are placed with distance $r + dr$ between the centers of mass.

(3) The decrease of the probability $dP_{link}(r)$ should be approximately proportional to the product of $P_{link}(r)$ (being linked) and the partial volume $4\pi r^2 dr$ of the configuration space:

$$dP_{link}(r) = -C_\ell \, P_{link}(r) \times 4\pi r^2 dr \,. \tag{43}$$

(4) Integrating the above differential equation we have

$$P_{link}(r) = P_{link}(0) \exp(-\alpha r^3) \tag{44}$$

where $\alpha = 4\pi C_\ell / 3$.

If we assume that constant C_ℓ depends on r as $C_\ell(r) = C_0 r^{\nu_1 - 3}$, then by integration we have

$$P_{link}(r) = P_{link}(0) \exp(-\alpha r^{\nu_1}) \tag{45}$$

where $\alpha = 4\pi C_0 / \nu_1$. If we simply assume that $C_\ell(r)$ is a constant with respect to r, we have $\exp(-\alpha r^3)$ in eq. (45).

The above intuitive argument gives the first term of formula (42). We can derive the second term if we consider contributions from other link conditions in the right hand side of eq. (43). The second terms can be considered as a correction to the first term of eq. (42).

We find that the fitting parameter C of formula (42) for $P_{link}(r)$ becomes very small for random polygons where $r_d = 0$ and N is large.[87] Furthermore, ν_1 becomes close to 3.0 as the number of nodes N increases. Thus, the above intuitive argument should be appropriate only for random polygons of very large N and small r_d.

By a similar intuitive argument we can derive formula (42) also for other link types. For instance, in the Hopf link case, we obtain the difference of two exponentials as follows:[86]

$$P_{2_1}(r) = B_1 \exp(-\beta_1 r^3) - B_2 \exp(-\beta_2 r^3) \,. \tag{46}$$

Let us recall the derivation of Formula (46).[86] We first assume the following:

(1) If a given pair of polygons with distance r between the centers of mass gives the trivial link, then it should also be trivial when the distance between the centers of mass is given by $r + dr$.

(2) If a given pair of polygons with distance r between the centers of mass gives the Hopf link, then it may become a different link when the distance between the centers of mass is given by $r + dr$.

(3) If the pair is neither the trivial or the Hopf link, it may become a Hopf link when the distance between the centers of mass is given by $r + dr$.

Then we have

$$dP_{2_1}(r) = -\gamma_1 P_{2_1}(r)\, dv + \gamma_2(P_{link}(r) - P_{2_1}(r))\, dv\,, \qquad (47)$$

where $v = 4\pi r^3/3$, and γ_1 and γ_2 are constants. By integrating eq. (47) we have eq. (46).

Similarly, if we assume a sequence of links from simple to complex links, through similar intuitive arguments, the linking probability can be expressed as a sum of several exponential terms of r^3.

6.4. *Anomalous osmotic pressure and topics related to the linking probability in polymer physics*

The linking probability $P_{link}(r)$ is related to the anomalous osmotic pressure observed for ring polymers in a θ solvent. Here, the θ temperature is given by that of the corresponding linear polymers. In the experiment,[17] the second virial coefficient of the ring polymer solution at the θ temperature, $A_2(\theta)$, did not vanish. It is therefore called "anomalous".

It was then addressed that the non-vanishing second virial coefficient at the θ temperature should be related to the topological interaction between each pair of ring polymers in solution.[18,19] The pairwise topological interaction was estimated by making use of the linking number.[85]

The $A_2(\theta)$ was calculated numerically by evaluating the linking probability $P_{link}(r, N, r_d = 0)$ through making use of two link invariants.[52]

$$A_2(\theta) = \frac{N_A R_{g,0_1}^3}{2M^2} \int_0^\infty P_{link}(r, N, r_d = 0)\, 4\pi r^2 dr\,. \qquad (48)$$

Here $M = M_{WT}$ denotes the molecular weight of the ring polymer, and N_A the Avogadro number. We recall that $(R_{g,0_1})^2$ denotes the mean square radius of gyration of the ring polymer, where it has N statistical segments with $N = M/M_0$ and has the trivial knot type 0_1.

The molecular-weight dependence of the second virial coefficient at the θ temperature, $A_2(\theta)$, is studied by evaluating the linking probability for some models of ring polymers.[52,89,90] The theoretical values of $A_2(\theta)$ for the Gaussian random polygons[52] are roughly consistent with the values of $A_2(\theta)$ obtained by Roovers and Toporowski.[17] It seems, however, that the theoretical estimates are not completely consistent with the recent experimental data[88] particularly with respect to the molecular-weight dependence. It is thus an interesting open problem to improve the theory to explain the N-dependence of $A_2(\theta)$.

Finally we remark that the linking probability has been successfully applied to show numerically the nonlinear rubber elasticity of topologically constrained polymer networks.[105] There should be many interesting problems in polymer physics which can be studied by applying the linking probability.

Acknowledgements

The authors would like to thank many colleagues and graduate students for their powerful and fruitful collaboration. They would also like to thank Prof. S.G. Whittington for useful comments.

A.1. Algorithm for generating SAP

Let us introduce the rod-bead model of SAP.[47] A conformation of the SAP is given by a sequence of N line segments of unit length. A solid ball of radius r_d is placed at each vertex of the polygon and the balls are not allowed to intersect each other for a valid SAP.

To construct random conformations of off-lattice SAP, we introduce the crank-shaft moves and the bond-interchange moves. Here we note that the former moves are similar to some parts of the pivot moves for generating self-avoiding polygons on a lattice.[122]

The crank-shaft algorithm consists of the following procedures. First we take a conformation of SAP in three dimensions. Second, we choose randomly two nodes of the SAP, and we rotate a subchain between the nodes around the axis passing through them. Here the rotation angle θ ($0 \leq \theta < 2\pi$) is chosen randomly with uniform probability. If there is an overlap among segments in the rotated conformation, then we throw it away and return to the initial conformation. If the rotated conformation has no overlaps among segments, we accept it as a valid SAP. We repeat the procedure until the chain reaches a state which is effectively independent of the initial one.

In addition to the crank-shaft moves, we also apply the *bond-interchange* moves, by which possible local correlations may decrease rapidly between the initial and the deformed conformations of SAP.[123] In every bond-interchange move, we choose two bonds randomly among all bonds of a given polygon and then interchange them, and search possible overlaps among the segments of the new conformation. If there is no overlap, we accept the interchanged conformation as a valid SAP. If they have an overlap, we throw it away and return to the initial conformation. Through the moves, the alignment of bonds is rearranged while the directions of the bonds are preserved.

In the simulation we perform the crank-shaft and the bond-interchange moves, for instance, $2N$ and N times, respectively, and we make a new SAP. In fact, according to previous studies, it is appropriate to perform the crank-shaft moves at least N times for obtaining such a new conformation that is effectively independent of the initial one.[57] For lattice polygons, it was shown that the correlation with the initial conformation decreases with respect to the number of applied moves.[57]

After constructing a large number of SAPs we calculate the second order Vassiliev invariant and the determinant of knots for the SAPs. We then select such SAPs that have the same set of values of the two knot invariants with the trivial knot. We thus practically select the SAPs of the trivial knot among the large number of generated SAPs.

References

1. *Cyclic Polymers*, ed. J.A. Semlyen (Elsevier Applied Science Publishers, New York, 1986); 2nd Ed. (Kluwer Academic Publ., Dordrecht, 2000).
2. A.D. Bates and A. Maxwell, *DNA Topology* (Oxford Univ. Press, 2005).
3. E. Orlandini and S. G. Whittington, *Rev. Mod. Phys.*, **79**, 611 (2007).
4. C. Micheletti, D. Marenduzzo and E. Orlandini, *Phys. Rep.*, **504**, 1 (2011).
5. W.R. Taylor, *Nature*, **406**, 916 (2000).
6. "Statistical physics and topology of polymers with ramifications to structure and function of DNA and proteins", eds. T. Deguchi et al., *Prog. Theor. Phys. Suppl.*, **191**, (2011).
7. "Knots and soft-matter physics", *Bussei-Kenkyu*, **92**-1, (2009).
8. D. Rolfsen, *Knots and Links* (Publish or Perish, Wilmington DE, 1976).
9. M. Doi and S.F. Edwards, *The Theorey of Polymer Dynamics* (Clarendon Press, Oxford, 1986).

10. P.G. de Gennes, *Scaling Concepts in Polymer Physics* (Cornell Univ. Press, Ithaka and New York, 1979).
11. M. Delbrück, in *Mathematical Problems in the Biological Sciences*, ed. by R.E. Bellman, *Proc. Symp. Appl. Math.*, **14**, 55 (1962).
12. H.L. Frisch and E. Wasserman, *J. Amer. Chem. Soc.*, **83**, 3789 (1961).
13. A.V. Vologodskii, A.V. Lukashin, M.D. Frank-Kamenetskii, and V.V. Anshelevich, *Sov. Phys. JETP*, **39**, 1059 (1974).
14. A.V. Vologodskii, A.V. Lukashin, and M.D. Frank-Kamenetskii, *Sov. Phys. JETP*, **40**, 932 (1975).
15. J. des Cloizeaux and M.L. Mehta, *J. Phys. (Paris)*, **40**, 655 (1979).
16. J. des Cloizeaux, *J. Phys. Let. (France)*, **42** , L433 (1981).
17. J. R. Roovers and P. M. Toporowski, *Macromolecules*, **16**, 843 (1983).
18. K. Iwata, *Macromolecules*, **18**, 115 (1985); **22**, 3702 (1989).
19. F. Tanaka, *J. Chem. Phys.*, **87**, 4201 (1987).
20. M.A. Krasnow, A. Stasiak, S.J. Spengler, F. Dean, T. Koller and N.R. Cozzarelli, *Nature*, **304**, 559 (1983).
21. F.B. Dean, A. Stasiak, T. Koller and N.R. Cozzarelli, *J. Biol. Chem.*, **260**, 4975 (1985).
22. K. Shishido, N. Komiyama, and S. Ikawa, *J. Mol. Biol.*, **193**, 215 (1987).
23. A.V. Vologodskii, N.J. Crisona, B. Laurie, P. Pieranski, V. Katritch, J. Dubochet and A. Stasiak, *J. Mol. Biol.*, **278**, 1 (1998).
24. R. M. Robertson, S. Laib and D. E. Smith, *Proc. Natl. Acad. Sci. USA*, **103**, 7310 (2006).
25. S. Araki, T. Nakai, K. Hizume, K. Takeyasu and K. Yoshikawa, *Chem. Phys. Lett.*, **418**, 255 (2006).
26. C. W. Bielawski, D. Benitez and R. H. Grubbs, *Science*, **297**, 2041 (2002).
27. D. Cho, K. Masuoka, K. Koguchi, T. Asari, D. Kawaguchi, A. Takano and Y. Matsushita, *Polym. J.*, **37**, 506 (2005).
28. A. Takano, Y. Kushida, K. Aoki, K. Masuoka, K. Hayashida, D. Cho, D. Kawaguchi and Y. Matsushita, *Macromolecules*, **40**, 679 (2007).
29. Y. Ohta, M. Nakamura, Y. Matsushita, A. Takano, *Polymer*, **53**, 466 (2012).
30. T. Yamamoto and Y. Tezuka, *Polym. Chem.*, **2**, 1930 (2011).
31. P. Virnau, L.A. Mirny and M. Kardar, *PLoS Comput. Biol.*, **2**, e122 (2006).
32. J. I. Sulkowska, P. Sulkowski, P. Szymczak and M. Cieplak, *Phys. Rev. Lett.*, **100**, 058106 (2008).
33. H.A. Kramers, *J. Chem. Phys.*, **14**, 415 (1946).
34. B. H. Zimm and W. H. Stockmayer, *J. Chem. Phys.* **17**, 1301 (1949).
35. E. F. Casassa, *J. Polym. Sci., Part A*, **3**, 605 (1965).
36. V. Bloomfield and B. H. Zimm, *J. Chem. Phys.*, **44**, 315 (1966).
37. G. ten Brinke and G. Hadziioannou, *Macromolecules*, **20**, 480 (1987).
38. D.W. Sumners and S. Whittington *J. Phys. A: Math. Gen.*, **21**, 1689 (1988).
39. N. Pippenger, *Discrete Appl. Math.*, **25**, 273 (1989).
40. S.Y. Shaw and J.C. Wang, *Science*, **260**, 533 (1993).
41. V.V. Ryubenkov, N.R. Cozzarelli and A. Vologodskii, *Proc. Natl. Acad. Sci. USA*, **90**, 5307 (1993).
42. A. Y. Grosberg, A. Feigel and Y. Rabin, *Phys. Rev. E*, **54**, 6618 (1996).

43. V. Katritch, J. Bednar, D. Michoud, R.G. Scharein, J. Dubochet and A. Stasiak, *Nature*, **384**, 142 (1996).
44. *Ideal Knots*, eds. A. Stasiak et al., (World Scientific Pub. Co. Inc., Singapore, 1999).
45. J.P.J. Michels and F.W. Wiegel, *Proc. R. Soc. London, Ser. A*, **403**, 269 (1986).
46. E.J. Janse van Rensburg and S.G. Whittington, *J. Phys. A: Math. Gen.*, **23**, 3573 (1990).
47. K. Koniaris and M. Muthukumar, *Phys. Rev. Lett.*, **66**, 2211 (1991).
48. T. Deguchi and K. Tsurusaki, *Phys. Lett. A*, **174**, 29 (1993).
49. T. Deguchi and K. Tsurusaki, *J. Phys. Soc. Jpn.*, **62**, 1411 (1993).
50. T. Deguchi and K. Tsurusaki, *J. Knot Theory Ramif.*, **3**, 321 (1994).
51. T. Deguchi and K. Tsurusaki, *Phys. Rev. E.*, **55**, 6245 (1997).
52. T. Deguchi and K. Tsurusaki, in *Lectures at Knots '96*, ed. S. Suzuki, (World Scientific, Singapore, 1997) p. 95.
53. E. Orlandini, M.C. Tesi, E.J. Janse van Rensburg, and S.G. Whittington, *J. Phys. A: Math. Gen.*, **29**, L299 (1996).
54. E. Orlandini, M.C. Tesi, E.J. Janse van Rensburg, and S.G. Whittington, *J. Phys. A: Math. Gen.*, **31**, 5953 (1998).
55. M. K. Shimamura, and T. Deguchi, *Phys. Lett. A*, **274**, 184 (2000).
56. M. K. Shimamura and T. Deguchi, *J. Phys. Soc. Jpn.*, **70**, 1523 (2001).
57. A. Yao, H. Matsuda, H. Tsukahara, M. K. Shimamura, and T. Deguchi, *J. Phys. A: Math. Gen.*, **34**, 7563 (2001).
58. M. K. Shimamura and T. Deguchi, *Phys. Rev. E*, **66**, R040801 (2002).
59. M. K. Shimamura and T. Deguchi, *Phys. Rev. E*, **68**, 061108 (2003).
60. K.C. Millett and E.J. Rawdon , in *Physical and Numerical Models in Knot Theory*, eds. J.A. Calvo et al., (World Scientific, Singapore, 2005) p. 247.
61. M. Baiesi, E. Orlandini, A.L. Stella, *J. Stat. Mech.*, (2010) P06012.
62. Y. Akita, Master Thesis, Ochanomizu University, March, 2010 (in Japanese).
63. E.J. Janse van Rensburg and A. Rechnitzer, *J. Phys. A: Math. Theor.*, **44**, 162002 (2011).
64. S. R. Quake, *Phys. Rev. Lett.*, **73**, 3317 (1994).
65. J. M. Deutsch, *Phys. Rev. E*, **59**, R2539 (1999).
66. A. Yu. Grosberg, *Phys. Rev. Lett.*, **85**, 3858 (2000).
67. M. K. Shimamura, and T. Deguchi, *Phys. Rev. E*, **64**, 020801(R) (2001).
68. M. K. Shimamura and T. Deguchi, *Phys. Rev. E*, **65**, 051802 (2002).
69. P. -Y Lai, *Phys. Rev. E*, **66**, 021805 (2002).
70. M. K. Shimamura and T. Deguchi, *J. Phys. A: Math. Gen.*, **35**, L241 (2002).
71. A. Dobay, J. Dubochet, K. Millett, P. E. Sottas and A. Stasiak, *Proc. Natl. Acad. Sci. USA*, **100**, 5611 (2003).
72. H. Matsuda, A. Yao, H. Tsukahara, T. Deguchi, K. Furuta and T. Inami, *Phys. Rev. E*, **68**, 011102 (2003).
73. N. T. Moore, R. C. Lua and A. Y. Grosberg, *Proc. Natl. Acad. Sci. USA*, **101**, 13431 (2004).
74. M.K. Shimamura and T. Deguchi, in *Physical and Numerical Models in Knot Theory*, eds. J.A. Calvo et al., (World Scientific, Singapore, 2005) p. 399.

75. A. Takano, Y. Ohta, K. Masuoka, K. Matsubara, T. Nakano, A. Hieno, M. Itakura,K. Takahashi, S. Kinugasa, D. Kawaguchi, Y. Takahashi, and Y. Matsushita, *Macromolecules*, **45**, 369 (2012).
76. A. Yao, H. Tsukahara, T. Deguchi and T. Inami, *J. Phys. A: Math. Gen.*, **37**, 7993 (2004).
77. M. K. Shimamura, K. Kamata, A. Yao and T. Deguchi, *Phys. Rev. E*, **72**, 041804 (2005).
78. T. Deguchi and A. Yao, *OCAMI Studies*, **1**, 165 (2007).
79. A. Yao, Y. Akita, E. Uehara and T. Deguchi, in preparation.
80. B. Dünweg, D. Reith, M. Steinhauser and K. Kremer, *J. Chem. Phys.*, **117**, 914 (2002).
81. B. Liu and B. Dünweg, *J. Chem. Phys.*, **118**, 8061 (2003).
82. N. Kanaeda and T. Deguchi, *J. Phys. A: Math. Theor.*, **41**, 145004 (2008).
83. N. Kanaeda and T. Deguchi, *Phys. Rev. E*, **79**, 021806 (2009).
84. N. Kanaeda and T. Deguchi, *Prog. Theor. Phys. Suppl.*, **191**, 146 (2011).
85. K. Iwata and T. Kimura, *J. Chem. Phys.*, **74**, 2039 (1981).
86. T. Deguchi, in *Physical and Numerical Models in Knot Theory*, eds. J.A. Calvo et al., (World Scientific, Singapore, 2005) p. 343.
87. N. Hirayama, K. Tsurusaki and T. Deguchi, *J. Phys. A: Math. Theor.*, **42**, 105001 (2009).
88. A. Takano, Y. Kushida, Y. Ohta, K. Masuoka, Y. Matsushita, *Polymer*, **50**, 1300 (2009).
89. D. Ida, D. Nakatomi and T. Yoshizaki, *Polym. J.*, **42**, 735 (2010).
90. N. Hirayama, K. Tsurusaki and T. Deguchi, *Prog. Theor. Phys. Suppl.*, **191**, 154 (2011).
91. B. Marcone, E. Orlandini, A. L. Stella and F. Zonta, *J. Phys. A: Math. Gen.*, **38**, L15 (2005).
92. E. Orlandini, A. L. Stella, C. Vanderzande and F. Zonta *J. Phys. A: Math. Theor.*, **41**, 122002 (2008).
93. R. Metzler, A. Hanke, P.G. Dommersnes, Y. Kantor and M. Kardar, *Phys. Rev. Lett.*, **88**, 188101 (2002).
94. E. Ercolini, F. Valle, J. Adamcik, G. Witz, R. Metzler, P. De Los Rios, J. Roca and G. Dietler, *Phys. Rev. Lett.*, **98**, 058102 (2007).
95. G.E. Cates and J. M. Deutsch, *J. Phys. (France)*, **47**, 2121 (1986).
96. J. Suzuki, A. Takano, and Y. Matsushita, *J. Chem. Phys.*, **129**, 034903 (2008).
97. J. Suzuki, A. Takano, T. Deguchi and Y. Matsushita, *J. Chem. Phys.*, **131**, 144902 (2009).
98. T. Vettorel, S.Y. Reigh, D.Y. Yoon, and K. Kremer, *Macromol. Rapid Commun.*, **30**, 345 (2009).
99. T. Vettorel, A.Y. Grosberg, and K. Kremer, *Phys. Biol.*, **6**, 025013 (2009).
100. T. Sakaue, Phys. Rev. Lett. **106**, 167802 (2011).
101. Y. Diao, *J. Knot Theory Ramif.*, **2**, 413 (1993).
102. R. Scharein, K. Ishihara, J. Arsuaga, Y. Diao, K. Shimokawa, and M. Vazquez, *J. Phys. A: Math. Theor.*, **42**, 475006 (2009).
103. S. Saka and H. Takano, *J. Phys. Soc. Jpn.*, **77**, 034001 (2008).

104. M. Baiesi, E. Orlandini and A.L. Stella, *Phys. Rev. Lett.*, **99**, 058301 (2007).
105. N. Hirayama and K. Tsurusaki, *Nihon Reoroji Gakkaishi*, **39**, 65 (2011).
106. N. Madras and G. Slade, *The Self-Avoiding Walk* (Birkhäuser, Boston, 1993).
107. K. Murasugi and B. Kurpita, *Knot Theory and Its Applications* (Birkhäuser, Boston, 1996).
108. M. Polyak and O. Viro, *Int. Math. Res. Not.*, **11**, 445 (1994).
109. Y. Oono and M. Kohmoto, *J. Chem. Phys.*, **78**, 520 (1983).
110. B. Schaub and D. B. Creamer, *Phys. Lett. A*, **121**, 435 (1987).
111. Y. Oono, *Adv. Chem. Phys.*, **61**, 301 (1985).
112. J.M. García Bernal and M.M. Tirado, J.J. Freire, and J. García de la Torre, *Macromolecules*, **23**, 3357 (1990).
113. J.M. García Bernal and M.M. Tirado, J.J. Freire, and J. García de la Torre, *Macromolecules*, **24**, 593 (1991).
114. C. Weber, P. De Los Rios, G. Dietler and A. Stasiak, *J. Phys. Condens. Matter*, **18**, S161 (2006).
115. C. Weber, P. De Los Rios, G. Dietler and A. Stasiak, *Biophys. J.*, **90**, 3100 (2006).
116. A. Ortega and J. García de la Torre, *Biomacromolecules*, **8**, 2464 (2007).
117. J. G. Hernández Cifre, R. Pamies, M. C. López Martinez, J. García de la Torre, *Polymer*, **46**, 267 (2005).
118. A. Rey, J. J. Freire and J. García de la Torre, *Macromolecules*, **20**, 342 (1987).
119. A. Iniesta and J. Garcia de la Torre, *J. Chem. Phys.*, **92**, 2015 (1990).
120. J. Rotne and S. Prager, *J. Chem. Phys.*, **50**, 4831 (1969).
121. H. Yamakawa, *J. Chem. Phys.*, **53**, 207 (1970).
122. N. Madras, A. Orlitsky and L.A. Shepp, *J. Stat. Phys.*, **58**, 159 (1990).
123. K. Koniaris, *J. Chem. Phys.*, **101**, 731 (1994).

CHAPTER 13

DYNAMICS OF CYCLIC POLYMERS REVEALED BY SINGLE-MOLECULE SPECTROSCOPY

Satoshi Habuchi[a]

Department of Organic and Polymeric Materials, Tokyo Institute of Technology
O-okayama 2-12-1-S8-44, Meguro-ku, Tokyo 152-8550, Japan
E-mail: habuchi.s.aa@m.titech.ac.jp

Dynamics of polymers in a dilute and semi-dilute solution are investigated by means of single-molecule fluorescence microscopy techniques. In addition to the direct demonstration of fundamental concepts in polymer physics such as reptation motion in a highly concentrated solution, the single-molecule techniques also reveal effects of topology of the polymer chains on their diffusion behavior. Threading of the chains plays an important role in the heterogeneous diffusion observed in the cyclic polymers.

1. Introduction

Polymer dynamics have been investigated theoretically as well as experimentally for decades. A microscopic view of polymer dynamics in a dilute and semi-dilute solution and in a polymer melt have been systematically investigated by means of variety of ensemble experimental methods. However, those methods are not able to unravel the dynamics of single chains due to disorders which are intrinsic characteristics for flexible polymers. Flexible polymer chains have conformational disorders as depicted in Fig. 1. First, the polymer chains have conformational distribution (static disorder). An average value obtained from an ensemble

[a] Current address: Chemical and Life Sciences and Engineering Division, King Abdullah University of Science and Technology, Bldg. 2 Level 4, 4233, 4700 King Abdullah University of Science and Technology, Thuwal 23955-6900, Kingdom of Saudi Arabia, *E-mail: satoshi.habuchi@kaust.edu.sa*

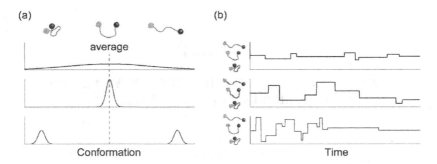

Fig. 1. Schematic illustrations of (a) conformational distributions (static disorder) and (b) conformational fluctuations (dynamic disorder) of individual flexible polymer chains.

measurement does not provide information about the conformation of each polymer chain (Fig. 1. (a)). Second, each chain shows conformation changes over time (dynamic disorder). The conformational dynamics of the individual chains cannot be revealed by a time-averaged ensemble measurement (Fig. 1. (b)). In order to unravel the static and dynamic disorders in polymer dynamics which are often hidden behind the ensemble averaging, the dynamics of individual chains have to be investigated.

This chapter describes methodologies to visualize and analyze diffusion of individual polymer chains. The first part of this chapter aims to give a general picture of sample preparation methods and experimental setup for visualizing single polymer chains, and analytical procedures of single polymer diffusion. After a discussion of a single-molecule view of polymer dynamics, the molecular level view of topology effects on polymer dynamics, especially for cyclic polymers, will be discussed in details. Disorders in polymer dynamics are discussed at the end of this chapter.

2. Methodologies to Visualize Diffusion of Single Polymer Chains

2.1. *Strategy of visualizing polymer dynamics of individual chains*

In order to follow dynamics of individual chains, a single chain has to be highlighted. Although a number of methods have been developed to

(a)   (b)

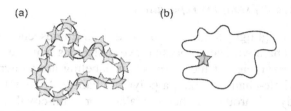

Fig. 2. Schematic illustrations of fluorescently labeled polymer chains; (a) whole chain labeled, and (b) site specifically labeled polymers.

visualize individual polymer chains, fluorescence microscopy has been a primary choice for studying dynamic behaviors of single polymer chains due to its ultra-high sensitivity and superior time-resolution.[1] Also fluorescence microscopy is a relatively non-invasive method. Especially, single-molecule fluorescence microscopy has been successfully applied to study dynamic behaviors of individual molecules in heterogeneous environments, such as meso-porous materials, lipid bilayers, and so on.[2,3] To visualize single chains in a dilute solution, the labeled chains are dispersed in the solvent at a very low concentration (pM–nM) such that each chain can be detected as a spatially isolated object (see section 3). The single chains in a melt or semi-dilute solution are visualized by mixing trace amounts of the labeled polymers with highly concentrated non-labeled chains.

One way to visualize a single polymer chain is to label the whole chain with fluorescent dye molecules (Fig. 2. (a)). The advantage of using this approach is that it allows visualizing directly both the shape and movement of the chain. One has to pay great attention to the effects of the labeling since the whole-chain labeling sometime alters drastically the elastic property of the chain.[4] Another strategy is to label a polymer chain at a specific site with a fluorescent dye molecule (Fig. 2. (b)). While the method does not provide directly the information about the shape of the chain, the site-specific labeling provides information about local polymer dynamics at a specific site. Also it offers an opportunity to measure polymer dynamics at the single-molecule level with a minimal impact of the labeling.

2.2. Procedures of sample preparation

A basic idea to highlight single polymer chains in fluorescence microscopy is to label the chain with small organic dye molecules which exhibit bright fluorescence. Examples of the whole-chain labeling are shown in Fig. 3. (a). For the whole-chain labeling, a polymer chain is densely labeled with fluorescent dyes through either covalent or non-covalent bonding. Biopolymers such as DNA are often labeled by using DNA intercalators which non-covalently bind to major and/or minor grooves of DNA tightly.

Frequently used DNA intercalators are listed in Fig. 3. (c). Since intercalating dyes potentially alter the elastic properties of the chain, one has to carefully control the labeling density such that the labeling does not have significant effects on the chain properties. While this method is often used for labeling long genomic DNA such as λ-phage DNA (48.5 kbp,

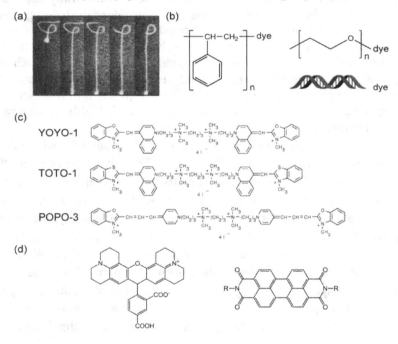

Fig. 3. Samples for single-chain fluorescence imaging experiments. (a) Example of the whole-chain labeled polymer. (b) Examples of site-specifically labeled polymers. DNA intercalators (c) and fluorophores (d) for single-chain fluorescence imaging. Reproduced with permission from Ref. 34 (Fig. 3. (a)). Copyright 1994, AAAS.

contour length of ~ 16.5 μm), it is also applicable to label shorter DNA as well as topologically unique DNA such as supercoiled and cyclic DNA. A covalent labeling of DNA with dye molecules is also possible although it has not been very popular.[5] At present, the whole-chain labeling of synthetic polymers are not frequently used approach to investigate polymer dynamics partly due to synthetic difficulties. So far, the whole-chain labeling of a polyisocyanide chain (up to a micrometer length) has been reported (Fig. 3. (b)) .[6]

The site-specific fluorescent labeling has often been done by covalently labeling chain ends. As shown in Fig. 3. (b), this approach has been used for both naturally occurring and synthetic polymers. The chain end labeled DNA of different lengths (2–23 kbp, contour length of ~ 0.7–7.8 μm) has been reported.[7] As for the synthetic polymers, a chain-end labeled polystyrene (PS) (M_n ~ 2,000–1,7000,000) as well as a chain-end labeled polyethylene oxide (PEO) have been reported.[8–10] In single-molecule fluorescence microscopy, spatio-temporal dynamics of individual chains can be followed only when the label is active. Therefore, it is of crucial importance to choose a highly photostable bright fluorophore as a fluorescent tag. Examples of the fluorescent dyes which meet the conditions are depicted in Fig. 3. (d). Rhodamine and perylenecarboxydiimide derivatives are well known bright and photostable dyes. Alexa and Atto dyes are also known as their superior fluorescent properties.

2.3. *Equipment for single-molecule fluorescence imaging*

As mentioned in the previous section, fluorescence microscopy is a primary tool for studying polymer dynamics at the single-molecule level. A typical experimental setup for the single-molecule fluorescence imaging is illustrated schematically in Fig. 4. The setup consists of an excitation laser, inverted optical microscope, and highly sensitive CCD camera. A diode-pumped solid-state laser or Ar ion laser with output power of around 50–500 mW are frequently used in single-molecule imaging experiments. The excitation laser is passed through an excitation filter, reflected by a dichroic mirror, and introduced into the microscope through a high numerical aperture (N.A.) microscope objective (N.A. ~ 1.2–1.65). This excitation configuration is called epi-illumination. A fluorescently labeled

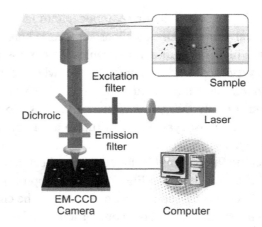

Fig. 4. Experimental setup for the single-molecule fluorescence imaging experiment.

polymer is optically excited into an electronically excited state by illuminating the excitation laser. Fluorescence from the labeled polymer chain is collected by the same objective, passes through the dichroic and emission filter, and detected by an electron multiplying (EM) CCD camera.

Spatial resolution of an optical microscope (R) is expressed as

$$R = 0.61\lambda/\text{N.A.} \tag{1}$$

where λ is a wavelength of the light.[11] The resolution ranges between 200–300 nm under the typical experimental conditions (N.A. ~ 1.3–1.45 and λ ~ 400–700 nm). Therefore, information about the shape of the chain can be obtained only when the chain has a structural dimension larger than hundreds of nanometer. Also each chain has to be spatially separated by at least several hundreds of nanometers to detect them as an isolated object. It should be noted that more sophisticated fluorescence microscopy methods have recently been developed which have a spatial resolution of nanometers to tens of nanometers.[12–14]

Time resolution of the measurement is determined by the brightness of the fluorescence obtained from a single polymer chain and the frame rate of the EM-CCD camera. Typical frame rate of the EM-CCD camera is in the range of 30–500 frames per second (fps). In reality, the time resolution is often improved in compensation for the spatial resolution of the measurement. The time resolution of maximum a few millisecond limits the detection of rapidly diffusing polymer chain. Diffusion coefficient (D) up to about 10 μm s^{-2} can be followed by the typical single-molecule fluorescence microscopy setup. Much faster frame rate (40,000 fps) has been achieved by using a high-speed camera. Since the sensitivity of this type of camera is much lower than that of the EM-CCD, the detected signal in this case is a scattered light rather than fluorescence.[15]

A depth of field of the typical fluorescence microscopy setup is in the range of ~ 0.5–1 μm. Because one can obtain a focused fluorescence image only when the molecules exist within the depth of field, a diffusion trajectory of the single polymer chain recorded by the EM-CCD corresponds to the 2 dimensional projection of the 3 dimensional motion (within ~ 0.5–1 μm depth) of the molecule. The movement of the chain beyond the depth of field cannot be followed by the conventional fluorescence microscopy setup. For 3 dimensional tracking/imaging of single molecules, more complex experimental systems are required.[16,17]

2.4. *Analysis of single-molecule diffusion*

A first step to analyze diffusion of a single chain is to localize the position of the chain in each image to generate diffusion trajectory as accurate as possible. As mentioned above, the spatial resolution of fluorescence microscopy is on the order of several hundreds of nanometers. Therefore,

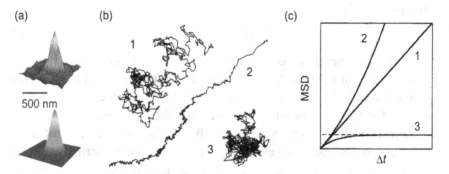

Fig. 5. Analysis of single-molecule diffusion. (a) Single-molecule fluorescence image (top) and 2 dimensional Gaussian fit of the raw image (bottom). (b) Examples of single-particle diffusion trajectories. The trajectories represent (1) random diffusion, (2) directed diffusion, and (3) restricted diffusion. (c) MSD-Δt plots for (1) random diffusion, (2) directed diffusion, and (3) restricted diffusion. Reproduced with permission from Ref. 20, Copyright 1993, Elsevier (Figs. 5. (b), 0.5. (c)).

the fluorescence image of a single molecule recorded by the camera has a spot size of at least several hundreds of nanometers even when the chain dimension is much smaller than this length scale (Fig. 5. (a) top). To localize the position of the molecule more accurately, the raw images are fit with an appropriate 2 dimensional function, typically Gaussian (Fig. 5. (a) bottom).[3] The fitting allows one to determine the spatial location of the molecule with accuracy of nanometer to tens of nanometers.[18,19] Using the fitting procedure, the positions of the molecule are determined frame by frame, and the diffusion trajectory of the single molecule/particle can be obtained by connecting those positions. Examples of single particle diffusion trajectories are shown in Fig. 5. (b).

Those trajectories provide information about diffusion modes of the molecule. While the trajectory 1 in Fig. 5. (a) represents a simple Brownian (random) diffusion, the trajectory 2 and 3 represent a direct and a restricted diffusion, respectively.[20] The different diffusion modes can be quantitatively characterized by calculating the mean square displacement (MSD) for every time interval (Δt) by using the formula,[21]

$$MSD(\Delta t) = \langle (x_{i+n} - x_i)^2 + (y_{i+n} - y_i)^2 \rangle \qquad (2)$$

where x_{i+n} and y_{i+n} describe the molecular position following a time interval Δt, given by n frame rate after starting at position x_i and y_i. i ranges from 1 to $N - n$, where N is the total number of positions recorded, and n takes on values 1, 2, 3 . . . $N - 1$. For a simple 2 dimensional Brownian diffusion,

$$MSD(\Delta t) = 4D\Delta t \qquad (3)$$

where D is the lateral diffusion coefficient. For simple Brownian motion, a linear increase of MSD with Δt is expected from the Eq. (3) (Fig. 5. (c) line 1), and D is calculated from the slope of the MSD–Δt plot. Typical MSD–Δt plots for other diffusion modes are also shown in Fig. 5. (c).[20, 22] A theoretical model which describes diffusion in the presence of fixed obstacles has also been developed.[23] We will examine distributions of D values determined for each molecule to investigate static disorders (heterogeneities) of a semi-dilute polymer solution in section 6.1. We will also refer to an alternative method to characterize multi-component diffusion in section 6.2.

In addition to the analysis of lateral diffusion of the chains, the MSD analysis can also be used to analyze modes of intrachain dynamics when a specific site on the chain is labeled fluorescently. The intrachain segment motion is described by either MSD $(t) \sim t^{2/3}$ or MSD $(t) \sim t^{1/2}$. The former formula describes the chain motion in the Zimm model, and the later describes the motion in the Rouse model (see below).

3. Single Molecule View of Polymer Dynamics

Polymer chain dynamics has been investigated by using light scattering, neutron scattering, viscosity, dielectric relaxation, and birefringence measurements. Although these traditional experimental methods have brought deep insight, they have not been able to provide direct information about molecular level dynamics of the chains. The motion of a single chain cannot be measured by those methods, and therefore, the dynamics of a single chain must be inferred from indirect measurements averaged over a large number of chains. Furthermore, the single-molecule fluorescence imaging provides most direct way of determining the diffusion behavior of isolated polymers in the limit of infinite dilution. Ensemble methods usually require higher sample concentrations, and therefore, extrapolating to the infinite dilution is necessary to determine the behavior of isolated polymers. In this section, I briefly overview the single-molecule studies on polymer dynamics.

3.1. *Polymer dynamics in a dilute solution*

The first successful molecular model of polymer dynamics was developed by Rouse, who simply ignored hydrodynamic interaction forces. In dilute solutions, hydrodynamic interactions between the monomers play a major role in the polymer dynamics. These effects were incorporated by Zimm, whose theory is generally believed to best describe polymer dynamics in the dilute concentration regime. In the Zimm model, the diffusion coefficient of a polymer chain is described by

$$D_{\text{Zimm}} = 0.196 \frac{k_B T}{\sqrt{6}\eta R_g} \qquad (4)$$

$$R_g \sim N^\nu \qquad (5)$$

where η is the solvent viscosity, k_BT is the thermal energy, R_g is the gyration radius, and N is the chain length. The scaling exponent v is $v = 3/5$ for a linear chain in good solvent conditions.

The scaling law of the polymer dynamics in a dilute solution was investigated at the single chain level using DNA as a model polymer.[24] The shape and diffusion of single DNA molecules can be visualized by fluorescently labeling them with DNA intercalator dyes (Fig. 6. (a)).[25] Smith *et al.* prepared DNA molecules ranging in length from 2 to 140 μm which were fluorescently labeled by the DNA intercalator TOTO-1, and the diffusion coefficients of individual chains were determined by the MSD analysis assuming the Brownian diffusion.[24] Fig. 6 (b). shows chain length dependent diffusion coefficient. The scaling exponent determined by a linear fit to log D vs log N is $v = 0.611$ which agrees with the prediction of the Zimm model. For shorter DNA molecules in the range of 0.03 to 7 μm, the scaling exponent of $v \sim 0.667$ was observed. This result suggests that DNA behaves as a semi-flexible polymer in this length range.

Intramolecular polymer relaxation time (τ_r) is expressed as $\tau_r \sim N^{3v}$ and $\tau_r \sim N^2$ for the Zimm and Rouse model, respectively. The scaling law of the

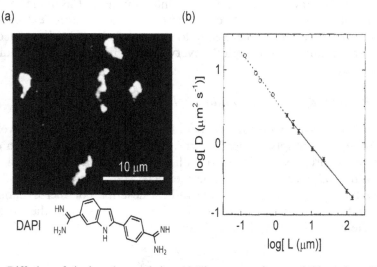

Fig. 6. Diffusion of single polymer chains. (a) Fluorescence image of 56 μm long DNAs labeled by the DNA intercalator DAPI. (b) Chain length dependence of the diffusion coefficient of the isolated DNA molecules determined by the MSD analysis of the fluorescence images. Reproduced with permission from Ref. 25, Copyright 1992, Elsevier (Fig. 6. (a)), and Ref. 24, Copyright 1996, American Chemical Society (Fig. 6. (b)).

(a)

(b)

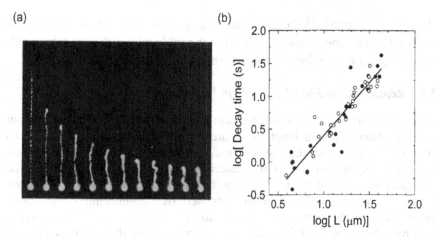

Fig. 7. Relaxation of single polymer chain. (a) Fluorescence image montages of single relaxing DNA molecule. Initially, the DNA is stretched to its full length (left image). Frames are spaced at 4.5-s intervals. (b) log-log plot of the decay time versus chain length. The data were taken in a sucrose solution (filled circles) and a glycerol solution (open circles). The solid line is a linear fit to the data. Reproduced with permission from Ref. 27. Copyright 1994, AAAS.

relaxation time was first studied by measuring rotational motion of freely diffusing fluorescently labeled DNA.[25] The experimentally obtained chain length dependent relaxation time was well described by the Zimm model. Intramolecular dynamics of a partially stretched DNA was also consistent with a semiflexible polymer with strong hydrodynamic interactions.[26] It is interesting to note that the single-molecule imaging experiment makes it possible to see polymer chain relaxation at conditions which deviate significantly from equilibrium. The chain relaxation of a fully stretched DNA displayed multiple relaxation components (Fig. 7. (a)), and the longest relaxation times followed a scaling law with chain length $\tau_r \sim N^{3\nu}$. The scaling exponent was measured to be $3\nu = 1.66$ (Fig. 7. (b)).[27] The scaling exponent determined from the extended chain is closer to that predicted by the Zimm model although the model describes small fluctuations of the chain about equilibrium.

The intramolecular motion about equilibrium at shorter time-scale was investigated by means of fluorescence correlation spectroscopy (FCS) technique.[7,28–30] Those studies reported very different results, and interpretation of the data is still controversial. A deviation from the Zimm model was observed for a single polymer chain immobilized by an electrokinetic trap.[31,32] Recently an attempt has been made to separate intramolecular motion completely from the molecular diffusion which will provide greater precision for characterizing the intramolecular motion.[33]

3.2. *Reptation motion of a single polymer chain*

Polymer dynamics in a semi-dilute solution above an entanglement concentration (C_e) and length (N_e) are governed by interactions between entangled polymer chains, which lead to a variety of unusual properties. The most successful model for describing polymer dynamics above the entanglement concentration is the reptation model. In the reptation model, the topological constrains imposed by neighboring chains on a given chain restrict its motion to a tube-like region. As a consequence, the chain behaves as if it were trapped in the tube that follows its own contour.

Single-molecule imaging is a perfect tool to demonstrate the model as it provides direct view of the detailed motion of a single chain. Perkins *et al.* manipulated one end of a fluorescently labeled long DNA chain in a concentrated polymer solution.[34] As the one end was pulled through the solution, it closely followed the path defined by its initial contour (Fig. 8. (a)), which clearly demonstrated the tube-like motion of the polymer. A tube-like motion of semirigid actin filaments has also been reported.[35] These experiments directly demonstrated the reality of the tube assumed in the reptation model. If the reptation model is applicable, the diffusion coefficient of the polymer chain follows the equation

$$D_{rep} \sim N^{-2} C^{-1.75} \qquad (6)$$

where C is the concentration of the polymer chain. While experimentally obtained scaling exponent ($N^{-1.8}$) of a concentrated solution (0.63 mg ml^{-1}) is close to the prediction of the reptation model, the scaling exponent ($N^{-1.27}$) of a less concentrated solution (0.40 mg ml^{-1}) did not agree with the prediction (Fig. 8. (b)).[36] The dependence of D on the concentration suggested that the long chain follows the scaling law predicted for the reptation, $D_{rep} = C^{-1.75}$. However, a shorter chain did not agree with the scaling prediction. Those results suggest that for molecules above a certain concentration and length satisfies the assumptions made in the reptation model. Chain mobility in a semi-dilute solution has also been studied by using FCS technique.[8,9]

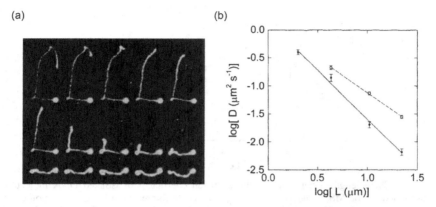

Fig. 8. Reptation motion of single polymer chain. (a) A series of fluorescence images showing the tube-like motion of a relaxing DNA molecule in a concentrated polymer solution. Frames are spaced at 2.3-s (row 1) and 13-s (row 2 and 3) intervals. (b) Chain length dependence of the diffusion coefficient of λ-DNA in a 0.63 mg/ml (filled circles) and a 0.40 mg/ml solution (open circles) determined by the MSD analysis of the fluorescence images. The lines are linear fits to the data. Reproduced with permission from Ref. 34, Copyright 1994, AAAS (Fig. 8. (a)), and Ref. 36, Copyright 1995, American Physical Society (Fig. 8. (b)).

4. Topology Effects on Polymer Dynamics: Isolated Polymers

Static and dynamic properties of cyclic polymers have been investigated theoretically and experimentally. Let me start with the ring polymer dynamics in a diluted solution. Isolated cyclic polymers in a dilute solution are characterized by smaller gyration radius, smaller intrinsic viscosity, and larger diffusion coefficient as compared with their linear counterparts.[37-39]

Topology effects on the diffusion of isolated polymers were investigated at the single-molecule level by using DNA molecules with different topologies in an aqueous buffer solution: namely cyclic, supercoiled, and linear DNA.[40] The chain length dependent diffusion coefficients determined by the MSD analysis are shown in Fig. 9. The scaling law $D \sim N^{-\nu}$ with nearly identical exponent ν was observed which is almost independent of the topology; 0.571, 0.589, and 0.571 for the linear, cyclic, and supercoiled DNA, respectively. Those are in good agreement with the predicted value of $\nu = 3/5$ for long linear polymers in a good solvent where excluded volume effects are appreciable. The ν values

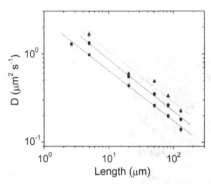

Fig. 9. Chain length dependence of the diffusion coefficient of isolated DNA molecules determined by the MSD analysis of the single-molecule fluorescence images. The points are the data obtained by linear (squares), circular (circles), and supercoiled circular (triangles) DNA. The lines are power-law fits to the data. Reproduced with permission from Ref. 40. Copyright 2006, National Academy of Science, U.S.A.

obtained in this study are higher than those reported for synthetic cyclic polymers ($v = 0.52$).[37] The smaller values observed in the studies could be due to the relatively narrow range of the molecular weight from which the v values were determined.

The D values obtained from the cyclic molecules are always larger than those obtained from the linear molecules of the same molecular weights (Fig. 9.). The ratios of D values of the linear (D_L) and cyclic molecules (D_C) were similar for all the molecules (mean value of $D_L/D_C = 1.32$). Ensemble measurements on synthetic polymers reported the ratio between $D_L/D_C = 1.11–1.36$.[37, 41] While the reported D_L/D_C ratios are not perfectly consistent with each other, the ratios agree with the values predicted theoretically ($D_L/D_C = 1.45$ [42]).

5. Topology Effects on Polymer Dynamics: Entangled Polymers in Semi-Dilute Solution

Dynamics of a linear polymer in a semi-dilute solution or in a melt can be described by the reptation model (see section 3.2.). In this model, the relaxation of constrains occurs mainly at the chain ends. Since rings have no ends, they cannot reptate like linear chains. The diffusion mechanism of cyclic polymers has been suggested theoretically as amoeba-like motion.[43,44] Despite a large number of experimental, theoretical, and simulation works, many controversies still remain. Furthermore, dynamics of blends of linear and cyclic polymers are suggested to be very different from the average properties of the two components.[45]

5.1. *Diffusion of linear and cyclic polymers in a semi-dilute solution*

Topology effects on polymer dynamics in a semi-dilute solution or in a polymer melt have been investigated by using either synthetic or naturally occurring polymers.[46-50] At the single-molecule level, the diffusion of linear and cyclic DNA in a semi-dilute solution was reported by Robertson *et al.*.[51,52] The topological combinations of the linear and cyclic DNA are shown schematically in Fig. 10. (a). The diffusion of single fluorescently-labeled linear and cyclic molecules (tracer molecules) in a matrix polymer solution which contain either linear or cyclic DNA at a semi-dilute concentration was recorded by means of single-molecule fluorescence imaging. The diffusion coefficients of the tracer molecules were relatively insensitive to their topology as well as the topology of the matrix polymers when they have a short chain length (Fig. 10. (b)). On the other hand, the diffusion coefficients of the longer chains depend considerably on the chain length (Fig. 10. (b)).

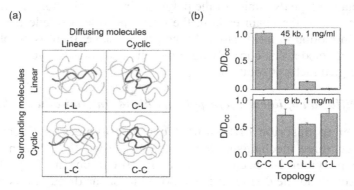

Fig. 10.   Topology effect on the diffusion coefficient of DNA molecules. (a) Schematic illustration of the topological combinations of entangled linear and cyclic DNA molecules. The four cases depicted are: a linear diffusing molecule surrounded by linear molecules (*L-L*), cyclic molecule surrounded by linear mlecules (*C-L*), linear molecule surrounded by cyclic molecules (*L-C*), and cyclic molecule surrounded by cyclic molecules (*C-C*). (b) Dependence of DNA self-diffusion coefficients (*D*) on molecular topology. Bar graphs display normalized *D* values versus topological case. All *D* values are normalized by the corresponding measured diffusion coefficient for a cyclic DNA surrounded by cyclic DNA. Reproduced with permission from Ref. 52. Copyright 2007, National Academy of Science, U.S.A.

The D values of the cyclic tracer molecules in the cyclic matrices (C-C) were always larger than those of the linear tracer molecules in the linear matrices (L-L). This is consistent with the results of simulation works.[53–55] The simulation works suggested more compact conformation and less interpenetration of the cyclic polymer in a melt state which leads to faster diffusion of the cyclic polymer.[53,56] In contrast, experimental studies at the ensemble level on synthetic polymers reported inconsistent results on the D values of C-C and L-L.[50,57,58]

While diffusion of linear polymers above C_e and L_e is described by the reptation model ((see Eq. (6)), the polymer diffusion below C_e and L_e (above critical concentration) is described by the Rouse model.

$$D_{Rouse} \sim N^{-1} C^{-0.5} \tag{7}$$

Both concentration and length dependence of the D values of the short chains were consistent with the Rouse model irrespective to the chain topology (Fig. 11. (a), Fig. 11. (b)). As expected, L-L showed transition from the Rouse to reptation type dynamics above C_e and L_e. C-C also showed similar transition (Fig. 11. (a), Fig. 11. (b)), partly consistent with the results of simulation studies.[59,60] While the results appear to indicate the reptation type dynamics of the polymers in the cyclic polymer matrices, the absence of the chain ends in the cyclic polymer exclude the normal reptation motion in these cases. At the present stage, there is no plausible theoretical model that correctly describes the dynamics of cyclic polymers in entangled melts or solution.

5.2. *Diffusion of linear-cyclic polymer blends in semi-dilute solution*

In order to describe diffusion of cyclic polymers in linear chain matrices, three mechanisms have been suggested [61–63]. When a large cyclic polymer is threaded by many linear chains, reptation is impossible and the diffusion is supposed to occur through a constraint release (CR) mechanism (Fig. 12. (a)). A small cyclic polymer could be unthreaded and the diffusion is supposed to occur through reptation (similar to a linear chain of half of its length), which is referred to as the restricted reptation (RR)mechanism (Fig. 12. (b)). Another mechanism referred to as once-threaded (R1) model, in which a cyclic polymer threaded once by a linear chain and diffuses along the contour of the linear chain (Fig. 12. (c)). When the linear and cyclic polymers

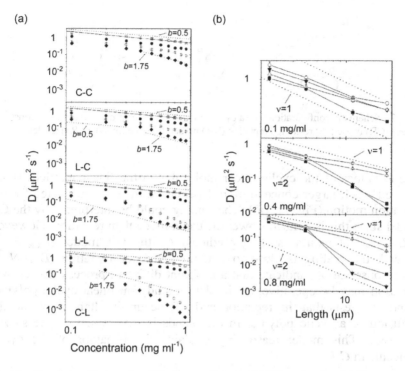

Fig. 11. Topology effect on the diffusion coefficient of DNA molecules. (a) Concentration dependence of the diffusion coefficient of the four cases of the topological combinations. The different point styles indicate the four different construct lengths: 5.9 kbp (open squares), 11.1 kbp (filled circles), 25 kbp (open hexagons), and 45 kbp (filled diamonds). The solid lines are power law fits to the 5.9 kbp data. The dotted lines indicate the scaling laws predicted by the Rouse ($b = 0.5$) and reptation ($b = 1.75$) models for linear molecules. (b) Chain length dependence of the diffusion coefficient of the three different concentrations. The different point styles indicate the four different topological cases: *C-C* (open circles), *L-C* (open triangles), *L-L* (filled squares), and *C-L* (filled triangles). The dotted lines indicate the scaling laws predicted by Rouse ($v = 1$) and reptation ($v = 2$) models for linear molecules. Reproduced with permission from Ref. 51. Copyright 2007, American Chemical Society.

have an identical chain length, the diffusion coefficient of the cyclic polymer in the CR and R1 models are expressed by the following equations.

$$D_{CR} \sim N^{-4} \qquad (8)$$

$$D_{R1} \sim N^{-2} \qquad (9)$$

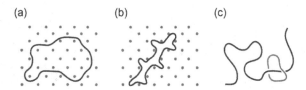

Fig. 12. Schematic configurations for a cyclic polymer in a linear chain matrix. (a) constraint release (CR) model; (b) restricted reptation (RR) model; (c) once-threaded (R1) model.

The D values of the cyclic tracer molecules in the linear matrices (C-L) showed much larger concentration dependence than that predicted for the reptation motion (Fig. 11. (a)). The finding could be interpreted by the CR model since this is a much slower diffusion mode than reptation. However, C-L does not exhibit scaling predicted by the CR model ($D_{CR} \sim N^{-4}$). Instead, the scaling is close to that predicted for reptation ($D_{rep} \sim N^{-2}$). Those observations indicate that a different diffusive process slower than reptation may be operative. A simulation study on linear-cyclic polymer blends suggests that the reptation and CR occur simultaneously and the diffusion of a cyclic polymer in the blends is determined by the slower process.[64] This model reasonably explains the behavior of the cyclic molecule in C-L.

6. Heterogeneities in Polymer Dynamics: The Role of Polymer Topology

Many diffusion mechanisms have been proposed to explain unique properties of cyclic polymers in a semi-dilute solution and in a melt observed in experimental, simulation, and theoretical studies. In spite of those efforts, the diffusion modes of cyclic polymers are not entirely clear yet. Especially, the diffusion mechanism of linear-cyclic blends is puzzling. This is partly due to the intrinsic heterogeneity that the polymer solution and melt possess. As discussed in the previous section, multiple diffusion modes could be operative simultaneously in the linear-cyclic melt. Unfortunately, these heterogeneities are hard to detect in ensemble experiments because they only provide an average value. In contrast, single-molecule experiments can eliminate ensemble averaging and provide an opportunity to directly detect the multiple diffusion modes.

6.1. *Direct detection of multi-mode diffusion*

In order to detect the multiple diffusion modes, a distribution of the diffusion coefficient obtained from individual molecules has to be carefully examined. Linear and cyclic polytetrahydrofuran (poly(THF)) incorporating a fluorophore, perylenediimide (PI) derivative (M_n = 4,200 and 3,800 for the linear and cyclic polymers, respectively) were synthesized for this purpose (Fig. 13. (a)).[65] The PI-labeled linear and cyclic poly(THF)s (tracer molecules) are mixed with linear matrix poly(THF)s (M_n = 3,000) dissolved in toluene at a semi-dilute condition (denoted by *L-L* and *C-L*, respectively) (Fig. 13. (b)), and fluorescence images of the labeled molecules were recorded (Fig. 14. (a), 14. (b)). While N_e of poly(THF) is unknown, the molecular weights of the polymers are about twice larger than N_e of a related polymer, PEO, indicating that the effective entanglement between the chains occurs.

Diffusion coefficients of the tracer molecules were obtained by analyzing the diffusion trajectories (Fig. 14. (c), 14. (d)) using the MSD analysis. Frequency histograms of the D values obtained for the *L-L* and *C-L* are depicted in Fig. 15. (bars). The difference in the mean D value between *L-L* and *C-L* is less than 15% (2.0 and 1.74 $\mu m^2\ s^{-1}$ for *L-L* and *C-L*, respectively). However, the distribution of the D values showed a significant difference.

(a)

(b)

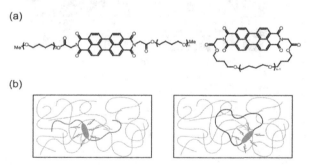

Fig. 13. Topology effect on the diffusion of poly(THF) molecules. (a) Chemical structures of the linear and cyclic poly(THF) molecules containing a perylenetetracarboxydiimide (PI) moiety. (b) Schematic illustration of the topological combinations of entangled linear and cyclic poly(THF) molecules. Left: a linear PI-labeled poly(THF) molecule surrounded by non-labeled linear poly(THF) molecules (*L-L*). Right: a cyclic PI-labeled poly(THF) molecule surrounded by non-labeled linear poly(THF) molecules (*C-L*). Reproduced with permission from Ref. 65. Copyright 2010, Wiley-Blackwell.

Generally, a distribution of D values measured by a single-molecule tracking experiment consists of a statistical contribution owing to a limited number of measured positions in a single trajectory,[66] and of a contribution due to the physical heterogeneity of the system. Statistical distributions corresponding to the theoretical diffusion of single molecules in a homogeneous environment are described by

$$p(D)\mathrm{d}D = \frac{1}{(N-1)!} \cdot \left(\frac{N_D}{D_0}\right)^{N_D} \cdot (D)^{N-1} \cdot \exp\left(\frac{-N_D D}{D_0}\right) \cdot \mathrm{d}D \qquad (10)$$

where N_D is the number of displacements in the diffusion trajectory, D_0 is the mean diffusion coefficient.[67,68] The calculated statistical distributions are shown in Fig. 15. (a) and Fig. 15. (b) as solid lines. The measured distribution of the D values for L-L is reproduced reasonably well by the calculated distribution (Pearson's correlation coefficient $r = 0.82$). This result demonstrates that the linear chains diffuse in a homogeneous environment with a single diffusion mode. On the other hand, the measured histogram of the D values for C-L deviates significantly from the calculated homogeneous statistical distribution ($r = 0.55$). Indeed, the D values for C-L appears to have bimodal distribution, peaking at around 1.0 and 4.5 μm^2 s^{-1}. The results indicate the presence of multiple diffusion modes in C-L.

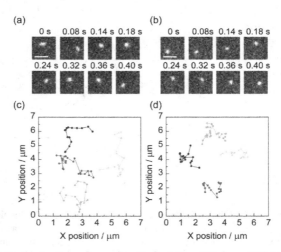

Fig. 14.   Single-molecule fluorescence imaging of the diffusion of PI-labeled poly(THF) molecules. Fluorescence images of (a) PI-labeled linear poly(THF) and (b) PI-labeled cyclic poly(THF) mixed with linear poly(THF) in toluene. Scale bar = 2 μm. 2D trajectories of diffusing molecules of (c) PI-labeled linear poly(THF) and (d) PI-labeled cyclic poly(THF). Reproduced with permission from [Habuchi et al., 2010]. Copyright 2010, Wiley-Blackwell.

6.2. *Quantitative analysis of the multi-mode diffusion*

Multi-component diffusion observed in the single-molecule tracking measurement can be analyzed in a quantitative fashion using a cumulative distribution function (CDF), P, which is the cumulative probability of finding a diffusion molecule within a radius r from the origin at a given time lag $i\Delta t$,[69]

$$P(r^2, i\Delta t) = \int_0^r p(r^{2'}, i\Delta t)dr' = 1 - \exp\left[-\frac{r^2}{4D(i\Delta t)}\right] \quad (11)$$

$$1 - P = \exp\left[-\frac{r^2}{4D(i\Delta t)}\right] \quad (12)$$

For a two-component model, the Eqs. are modified to,

$$P(r^2, i\Delta t) = \int_0^r p(r^{2'}, i\Delta t)dr' = 1 - A_1\exp\left[-\frac{r^2}{4D_1(i\Delta t)}\right] -$$

$$A_2\exp\left[-\frac{r^2}{4D_2(i\Delta t)}\right] \quad (13)$$

$$1 - P = A_1\exp\left[-\frac{r^2}{4D_1(i\Delta t)}\right] + A_2\exp\left[-\frac{r^2}{4D_2(i\Delta t)}\right] \quad (14)$$

where A_1 and A_2 are the fraction of the components. The CDF analysis and plots of $1-P$ versus r^2 are useful tools to extract the values of diffusion

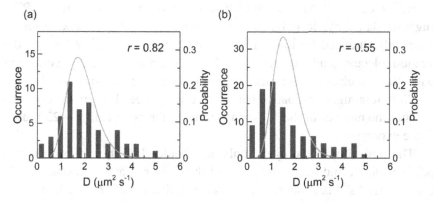

Fig. 15. Frequency histograms (bars) of the diffusion coefficient determined for (a) PI-labeled linear poly(THF) and (b) PI-labeled cyclic poly(THF). The solid lines show calculated theoretical statistical distributions corresponding to diffusion of single molecules in a homogeneous environment, with the diffusion coefficient given by means of the respective histograms. Reproduced with permission from Ref. 65. Copyright 2010, Wiley-Blackwell.

coefficients from trajectories that contain multiple diffusion modes (Fig. 16. (b)). Fast decay of 1–P reflects a rapidly decreasing probability that a molecule diffuses into larger distances and thus corresponds to slow diffusion (Fig. 16. (a)).

The CDF shows a single-exponential behavior for L-L, which suggests the presence of single diffusion mode (Fig. 16. (c)). The D value calculated from the CDF analysis is 2.9 $\mu m^2 s^{-1}$ (Fig. 16. (e)), which is consistent with the mean value obtained from the MSD analysis. In contrast, the CDF of C–L shows a multi-exponential behavior (Fig. 16. (d)), indicating the presence of multimode diffusion modes. The two diffusion coefficients (1.1 and 4.9 $\mu m^2 s^{-1}$, with fractions of 0.66 and 0.34) were obtained from a double-exponential fitting (Fig. 16. (f)). These results are also consistent with the inhomogeneous diffusion suggested by the D value distribution (Fig. 15. (b)). Indeed, the D values obtained from the CDF analysis agree well with the two peaks in the D value distribution in the MSD analysis.

Both the MSD and CDF analyses point to the heterogeneous multimode diffusion of the cyclic tracer polymer molecule in the linear matrix polymers. The observed heterogeneity can be interpreted as partial threading of the cyclic chains with the linear matrix chains. The heterogeneity in the D value is suggested to become larger when (partial) threading occurs.[63] Also a simulation study on a related polymer PEO suggests that a cyclic polymer which consists of 60 backbone atoms is efficiently threaded by linear matrix polymers in a melt.[70] Although the single-molecule study on the poly(THF)s cannot be directly compared with the simulation since the experiments are done in a semi-dilute solution, it is highly probable that the threading indeed occurs, taking into account the number of the backbone atoms of the poly(THF)s (~230) used in the experiments.

The threading of the cyclic polymer should slow the diffusion of the polymer down. Indeed, the D value of the slower component observed in C-L (1.1 $\mu m^2 s^{-1}$) is about twice smaller than that observed in L-L (2.0 $\mu m^2 s^{-1}$). Thus, the slower component can be reasonably attributed to the diffusion of the threaded cyclic polymer. It is not entirely clear from the experiment that the threading mode can be described by the R1 model (see Fig. 12. (c)). Relatively large fraction of the slow diffusion component

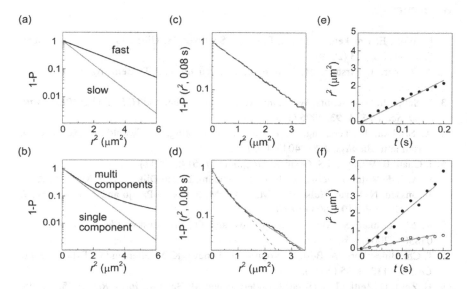

Fig. 16. Analysis of multi-mode diffusion of the poly(THF) molecules. (a, b) Examples of cumulative distribution functions (CDFs); (a) diffusion rate dependence, (b) component number dependence. Experimentally obtained CDFs in the form of 1-P (dots) for (c) PI-labeled linear poly(THF) and (d) PI-labeled cyclic poly(THF). The broken and solid lines show single- and double-exponential fittings. CDF coefficients at different time lags for (e) PI-labeled linear poly(THF) and (f) PI-labeled cyclic poly(THF). The solid lines show linear fittings. Reproduced with permission from Ref. 65. Copyright 2010, Wiley-Blackwell (Figs. 16. (c), 16. (d), 16. (e), 16. (f)).

suggests that the majority of the cyclic polymers are threaded by the linear matrix chains.

The faster component observed in the D value distribution can be attributed to unthreaded cyclic polymers. Diffusion of an unthreaded cyclic polymer in a linear matrix is suggested to be described by the RR model (see Fig. 12. (b)).[62,71] Because the effective chain length of the cyclic polymer is half of its linear counterpart in this model, the Eq. (6) predicts that the unthreaded cyclic polymer diffuses four times as fast as that of the linear polymer. The fast diffusing component observed in *C-L* ($D = 4.9$ μm^2 s^{-1}) and the D value observed in *L-L* ($D = 2.0$ μm^2 s^{-1}) are roughly consistent with the prediction of the RR model.

References

1. D. Woll, E. Braeken, A. Deres, F. C. De Schryver, H. Uji-i and J. Hofkens, *Chem. Soc. Rev.* **38**, 313 (2009).
2. A. Zurner, J. Kirstein, M. Doblinger, C. Brauchle and T. Bein, *Nature* **450**, 705 (2007).
3. T. Schmidt, G. J. Schutz, W. Baumgartner, H. J. Gruber and H. Schindler, *Proc. Natl. Acad. Sci. U. S. A.* **93**, 2926 (1996).
4. A. Sischka, K. Toensing, R. Eckel, S. D. Wilking, N. Sewald, R. Ros and D. Anselmetti, *Biophys. J.* **88**, 404 (2005).
5. J. Guan, B. Wang and S. Granick, *Langmuir* **27**, 6149 (2011).
6. P. A. J. de Witte, J. Hernando, E. E. Neuteboom, E. van Dijk, S. C. J. Meskers, R. A. J. Janssen, N. F. van Hulst, R. J. M. Nolte, M. F. Garcia-Parajo and A. E. Rowan, *J. Phys. Chem. B* **110**, 7803 (2006).
7. R. Shusterman, S. Alon, T. Gavrinyov and O. Krichevsky, *Phys. Rev. Lett.* **92**, 048303 (2004).
8. T. Cherdhirankorn, A. Best, K. Koynov, K. Peneva, K. Muellen and G. Fytas, *J. Phys. Chem. B* **113**, 3355 (2009).
9. H. Zettl, U. Zettl, G. Krausch, J. Enderlein and M. Ballauff, *Phys. Rev. E* **75**, 061804 (2007).
10. R. Chen, L. Li and J. Zhao, *Langmuir* **26**, 5951 (2010).
11. D. B. Murphy, *Fundamentals of light microscopy and electronic imaging* (Wiley-Liss, Inc., 2001).
12. E. Betzig, G. H. Patterson, R. Sougrat, O. W. Lindwasser, S. Olenych, J. S. Bonifacino, M. W. Davidson, J. Lippincott-Schwartz and H. F. Hess, *Science* **313**, 1642 (2006).
13. M. J. Rust, M. Bates and X. W. Zhuang, *Nat. Methods* **3**, 793 (2006).
14. S. W. Hell, R. Schmidt and A. Egner, *Nat. Photonics* **3**, 381 (2009).
15. A. Kusumi, C. Nakada, K. Ritchie, K. Murase, K. Suzuki, H. Murakoshi, R. S. Kasai, J. Kondo and T. Fujiwara, *Ann. Rev. Biophys. Biomol. Struct.* **34**, 351 (2005).
16. L. Holtzer, T. Meckel and T. Schmidt, *Appl. Phys. Lett.* **90**, 053902 (2007).
17. G. A. Lessard, P. M. Goodwin and J. H. Werner, *Appl. Phys. Lett.* **91**, 224106 (2007).
18. R. E. Thompson, D. R. Larson and W. W. Webb, *Biophys. J.* **82**, 2775 (2002).
19. A. Yildiz, J. N. Forkey, S. A. McKinney, T. Ha, Y. E. Goldman and P. R. Selvin, *Science* **300**, 2061 (2003).
20. A. Kusumi, Y. Sako and M. Yamamoto, *Biophys. J.* **65**, 2021 (1993).
21. H. Qian, M. P. Sheetz and E. L. Elson, *Biophys. J.* **60**, 910 (1991).
22. M. J. Saxton and K. Jacobson, *Ann. Rev. Biophys. Biomol. Struct.* **26**, 373 (1997).
23. M. J. Saxton, *Biophys. J.* **66**, 394 (1994).
24. D. E. Smith, T. T. Perkins and S. Chu, *Macromolecules* **29**, 1372 (1996).

25. M. Matsumoto, T. Sakaguchi, H. Kimura, M. Doi, K. Minagawa, Y. Matsuzawa and K. Yoshikawa, *J. Polym. Sci., Part B: Polym. Phys.* **30**, 779 (1992).
26. S. R. Quake, H. Babcock and S. Chu, *Nature* **388**, 151 (1997).
27. T. T. Perkins, S. R. Quake, D. E. Smith and S. Chu, *Science* **264**, 822 (1994).
28. D. Lumma, S. Keller, T. Vilgis and J. O. Radler, *Phys. Rev. Lett.* **90**, 218301 (2003).
29. E. P. Petrov, T. Ohrt, R. G. Winkler and P. Schwille, *Phys. Rev. Lett.* **97**, 258101 (2006).
30. R. G. Winkler, S. Keller and J. O. Radler, *Phys. Rev. E* **73**, 041919 (2006).
31. A. E. Cohen and W. E. Moerner, *Proc. Natl. Acad. Sci. U. S. A.* **104**, 12622 (2007).
32. A. E. Cohen and W. E. Moerner, *Phys. Rev. Lett.* **98**, 116001 (2007).
33. K. McHale and H. Mabuchi, *J. Am. Chem. Soc.* **131**, 17901 (2009).
34. T. T. Perkins, D. E. Smith and S. Chu, *Science* **264**, 819 (1994).
35. J. Kas, H. Strey and E. Sackmann, *Nature* **368**, 226 (1994).
36. D. E. Smith, T. T. Perkins and S. Chu, *Phys. Rev. Lett.* **75**, 4146 (1995).
37. D. F. Hodgson and E. J. Amis, *J. Chem. Phys.* **95**, 7653 (1991).
38. S. S. Jang, T. Cagin and W. A. Goddard, *J. Chem. Phys.* **119**, 1843 (2003).
39. M. Muller, J. P. Wittmer and M. E. Cates, *Phys. Rev. E* **61**, 4078 (2000).
40. R. M. Robertson, S. Laib and D. E. Smith, *Proc. Natl. Acad. Sci. U. S. A.* **103**, 7310 (2006).
41. P. C. Griffiths, P. Stilbs, G. E. Yu and C. Booth, *J. Phys. Chem.* **99**, 16752 (1995).
42. H. Johannesson, D. B. Creamer and B. Schaub, *J. Phys. A: Math. Gen.* **20**, 5071 (1987).
43. T. McLeish, *Science* **297**, 2005 (2002).
44. S. P. Obukhov, M. Rubinstein and T. Duke, *Phys. Rev. Lett.* **73**, 1263 (1994).
45. S. Nam, J. Leisen, V. Breedveld and H. W. Beckham, *Macromolecules* **42**, 3121 (2009).
46. M. Kapnistos, M. Lang, D. Vlassopoulos, W. Pyckhout-Hintzen, D. Richter, D. Cho, T. Chang and M. Rubinstein, *Nat. Mater.* **7**, 997 (2008).
47. G. B. McKenna, B. J. Hostetter, N. Hadjichristidis, L. J. Fetters and D. J. Plazek, *Macromolecules* **22**, 1834 (1989).
48. S. Nam, J. Leisen, V. Breedveld and H. W. Beckham, *Polymer* **49**, 5467 (2008).
49. D. Richter, B. Ewen, B. Farago and T. Wagner, *Phys. Rev. Lett.* **62**, 2140 (1989).
50. E. von Meerwall, R. Ozisik, W. L. Mattice and P. M. Pfister, *J. Chem. Phys.* **118**, 3867 (2003).
51. R. M. Robertson and D. E. Smith, *Macromolecules* **40**, 3373 (2007).
52. R. M. Robertson and D. E. Smith, *Proc. Natl. Acad. Sci. U. S. A.* **104**, 4824 (2007).
53. S. Brown and G. Szamel, *J. Chem. Phys.* **109**, 6184 (1998).
54. J. D. Halverson, W. B. Lee, G. S. Grest, A. Y. Grosberg and K. Kremer, *J. Chem. Phys.* **134**, 204904 (2011).
55. M. Muller, J. P. Wittmer and M. E. Cates, *Phys. Rev. E* **53**, 5063 (1996).
56. J. Suzuki, A. Takano, T. Deguchi and Y. Matsushita, *J. Chem. Phys.* **131**, 144902 (2009).

57. T. Cosgrove, P. C. Griffiths, J. Hollingshurst, R. D. C. Richards and J. A. Semlyen, *Macromolecules* **25**, 6761 (1992).
58. T. Cosgrove, M. J. Turner, P. C. Griffiths, J. Hollingshurst, M. J. Shenton and J. A. Semlyen, *Polymer* **37**, 1535 (1996).
59. K. Hur, C. Jeong, R. G. Winkler, N. Lacevic, R. H. Gee and D. Y. Yoon, *Macromolecules* **44**, 2311 (2011).
60. B. V. S. Iyer, S. Shanbhag, V. A. Juvekar and A. K. Lele, *J. Polym. Sci., Part B: Polym. Phys.* **46**, 2370 (2008).
61. J. Klein, *Macromolecules* **19**, 105 (1986).
62. P. J. Mills, J. W. Mayer, E. J. Kramer, G. Hadziioannou, P. Lutz, C. Strazielle, P. Rempp and A. J. Kovacs, *Macromolecules* **20**, 513 (1987).
63. Y. B. Yang, Z. Y. Sun, C. L. Fu, L. J. An and Z. G. Wang, *J. Chem. Phys.* **133**, 064901 (2010).
64. G. Subramanian and S. Shanbhag, *Macromolecules* **41**, 7239 (2008).
65. S. Habuchi, N. Satoh, T. Yamamoto, Y. Tezuka and M. Vacha, *Angew. Chem. Int. Ed.* **49**, 1418 (2010).
66. M. J. Saxton, *Biophys. J.* **72**, 1744 (1997).
67. S. Y. Nishimura, S. J. Lord, L. O. Klein, K. A. Willets, M. He, Z. K. Lu, R. J. Twieg and W. E. Moerner, *J. Phys. Chem. B* **110**, 8151 (2006).
68. M. Vrljic, S. Y. Nishimura, S. Brasselet, W. E. Moerner and H. M. McConnell, *Biophys. J.* **83**, 2681 (2002).
69. G. J. Schutz, H. Schindler and T. Schmidt, *Biophys. J.* **73**, 1073 (1997).
70. C. A. Helfer, G. Q. Xu, W. L. Mattice and C. Pugh, *Macromolecules* **36**, 10071 (2003).
71. S. F. Tead, E. J. Kramer, G. Hadziioannou, M. Antonietti, H. Sillescu, P. Lutz and C. Strazielle, *Macromolecules* **25**, 3942 (1992).

CHAPTER 14

PROGRESS IN THE RHEOLOGY OF CYCLIC POLYMERS

Dimitris Vlassopoulos, Rossana Pasquino and Frank Snijkers

Institute of Electronic Structure & Laser, FORTH and Department of Materials Science & Technology, University of Crete GR 71110 Heraklion, Crete, Greece E-mail: dvlasso@iesl.forth.gr

The ability to purify cyclic polymers with liquid chromatography at the critical condition and the recent advances in theoretical modeling and advanced simulations have opened the route for determining and understanding their rheological properties, and hence resolving a nearly 30-years old controversy. In the unentangled regime, the zero-shear viscosity of cyclic polymers is half that of their linear counterparts and their rheology conforms to the Rouse model. The behavior of entangled rings conforms to the lattice animal configuration and is characterized by a self-similar stress relaxation with a power law exponent of about 0.4 extending over 3 decades in time. No entanglement plateau is observed. Their viscosity remains much smaller in comparison to linear polymers. Tiny amounts of linear chain contaminants (of the order of 0.1%, below their overlap concentration) dramatically alter the rheology of rings as they entropically penetrate them, forming a transient network of linear chains bridged by rings. Despite the significant progress, many outstanding challenges lie ahead and make this topic fascinating.

1. Introduction

The rheological properties of polymers are of central importance for the design, fabrication, processing and the performance of final products. In fact, one of the grand challenges in polymer science has been the understanding and control of macromolecular motion. This reflects the viscoelasticity and the structure-property relationships of polymers, which extend over a broad range of length and time scales, depend on macromolecular architecture and composition, and are responsible for their processing and final properties. The fundamental material function governing macromolecular thermal motion is the viscoelastic stress relaxation modulus, $G(t)$, which describes the stress response to a sudden small strain imposed on the material at rest (the small strain guarantees that the response is linear). It can also be obtained from the inverse Fourier transformation of the frequency-dependent storage and loss moduli, G' and G'', respectively. The latter are measured in a popular small amplitude oscillatory shear experiment, where typically a small strain $\gamma = \gamma_0 \sin(\omega t)$ is imposed on the test material (with γ_0 the strain amplitude and ω the frequency), and the stress response is probed:[1]

$$\sigma = \sigma_0 \sin(\omega t + \delta) = \gamma_0 [G' \sin(\omega t) + G'' \cos(\omega t)]$$

Here σ_0 is the stress amplitude and $\delta = \arctan(G''/G')$ the phase angle. For simple, short unentangled linear polymer melts (with molar mass M smaller than the molar mass of an entanglement segment, M_e), $G(t)$ exhibits a self-similar power-law-like dynamics as described by the Rouse model.[2] When $M > M_e$, linear polymers become entangled and their dynamics are severely hindered. The key concept developed and advanced over the past four decades to describe the dynamics of entanglements is that of the tube as a single-chain approximation to uncrossability constraints by neighbouring chains.[3] Originally conceived in the 1970s by Edwards, de Gennes and Doi, the tube model has led to accurate effective-medium theoretical predictions of stress relaxation.[3] Each chain relaxes its stress by diffusing along the tube. Note that only the two free end-segments of every chain can randomize their direction. By diffusing back-and-forth curvilinearly along the tube, the chain eventually loses all memory of the original tube and escapes into a new configuration. This is the mechanism of reptation, which when coupled to contour length fluctuations and constraint release provides quantitative predictions of rheological experiments. As a consequence, $G(t)$ (or G') exhibits a plateau at intermediate times (frequencies) and exponential decay at longer times.[4,5] Stress relaxes in proportion to the amount of polymer restored to random-walk configurations.

Branched polymers are different, while they can still be described by the tube model. Stars, the simplest branched polymers, have arms with only one free end while the other end is fixed at the core (acting as a solid wall). In such a case reptation does not operate, but instead stars exhibit a logarithmic $G(t)$ plateau and relax stress via arm retraction.[3,5] Macromolecules with complex branched architectures (such as H-polymers, combs, Cayley-trees and pom-pom polymers) relax via the so-called hierarchy of modes:[6-9] different branching generations relax at different time scales in a sequential fashion starting from the outermost dangling ends and moving inwards.

The challenge with understanding the rheology of cyclic polymers (also called rings hereafter) stems from their lack of free ends. The history of unknotted entangled cyclic polymer melts is long and fascinating. Starting with their conformation, the prevailing idea has been that a ring in an array of fixed obstacles forms a "lattice-animal"—like conformation comprising double-folded loops, which relax stress by sliding along the contour of other loops.[10-12] Note that the lattice-animal model is two-dimensional. Molecular simulations have confirmed the reduced size of cyclic polymers as compared to their linear counterparts of the same molar mass, due to their topology.[13,14] Moreover, recent findings have suggested an overall conformation of a crumpled globule and strong ring-ring interactions.[14] Concerning the latter, note that the possibilities that double folds of neighboring rings in melts can move and free space or penetrate each other, open-up the folded structure, and temporarily block simple sliding motions of loops, are not considered in this two-dimensional lattice-animal model.

Turning to experimental evidence, nearly the bulk of the rings considered have been synthesized from linear chains with two functional end groups which eventually bind and close the loop (see Chapter 10). The main challenges have been (and remain) to ensure that rings are unknotted, unlinked (unconcatenated) and pure. The first has been promoted by carrying out the ring closure in good solvents.[15,16] The second has been met by carrying out the synthesis in very dilute solutions.[15,16] The third has been the subject of a long controversy, which appears to have been resolved only recently.

In this Chapter, we present the developments on the rheology of cyclic polymers. The rheological results of the 1980s, mainly from Roovers and McKenna and co-workers, are critically discussed in the next section. The outstanding problem is the lack of consensus on several controversial issues which are mentioned. We then discuss recent work which has demonstrated that ring contamination with linear polymers is at the origin of this controversy. In particular, liquid chromatography at the critical condition (LCCC),[17] a mixed mode of separation mechanisms compensating the entropic size exclusion and enthalpic interaction of polymers with the porous packing materials, has opened the route for separating linear from ring polymers. Rheological data have been collected with "as-pure-as-possible" ring polymers obtained with this technique and are presented and discussed in view of theoretical predictions in section 3. Then, section 4 includes a summary of recent results from atomistic and molecular simulations on pure rings, while a critical comparison between results from experiments and simulations is presented in section 5. Blending pure cyclic and linear polymers is the subject of section 6. It has important consequences in understanding the effects of linear contaminants as well as the physics of rings, and the experimental results are discussed in view of developments in both theoretical analysis and simulations. Emerging experimental evidence confirming the key recent findings is presented in section 7. Finally, the current status of the field and the perspectives are summarized in section 8.

2.   Literature Review: The Early Controversial Results

A review of the rheological findings of the 1980s is already available.[16] Whereas the linear chain contamination has been known, these early works addressed the issue by using cyclic polymers fractionated via standard means, i.e., SEC[18] and often refractionated.[19] Unfortunately, this has not resolved the problem, which is more pronounced the higher the molar mass. However, SEC fractionation was the only approach used at that time. At this point, a short clarification of the fractionation process used in the past[16,18,19] for obtaining ring polymers is in order. The standard means of purifying ring polymers were always by solvent–non-solvent fractional precipitation. Refractionation was so common that it is not always reported. The fractionation involves two different steps. The first stage is the separation of the low-M fraction from the high-M polycondensation material formed in the cyclization step. This can be reasonably well ascertained by SEC analysis. The second step would involve the separation of linear and ring of the same M. This was often omitted on the assumption that there was no linear contamination. In any case this separation proved very inefficient. The other reason why the older samples are not trustworthy is that the standard SEC analysis has insufficient resolution to clearly separate linear and ring in order to determine the fraction of linear contamination. As a result one relied on the maximum difference between ring and its linear parent expressed by the apparent M of the ring in function of the M of the linear parent, e.g. $M_{ring} = 0.71\ M_{linear}$ (delayed elution). It was assumed that the lower the prefactor the purer the ring. It should be a constant. But there is evidence that it increases for decreasing M. The width and skewing of the elution band of the ring were seldom compared to the linear. Other dilute solution properties were measured in conjunction with the M of linear and ring[16] (see also Chapter 11). These again provide a single number giving an "average with associated uncertainty". There were various theoretical ratios of ring and linear to assess the quality of the ring. Preparative SEC was rarely attempted then because the resolution is usually not better than in an analytical SEC set-up.

In general, the rheology of unentangled rings appears to conform to the Rouse model and their zero-shear viscosity was found to be half that of the corresponding linear polymer and follow $\eta_0 \propto M$.[18–20] The only exception appears to be the work by Semlyen *et al.* on cyclic and linear poly(dimethylsiloxane), PDMS.[21] They found that, at low molar masses (below the critical value M_e marking entanglements) the viscosity of the cyclic polymer is higher than the viscosity of the linear counterpart. Note that the experimental results[18,19,21] were reported at iso-frictional conditions, i.e. same distance from the glass transition temperature. Such a comparison is imperative, especially for low molar masses M where the glass transition temperature (T_g) depends on M and the difference in T_g between cyclic and linear polymers is large.[19,21] However, the M-dependent T_g measurements gave contradictory results: for small PS rings,[19,20] T_g increased with increasing M, whereas the opposite was found for the PDMS rings.[21–23] In both cases a constant T_g was reached at higher M and the respective values of the linear chains where nearly identical in that regime. Moreover, in the low-M regime, the T_g of the linear chains decreased with decreasing M and compared to that of rings it remained lower for PS.[18-20] The PDMS results of increasing T_g as M decreases were explained on the basis of different configuration entropy of the rings by DiMarzio and Guttman.[24] In addition, for PDMS rings the zero-shear viscosity was found to follow $\eta_0 \propto M^{0.6}$.[21,25]

For higher molar masses, there is consensus that the zero-shear viscosity of entangled polymer rings is smaller that of linear polymers of the same molar mass.[18-20,26] Roovers suggested that based on his 1,4-polybutadiene (PBD) ring data, the ring zero-shear viscosity was about 10 times smaller than that of the respective linear.[26] He also observed that at very high molar masses of PS (typically above 20 entanglements) the ring viscosity equals and also surpasses that of the linear polymer.[18] Moreover, in the entanglement regime (based on the well-defined M_e of linear chains) a plateau storage modulus was extracted from dynamic oscillatory[18,26] and creep[19] measurements. Based on experiments with different rings from different sources, its reported value varied between 1/6 and 1/2 of the corresponding linear polymer.[18,19,26] The discussion about ring entanglements was further triggered by forward recoil spectroscopy diffusion measurements of rings in deuterated linear matrices,[27,28] and it was suggested that rings form double-folded linear conformations[29] and undergo some kind of restricted reptation. In most but not all of the dynamic frequency measurements the terminal slopes $(G' \propto \omega^2, G'' \propto \omega)$ were reached for the rings.[18,26] In addition, the extracted values of the zero-shear viscosity and the recoverable compliance were often inconsistent when rings with the same number of entanglements were compared.[18-20] Further controversy was created by the molar mass dependence of the zero-shear viscosity in the entangled regime since McKenna and co-workers favored a relation similar to linear chains $(\eta_0 \propto M^{3.9})$[19,20] whereas Roovers proposed an exponential dependence $(\eta_0 \propto \exp(aM)$ with a being a prefactor).[18,26,30] The debates on these findings did not resolve the problem, although it was recognized that the issue of linear chain contamination is crucial. In addition, focusing on PS polymers which were prepared in different laboratories and samples were available over a wide range of molar masses and the results compared, those studied by McKenna were cyclized by Rempp and co-workers in good solvent conditions[20,31,32] whereas those of Roovers in near-theta solvent conditions.[33] The latter were questioned due to the high probability of having knots.[19,20] Whereas Roovers[26] argued that knotting should not be expected in cyclics for molecular weights below 10^6, this issue remains an important one. In fact, knots formation is more likely in theta than in good solvent conditions,[34] and their presence reduces the ring size slightly,[35,36] and hence one may expect changes in rheology. It should be also mentioned that from the three major rheological data sets available, McKenna and Plazek performed creep and recovery measurements,[18–20] Roovers small amplitude oscillatory shear (dynamic frequency sweeps)[18,26,30] and Semlyen and co-workers steady shear reporting only the viscosity.[21] Hence direct comparison is possible only for the zero-shear viscosity.

In an effort to resolve the controversies arising from the properties of rings of different origin and to further explore their viscoelasticity, different fractions of ring polymers were mixed with their linear counterparts.[26,37] The experiments revealed that, high concentrations of the latter greatly enhanced the zero-shear viscosities of the cyclic fractions. In fact, a consensus was reached on the dependence of blend's relative zero-shear viscosity (normalized to the linear polymer viscosity $\eta_{0,linear}$) on the volume fraction of linear polymer ϕ_{linear}, $\eta_{0,blend} / \eta_{0,linear} \propto \phi_{linear}^{5.6}$, for values of $0.05 \leq \phi_{linear} \leq 0.30$. Moreover, Roovers reported for a fraction of ring $0 \leq \phi_{ring} \leq 0.8$ the viscosity of the blend was higher than that of either pure component, reaching a maximum ratio $\eta_{0,blend} / \eta_{0,ring}$ of about 2.3 at $\phi_{linear} \approx 0.5$.[26,30] At the time this result was considered intriguing for architecturally different polymer blends and was discussed in the context of tube-model mechanisms of relaxation,[16,26] but remained unexplained. Finally, the extracted plateau modulus of the rings from dynamic oscillatory measurements was also affected by the presence of linear polymers. Its value was reported to shift from about 1/6 that of the linear polymer of the same molar mass to about 1/2, when the linear fraction was 20% to 25%. In that range, the melt viscosities of blend and linear polymer were nearly identical.[26]

3.   Linear Rheology of Experimentally Pure Cyclic Polymers

Currently, the purest possible cyclic polymers are obtained via liquid chromatography at the critical condition (LCCC).[17,38] We show here results from rheological experiments with two old polystyrene (PS) rings from Roovers[18] which were reported to exhibit entanglement plateau (see also Fig. 1b below). These samples have been recently fractionated by LCCC[38–40] and their linear viscoelastic properties measured at different temperatures by small amplitude oscillatory shear.[41] To date, this rheological dataset remains the only one with LCCC-purified entangled rings. More specifically, two nearly monodisperse samples were investigated, with molar masses 160 and 198 kg/mol (considering $M_e = 17.5$ kg/mol for polystyrene, this corresponds roughly to $Z = 10$ and $Z = 12$ entanglements, respectively).[18,33,42] The amounts of fractionated samples[38] were of the order of 10 mg, which poses problems when measuring bulk rheological properties. This challenge was met by using specially designed parallel plate geometries with a small diameter of 4 mm.

Master curves were obtained using time-temperature superposition. The vertical shifting reflects the variation of density with temperature, the horizontal shift factors conform to the temperature dependence of the monomeric friction coefficient and follow the well-known WLF-equation with nearly the same coefficients as linear chains. The temperature-dependent shift factors were nearly identical for polystyrenes of different architectures (linear, cyclic and various branched).[41] The master curves for the storage and loss moduli as function of frequency are shown in Fig. 1a at a reference temperature $T_{ref} = 170°C$ for one purified cyclic polymer ($Z = 12$) along with its linear precursor. The linear polymer displays the usual viscoelastic behavior with the storage modulus G′ exhibiting a rubbery plateau at intermediate frequencies and terminal slopes of 2 and 1 for G′ and G″, respectively, at the lowest frequencies. On the other hand, for the pure cyclic polymer the terminal slopes are not reached in the investigated frequency range. The reason for the absence of a terminal regime remains elusive. Two tentative explanations relate to very slow diffusive processes due to ring-ring interpenetration (in 3D) and the possible tiny traces of remaining linear polymer "impurities", which cannot be removed even by critical fractionation and to which the long-time region is particularly sensitive.[41]

The most important observation in the linear viscoelastic response of the purified cyclic polymer is the absence of a plateau in G′, in sharp contrast to earlier data[18–20,26,30]. Instead, G′ displays a weak frequency dependence with a power-law exponent close to 0.4 for about three decades in frequency. This result conforms to the scaling predictions of Rubinstein[5,41] and, more recently Milner,[43] both based on a lattice animal configuration of the rings. In Fig. 1a the experimental data are compared against Rubinstein's model. The latter predicts the following expression for the stress relaxation modulus:[41]

$$G(t) = G_N^0 \left(\frac{t}{\tau_e}\right)^{-\frac{2}{5}} \exp\left(-\frac{t}{\tau_{ring}}\right) \text{ with } \tau_{ring} = \tau_e \left(\frac{M}{M_e}\right)^{\frac{5}{2}}, \text{ for } t > \tau_e$$

with G_N^0 the plateau modulus of the linear precursor, τ_e the relaxation time of an entanglement segment and τ_{ring} the longest relaxation time of the ring. This expression yields G' and G'' when converted to the frequency regime:

$$G'(\omega) = G_N^0 \frac{M_e}{M} \left(\frac{\omega}{\beta}\right)^{\frac{2}{5}} \frac{\Gamma\left(\frac{3}{5}\right)}{\left[\left(\frac{\beta}{\omega}\right)^2 + 1\right]^{\frac{3}{10}}} \sin\left[\frac{3}{5}\arctan\left(\frac{\omega}{\beta}\right)\right]$$

$$G''(\omega) = G_N^0 \frac{M_e}{M} \left(\frac{\omega}{\beta}\right)^{\frac{2}{5}} \frac{\Gamma\left(\frac{3}{5}\right)}{\left[\left(\frac{\beta}{\omega}\right)^2 + 1\right]^{\frac{3}{10}}} \cos\left[\frac{3}{5}\arctan\left(\frac{\omega}{\beta}\right)\right]$$

with $\beta = (M_e / M)^{2.5} / \tau_e$ and Γ the gamma function. The lines in Fig. 1a represent these predictions. The molecular parameters were taken from linear polystyrene data:[41] $\tau_e = 4.10^{-4}$ s, $M_e = 17.5$ kg/mol, and $G_N^0 = 0.18$ MPa ($G_N^0 = \rho RT/M_e$ with ρ the density and R the universal gas constant[5]). The agreement between experimental data and predictions in this power-law region is remarkable and confirms the absence of rubbery plateau in pure entangled cyclic polymers.

To appreciate the significance of this result in view of the early controversies, the experimental data of Kapnistos *et al.*[41] are compared against the original data of Roovers on the same polystyrene.[18] In fact, Fig. 1b demonstrates the power of LCCC by comparing the dynamic rheological data of the same cyclic PS with $M = 198$ kg/mol before and after critical fractionation. The old data clearly display a plateau in G' reflecting the presence of unlinked linear chains (contaminants) which could not be completely removed by consecutive conventional size-exclusion chromatography (SEC) separations. When the rings were treated with LCCC and remeasured, the observed rheological behavior was drastically different. Hence, there is no doubt that the presence of impurities renders the old data ambiguous, and this explains the early controversies concerning the rheology of rings (i.e. large differences in the observed plateau moduli for the rings, differences in the extracted values for the zero shear viscosity and recoverable compliance).

From the above evidence, it is established that cyclic polymers do not relax their stress by reptation or by retraction, but by self-similar local motions giving rise to a power-law response. It is also clear that the traditional fractionation methods are not so successful in efficiently fractionating cyclic polymers. LCCC instead can yield 99.5% or more pure cyclic polymers.[38-40] The inset of Fig. 1b depicts the chromatographs before and after critical fractionation: the removal of linear contaminant (the peak at low times) is evident.[41]

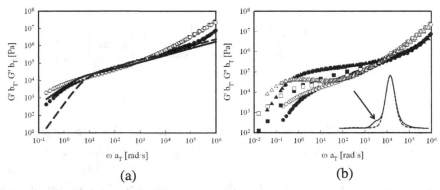

Fig. 1. (a) Master curves for the storage G' and loss G'' moduli as function of shifted frequency at $T_{ref} = 170°C$ of the purified ring 198 kg/mol (G'' ○, G'●), taken from Kapnistos *et al.*[41] and the model predictions (see text) $G'(--)$, $G''(-)$. (b) Master curves for G' and G'' as function of frequency at $T_{ref} = 170°C$ of the PS linear 198 kg/mol (G''△, G'▲), PS 198 kg/mol ring before[18] (G''□, G'■) and after LCCC-fractionation[41] (G'' ○, G'●). Inset: Chromatograms (refractive index vs. elution volume) of the ring before (-) and after (--) fractionation by LCCC. The arrow indicates the peak related to the presence of linear contamination.

Despite the complexity of the terminal regime, it is also clear that pure cyclic polymers relax faster than their linear counterparts. This important finding corroborates earlier suggestions from interdiffusion studies of labeled polystyrene rings. Recently, Takano and co-workers have reported on the synthesis and detailed characterization of ring polystyrenes with high purity based on LCCC.[44] Using protonated and deuterated rings with $Z \approx 6$, they measured the evolution of the interfacial thickness of films with neutron reflectivity and dynamic secondary ion mass spectroscopy during annealing.[45] Their results clearly indicated that cyclic polystyrene diffused much faster than the corresponding linear.

It is instructive to make another comparison with the old data set from McKenna *et al.*,[19] on a PS ring sample with molar mass $M = 185$ kg/mol, i.e., only slightly different from the sample of Roovers discussed before. At that time, in order to obtain pure rings, the sample was refractionated with conventional SEC. The time-dependent recoverable compliance extracted from creep and recovery measurements showed that indeed the refractionated sample was different from the single-time fractionated one.[19] Hence, a comparison with the LCCC-fractionated samples is in order. To do so, we transformed the dynamic ata of the purified PS rings from Kapnistos *et al.*[41] (see Fig. 1b) to creep compliance.

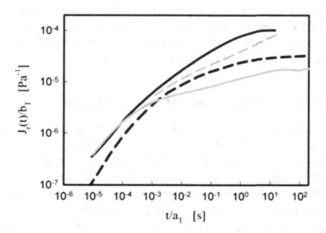

Fig. 2. Master curves for the recoverable compliance versus time at $T_{ref} = 170$ °C. Full black line: purified PS ring 198 kg/mol data from Kapnistos *et al.* data[41] (conversion of data in Fig. 1a). Dashed gray line: respective unfractionated PS ring 198 kg/mol data[18] (conversion of data in Fig. 1b). Continuous gray line: data of linear PS 200 kg/mol sample[41] (conversion of data in Fig. 1b). Dashed black line: McKenna's data with PS refractionated ring 185 kg/mol.[19] Note that the small deviation of this different data set with different sample from the rest may be attributed to the different temperature calibrations, shift factors (not available in Ref. 19) and errors in the conversion of the dynamic data of Ref. 41.

Figure 2 shows the results of this conversion along with the data from McKenna *et al.*[19], all shifted (horizontally and vertically) with respect to a reference temperature of 170°C. One can estimate the steady-state recoverable compliance (long-time steady value) from the data. The converted recoverable compliance data of the LCCC-purified ring of Fig. 1 exhibits an approximate steady value of about 7.10^{-5} Pa^{-1}. The original unpurified ring sample does not seem to reach any steady-state value in the experimental time window. Interestingly, also McKenna's recoverable compliance data of the refractionated ring of similar molar mass approaches a plateau, whereas his once-fractionated (via SEC) ring does not[19] (data not shown here). Finally, the linear PS data from Kapnistos *et al.*[41] also reaches a steady state, lower than that of the respective ring. Hence, the observation of steady-state recoverable compliance appears to conform to a high level of purity of the cyclic polymers, in agreement to McKenna's earlier claims,[19] although it cannot be considered as a proof. Given this result, the behavior of the linear chains (also reaching steady state recoverable compliance), the different origin of old samples, the data with the LCCC-purified PS rings[41] and the limitations in comparing directly different rheological data, we suggest that LCCC is the most powerful method to date to purify cyclic polymers.

4.   Results from Atomistic and Molecular Simulations

Given the formidable challenge of obtaining truly pure ring polymers experimentally, computer simulations become indispensable complementary tools, as they do not suffer from the same limitations and the issue of contamination is solved naturally. One can study 100% pure rings and their blends with linear polymers at any precisely defined ring concentration. Several simulation studies have been performed in the past two decades. Most are based on the lattice bond fluctuation model (BFM), coarse-grained Monte Carlo, atomistic and various molecular dynamics approaches. There is a general consensus about the restricted conformations of the rings in the melt,[13,14,46–49] conforming to the lattice-animal-like theoretical preditions.[10,11] Turning to dynamics, it very much depends on the size of the rings. Very small rings (with molar mass much smaller compared to the entanglement limit of linear chains) exhibit a complex behavior with smaller diffusion coefficient[13,50] and smaller viscosity,[50] compared to linear polymers, due to their larger density.[13] With increasing molar mass below the entanglement limit, both viscosity and self-diffusion suggest faster ring motion compared to linear chains.[13,50,51] For large entangled rings, despite some reports for comparable (to linear counterparts) dynamics,[52] different results indicate that they diffuse faster than their linear counterparts[13,48,50,51,53,54] and consequently they relax their stress faster.[51] For large entangled rings the ratio of the ring-to-linear zero-shear viscosities is much smaller than that of the unentangled regime (where Rouse theory predicts a ratio of 0.5).[50,51] In that regime, the dependence of the diffusion coefficient on molar mass was weaker for the simulated rings (with atomistic molecular dynamics) compared to the linear polymers, with a ratio of power law-exponents of about 1/2.[54] The rings approached the linear polymer scaling at the largest molar masses.[54]

5. Critical Comparison of Experiments and Simulations

To further assess the experimental results with purified cyclic polymers, we now focus on the dependence of the zero-shear viscosity (often obtained by extrapolation or fitting) on molar mass. Figure 3 depicts the viscosity ratio of ring and linear polymers having the same molar mass, as function of the number of entanglements of the linear polymer ($Z = M/M_e$). Selected experimental and simulations data taken from the literature are shown. The experimental data include recently purified rings as well as old rings from the 1980s. The plot also includes recent data on unentangled poly(ethylene oxide), PEO, which was prepared anionically and fractionated multiple times by SEC[55] and compared favorably with LCCC analysis[56] to ensure that linear contaminants were removed. Also included are recent data on unentangled cyclic poly(oxyethylenes), POE.[57] They were synthesized nearly monodisperse and thoroughly characterized. Removal of contaminants by inclusion complexation and fractional precipitation ensured high purity as confirmed by the absence of free end-groups from ^{1}H NMR.[57] As seen in Fig. 3, the experimental data follow closely the predictions from Rouse theory,[58] and the respective viscosity ratio from both experiments and simulations is found to be close to the predicted value of 0.5.[55,58] The ratio decreases upon increasing molar mass and at high Z-values recent molecular dynamics simulations predict $\eta_{0,ring}/\eta_{0,linear} \sim Z^{1.4}/Z^{3.4} \sim Z^{-2}$ (white stars in Fig. 3)[51]. The experimental data of Kapnistos *et al.*[41] on rings purified via LCCC are in excellent agreement with the simulation results, while the earlier data with rings (Roovers,[18] McKenna[19] and Semlyen[21] data) deviate from the predictions. Given the accuracy of the molecular dynamics simulations[51] and their unique ability in studying 100% pure rings, this result again points to the significance of the contamination problem, the importance of LCCC, and the good quality (i.e. high purity) of the recent experimental data.

Concerning the low-molar mass unentangled rings, a remark on the iso-frictional conditions is in order. In section 2 and in reference to this issue, the literature evidence on the dependence of T_g on molar mass for cyclic and linear polymers was briefly mentioned. It turns out that this has been controversial as well. There is only one reported measurement of LCCC-purified PS ring from Roovers[33,38] indicating that the ring T_g remains nearly identical to the high-molar mass plateau value.[59] For the PEO rings of Fig. 3, the experimental melting temperatures (by DSC) were indistinguishable from those of their linear counterparts.[56] Noteworthy is also the good agreement of the POE, PEO and atomistic simulations data on PE,[50] although the latter surprisingly fall slightly below the 0.5 value. On the other hand, the PDMS data of Semlyen and co-workers[21] deviate sharply. This can be attributed to impure rings, although in general linear contamination is expected to be lower at lower molar masses.[56]

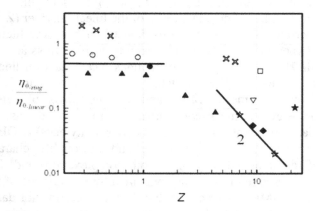

Fig. 3. Dependence of the ratio of zero-shear viscosities of cyclic and linear polymers having the same molar mass on the number of entanglements, Z. Experimental data with LCCC-purified PS from Ref. 41(◆); multiply-fractionated (SEC) PEO ring from Ref. 55 (●); POE rings from Ref. 57 (○); old data with rings fractionated via standard SEC, PS from Ref. 18 (▽); old PS from Ref. 19 (□); old PBD from Ref. 26 (★); old PDMS from Ref. 21 (✕). Simulations data: PE from Ref. 50 (▲); PS from Ref. 51 (☆). The POE data were made iso-frictional by shifting based on the reported Tg of cyclic and linear polymers in Ref. 57. Lines are drawn to guide the eye to the low-Z and high-Z scalings (0 and −2, respectively).

With the issue of ring purity addressed, the next grand challenge is the rheological determination of the degree of purity and, more importantly, understanding the role of contamination with linear chains and further quantifying. The motivation for this goes back to the work of McKenna and Plazek[37] and then Roovers,[26] whose results however are hard to assess as already discussed. Now, this challenge can be met by deliberately adding linear polymers to LCCC-purified cyclic polymers. This is the topic of the next section.

6. Mixtures of Cyclic and Linear Polymers

An experimental investigation of the rheological behavior of blends of purified cyclic with linear polymers has been reported by Kapnistos *et al.*[41] An intriguing finding in that work is the value of the minimum concentration of linear contaminant that could be detected in the rheological response of the blend.

Figure 4 shows the master curves for the storage modulus G' of the blends at a reference temperature of 170 °C for different concentrations of linear polymer.[41] For clarity, only a few selected concentrations are shown. Remarkably, even a fraction of added linear polymers as low as 0.07% leads to significant differences in the linear rheology (an increase of about 70% at low frequencies). This surprisingly low concentration is approximately 50 times below the overlap concentration $c*$ of the linear chains.[41] Moreover, the rheological response of a blend with 5% of linear polymers already displays a plateau in the elastic modulus G' at intermediate frequencies and the overall dynamic response virtually overlaps with that of the linear polymer. Hence, the rheology of the pure rings is extremely sensitive to the addition of minute amounts of linear chains, i.e. to contamination. This seems to be a plausible explanation of the controversies of the 1980s. The only meaningful comparison of the recent data with pure rings and the early ones concerning rheology[26,30,37] and diffusion[27,28] is qualitative, regarding the observations of slowing-down of the dynamics in ring-linear polymer blends compared to the pure rings. But no quantitative comparison can be made.

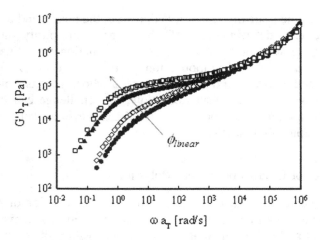

Fig. 4. Master curves for G' as function of shifted frequency for linear-ring polymer blends at $T_{ref} = 170$ °C for different concentrations of linear polymer φ_{linear}: 0% (●); 0.07% (◇); 5% (▲); 100% (□). Data are taken from Ref. 41.

At high concentrations of linear polymer instead, the early works of the 1980s showed that the zero-shear viscosity of the blends was higher than the zero-shear viscosity of the corresponding pure linear polymer.[26,30,37] It is important to note here that at high concentrations of linear polymer (typically above 1%), the issue of purity is not so crucial anymore, as it simply introduces a small error to the (contaminated) ring response. This is in sharp contrast to low concentrations where truly pure cyclic polymer samples cannot be obtained and the sensitivity for the rheology is high. In the recent data set of Kapnistos *et al.* with experimentally pure rings, the highest investigated concentration of added linear polymer was 20%,[41] clearly below the concentration region where the viscosity of the blend can be expected to overcome both components, based on the earlier works,[26,30,37] as well as recent molecular dynamics simulations.[60]

Figure 5 shows the ratio of the zero-shear viscosity of the blend to that of the linear polymer as function of concentration of linear polymer. The plot includes the recent data with LCCC-purified rings and selected old data. This way of plotting ensures iso-frictional conditions for these high-M polymers. This plot does not imply a universal scaling as it depends on the number of entanglements Z and the overlap concentration c^*. A detailed analysis and discussion of this graph is beyond the scope of his chapter. Nevertheless, it is useful to point out some key observations in the high concentration regime: when the concentration of linear polymer reaches a value of roughly 20% (the highest considered by Kapnistos *et al.*[41]), the viscosity ratio approaches a value of 1. This at first sight surprising result is confirmed by the recent molecular dynamics simulations,[60] in which the ring concentration is well-defined and not affected by experimental error. It also agrees with the data of Roovers on PBD ring-linear polymer blends who however observed the same effect at lower linear polymer concentrations.[26,30] Furthermore, the data of Kapnistos *et al.*[41] are in excellent agreement with the recent molecular dynamic simulations of Halverson *et al.*[51] concerning the zero-shear viscosity of the pure ring polymers of identical molar masses, suggesting that the experimental rings are indeed as pure as currently possible.

At higher linear fractions, the old experimental data[26,37] appear to conform to the overall trend of increasing $\eta_{0,blend} / \eta_{0,linear}$ with ϕ_{linear}. However, as seen clearly in Fig. 5, the experimental data (with a maximum ratio of about 2.5) are well above the simulations data (with a maximum ratio of about 1.3). In addition, the suggested scaling exponent of 5.6 is not supported by the available simulations results. The difference between the old experimental data and the recent simulations data points again to linear contamination problems associated with the early cyclic samples.

In the region of low linear polymer concentrations, the experimental blend viscosities clearly deviate from the respective simulations data.[60] The dependence of the blend viscosity on the linear contaminants is much stronger in the experimental data than in the simulations. Whereas in the former case, the effect of contaminant was detected at a linear polymer concentration close to $c^*/50$, the respective critical concentration in the latter case was around c^*.[60] To date, the origin of this discrepancy remains elusive. Clearly, the simulations have the important advantage of unambiguously and precisely controlling the concentration. Note that a recent simulation study[61] using the bond fluctuation model has also attempted at explaining the experimental data and found a critical concentration at which the linear chains started to affect the behavior of the rings in the range $c^*/30 - c^*/20$, i.e. closer to the experiments. However, the molecular dynamics simulations are considered more accurate, hence the issue remains open.

Despite the quantitative issues, the strong sensitivity of ring rheology to the presence of linear chains is an undisputed finding. Whereas the simulations have shed light onto this intriguing effect,[60-63] and clearly motivate further work, it can be explained in the context of entropy considerations. In a nutshell, when added to a ring melt a linear chain will thread as many rings as possible since in this way it maximizes the entropy of the system. The latter comes about by an opening of the local loops of the ring without however increasing its overall size substantially.[62,63] This arrangement also increases the linear chain's entropy as otherwise it would have to avoid the obstacles of the rings by shrinking. The schematic cartoon of Fig. 6 attempts at illustrating such a situation. The rings assume the lattice animal conformation (left) and the chain penetrates them as seen on the right-hand side. A linear chain will penetrate as many rings as possible and a reasonable estimate is Z rings per chain although a ring can be penetrated by more than one chain. From the number concentration of linear chains one can estimate their average distance (the inverse cubic root) which gives an idea of the thread formed by the chain penetrating rings. Even for the lowest concentrations considered here, the numbers suggest the formation of a transient network of linear chains bridged by rings.

Threading of multiple cyclic polymers by their linear counterparts has been investigated with simulations[64,65] and diffusion experiments using pulsed field gradient NMR.[66,67] The implication of this threading was the formation of a percolating transient network (entanglement-like) which was responsible for the suppression of the melt dynamics.[65,67] Incidentally, small cyclic alkanes were found to diffuse more slowly compared to the respective linear alkanes,[66] confirming the results on very small polymers,[13,50] as already discussed in section 4. This can be attributed to topological differences of the two species.

One can now appreciate and understand, at least qualitatively, the increase of viscosity of pure rings in the presence of linear chains as depicted in Fig. 5, or else the slowing-down of their viscoelastic relaxation. Since a linear chain threads rings, the terminal relaxation of the latter will take place after the chain has moved out. Hence, the relaxation of the rings is slowed down. At high enough concentrations of linear chains (typically 20% or higher) the slowing down is such that the blend terminal time (usually extracted by extrapolation experimentally) is slower than that of the linear chains as well. The data of Fig. 5 are revealing. A quantitative analysis of this process accounts for the linear chains percolation through the rings, the self-similar relaxation of the rings via sliding motion of local double folds, the relaxation of the linear chain's section at times up to the ring relaxation time and the slower relaxation of the rest. Details can be found in Ref. 41.

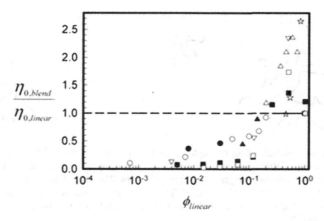

Fig. 5. Ratio of the zero-shear viscosity of the ring-linear polymer blend to that of the linear polymer as function of concentration of the linear polymer. Halverson *et al.* molecular dynamics simulations[60] for two different degrees of polymerization: □ and ■; Experimental data from Kapnistos *et al.*:[41] PS 198 kg/mol (○); PS 160 kg/mol (●); old experimental data from McKenna and Plazek[37] PS (☆); Roovers[18] PS (△); Roovers [26] PBD (▽); Roovers [26] PBD (▲). The horizontal dashed line through a ratio value of 1 is drawn to guide the eye as to the "excess" relative viscosity at high fractions of added linear polymer.

Therefore, trace amounts of linear chains in a melt of rings bridge between the rings and dramatically alter the rheology of the purified ring melt. This provides the framework for understanding the role of linear contamination on the rheology of cyclic polymers. In general, the so-called architectural dispersity, i.e. the presence of side-products resulting from the synthesis of targeted architecturally complex polymers, has received a great deal of attention. Even with the highest-precision synthetic approaches such as high-vacuum anionic synthesis, this type of contamination is inevitable. It appears that advanced chromatographic techniques are the key for purifying and analyzing the reaction products.[68,69] However, there is no doubt that, due to their topology, the problem with ring contamination is extreme in terms of the rheological consequences, and in this respect unique. On the other hand, unlike their drastic effects on the rheological behavior, small amounts of linear additives appear to have a much weaker effect on the diffusion of rings, apparently due to their different sensitivity at low fractions.[25,60,70] We note that the overall picture emerging from Fig. 5 has been recently confirmed with new experimental data.[71]

Fig. 6. Schematic sketch of an entangled ring polymer in an array of fixed obstacles (not shown) assuming a lattice-animal conformation (on the left side) and a linear chain which penetrates and threads two neighboring ring polymers. The dotted lines are relaxing chain sections up to the ring relaxation time.[41] Thus, the end-to-end distance of the unrelaxed section of the linear chain is smaller than the end-to-end distance of the linear chain in absence of rings.[41]

7.   Additional Experimental Evidence

It is now clear that obtaining pure cyclic polymers for experimental work remains a challenge. In particular for bulk rheology, and despite the recent advances in measuring tiny amounts of sample,[41] there is a need for having at least samples of the order of tens of milligrams for conducting thorough investigations. Recently more activity has been reported in this direction. More specifically, Takano and co-workers have synthesized and characterized rings, which were further purified by LCCC.[44,72] Their number of entanglements range from $Z \approx 1$ to about $Z \approx 33$. In that range the radius of gyration of the rings in theta solvents remained smaller than that of the respective linear polymer and with slightly stronger molar mass dependence (scaling exponent 0.53 vs. 0.49). The available rheological evidence[73] for pure polystyrenes up to $Z \approx 14$ (very similar to that of Ref. 41) confirmed the earlier experimental findings:[41,55] unentangled rings behave like Rouse chains with a viscosity about half that of the respective linear polymers; entangled rings do not exhibit a rubbery plateau but a power-law stress relaxation instead; their zero-shear viscosity has stronger molar mass dependence than that of unentangled rings, but still weaker than that of the respective linear chains; and finally, the addition of linear chain contaminants drastically affects the pure ring rheology. In that work, the lowest linear fraction investigated was 1%.

8. Conclusions and Outlook

Cyclic polymers hold the key for exploring new directions in polymer physics and especially rheology of entangled polymers, and the current state-of-the-art holds the premise to pursue this direction.[74] Furthermore, the topic has implications in chemical technology and beyond. Mitochondrial and plasmic DNA are cyclic and often have knotted structure; therefore ring polymers are ideal models for fundamental biophysics problems as well.[75–79] For example, the organization of chromatin in the cell nucleus can be understood by drawing analogies to the conformation and dynamics of densely packed rings. Chromatin fibers are packed *in vivo* at a rather high density, like a melt of linear chains. However, the different chromosomes in the nucleus do not intermix but instead segregate in different stable distinct "territories", in analogy with the conformation in a melt of non-concatenated rings. This type of segregated conformation appears typical for the cells of higher eukaryotes, including humans. In addition, the fundamental understanding of ring structure and dynamics is particularly relevant for applications ranging from DNA separation to enzymology and from protein structure stabilization to drug delivery and pharmacokinetics.[77,78,80]

Whereas obtaining 100% pure ring polymer experimentally remains a challenge, it is now established that as-pure-as-currently-possible rings can be obtained and studied systematically, and that liquid chromatography at the critical condition is an indispensable tool in this regard. Moderately entangled pure rings do not exhibit an entanglement plateau analogous to that of the respective linear polymers, but instead relax their stress via a power-law with an exponent very close to 0.4, in agreement with lattice-animal-based theoretical predictions and molecular dynamics simulations. Note that, whereas the theory was originally developed for entangled rings moving in an array of fixed obstacles, it does in fact account for constraint release effects and predicts the relaxation for rings in a two-dimensional array of moving obstacles. This is elaborated in Ref. 41.

Adding linear polymers to pure rings at concentrations far below the overlap concentration for linear chains induces distinct changes in the linear rheology. The exact value of the critical concentration for which there is a distinct change in the rheological response is still not clearly understood quantitatively. Nevertheless this unambiguous effect of entropic origin points to dramatic sensitivity of rings to linear polymer contamination.

These exciting developments have opened Pandora's box with many important challenges ahead. Despite the various emerging activities on block copolymer rings[81,82] (see also Chapter 11), ring/linear blends with chemically different components,[83,84] and semi-crystalline systems,[85–89] here we focus on open problems in the rheology of amorphous homopolymer rings: The validity of power-law relaxation at very high molar masses (with much more than 20 entanglements). The exact determination of the levels of impurity after critical fractionation and of the related issue of the critical concentration of added linear chains for affecting ring rheology. The elucidation of the terminal regime in rings. Here, one should note that the lattice-animal model is two dimensional and ring-ring interpenetration is not considered, though it may be important.[48,51,53] Recently, this model has been confirmed directly by atomic force microscopy measurements of DNA rings in constrained environments.[90] Furthermore, the rheology of binary blends of rings, the small being unentangled or entangled. The role of linear chain molar mass in the ring/linear blend rheology. The apparent strong difference in molar mass dependence of zero-shear viscosity and self-diffusion. The solution rheology of rings throughout the whole concentration range from dilute to concentrated. Related to the latter topic, recent fluorescence microscopy studies on circular DNA and synthetic poly(tetrahydrofuran) in solution suggest entanglement-like effects on the diffusion dynamics which may be multimodal.[91–94] Also, the slowing-down of self-diffusion in semidilute solution has been discussed on the basis of scaling arguments.[94]

The nonlinear rheological behavior of rings, both in shear and in extension, is virtually unexplored, both experimentally and theoretically. A few scarce theoretical and simulation investigations indicate that the properties of these distinct macromolecules (such as shear thinning, coil-stretch transition) can be different from their linear analogues, albeit not necessarily qualitatively.[95–97] Apart from a complete molecular understanding which seems still a big step, even a first assessment of the usefulness of simple empirical rules, such as Cox-Merz rule[1] which has proven to be reliable for linear entangled polymers for example, would shed some light on the nonlinear behavior of ring polymers and their possible very different behavior in strong flows. In this regard, rheo-physical experiments, providing information of structural changes under flow will be particularly enlightening, as has already been pointed out.[58]

Last but not least, the issue of knotting is a very important one and should be addressed. It was invoked as one of the reasons for the discrepancies in the rheological data of the 1980s. There were strong arguments calling for the crucial role of knots on the ring dynamics, given their influence on the ring size. However, recent investigations of LCCC-purified polystyrene rings suggested that knotting does not affect the rheology significantly. Nevertheless, the issue is far from settled. Recent work on the preparation of well-defined high purity knotted ring polystyrenes as large as 23 entanglements,[98] is therefore a very promising development.

Acknowledgments

D.V. is grateful to his longtime friend and collaborator J. Roovers for introducing him to the fascinating world of cyclic polymers, and for enlightening comments. The work of the authors has benefited a great deal from stimulating collaborations and discussions with M. Rubinstein, T. Chang, W. Pyckhout-Hintzen and T.C.B. McLeish. Partial support of the EU during preparation of this chapter is acknowledged (ITN-MC-DYNACOP, grant 214627).

References

1. R. G. Larson, *The Structure and Rheology of Complex Fluids* (Oxford University Press, New York, 1999).
2. P. E. Rouse, *J. Chem. Phys.* **21**, 1272 (1953).
3. M. Doi and S. F. Edwards, *The Theory of Polymer Dynamics* (Oxford University Press, New York, 1986).
4. A. E. Likhtman and T. C. B. McLeish, *Macromolecules* **35**, 6332 (2002).
5. M. Rubinstein and R. H. Colby, *Polymer Physics* (Oxford University Press, New York, 2003).
6. T. C. B. McLeish, *Adv. Phys.* **51**, 1379 (2002).
7. T. C. B. McLeish, J. Allgaier, D. K. Bick, G. Bishko, P. Biswas, R. Blackwell, B. Blottière, N. Clarke, B. Gibbs, D. J. Groves, A. Hakiki, R. K. Heenan, J. M. Johnson, R. Kant, D. J. Read and R. N. Young, *Macromolecules* **32**, 6734 (1999).
8. R. G. Larson, *Macromolecules* **34**, 4556 (2001).
9. M. Kapnistos, D. Vlassopoulos, J. Roovers and L. G. Leal, *Macromolecules* **38**, 7852 (2005).
10. M. Rubinstein, *Phys. Rev. Lett.* **24**, 3023 (1986).
11. M. E. Cates and J. M. Deutsch, *J. Phys. (Paris)* **47**, 2121 (1986).
12. S. P. Obukhov, M. Rubinstein and T. A. Duke, *Phys. Rev. Lett.* **73**, 1263 (1994).
13. K. Hur, R. G. Winkler and D. Y. Yoon, *Macromolecules* **39**, 3975 (2006).
14. J. D. Halverson, W. B. Lee, G. S. Grest, A. Y. Grosberg and K. Kremer, *J. Chem. Phys.* **134**, 204904 (2011).

15. Y. Ederle, K. S. Naraghi and P. J. Lutz, in *Materials Science and Technology. A comprehensive treatment*, ed. R. W. Cahn, P. Haasen and E. J. Kramer, *Synthesis of polymers* (ed. A. D. Schlütter), Chapter 19 "Synthesis of cyclic macromolecules" (Wiley-VCH, New York, 1999).

16. J. Roovers, in *Cyclic Polymers*, 2nd Ed., ed. J. A. Semlyen, Chapter 10 "Organic cyclic polymers" (Kluwer Academic Publishers, Dordrecht, The Netherlands, 2000).

17. T. Chang, *Adv. Polym. Sci.* **163**, 1 (2003).

18. J. Roovers, *Macromolecules* **18**, 1359 (1985).

19. G. B. McKenna, B. J. Hostetter, N. Hadjichristidis, L. J. Fetters and D. J. Plazek, *Macromolecules* **22**, 1834 (1989).

20. G. B. McKenna, G. Hadziioannou, P. Lutz, G. Hild, C. Strazielle, C. Straupe, P. Rempp and A.J. Kovacs, *Macromolecules* **20**, 498 (1987).

21. D. J. Orrah, J. A. Semlyen and S. B. Ross-Murphy, *Polymer* **29**, 1452 (1988).

22. S. J. Clarson, K. Dodgson and J. A. Semlyen, *Polymer* **26**, 930 (1985).

23. S. J. Clarson, J. A. Semlyen and K. Dodgson, *Polymer* **32**, 2823 (1991).

24. E. A. Di Marzio and C. M. Guttman, *Macromolecules* **20**, 1403 (1987).

25. T. Corgrove, T. C. Griffiths, J. Hollingshurst, R. D. C. Richards and J. A. Semlyen, *Macromolecules* **25**, 6761 (1992).

26. J. Roovers, *Macromolecules* **21**, 1517 (1988).

27. P. J. Mills, J. W. Mayer, E. J. Kramer, G. Hadziioannou, P. Lutz, C. Strazielle, P. Rempp and A. J. Kovacs, *Macromolecules* **20**, 513 (1987).

28. S. F. Tead, E. J. Kramer, G. Hadziioannou, M. Antonietti, H. Sillescu, P. Lutz and C. Strazielle, *Macromolecules* **25**, 3942 (1992).

29. J. Klein, *Macromolecules* **19**, 105 (1986).

30. J. Roovers, *Rubber Chem. Technol.* **62**, 33 (1989).

31. G. Hild, A. Kohler and P. Rempp, *Eur. Polym. J.* **16**, 525 (1980).

32. G. Hild, C. Strazielle and P. Rempp, *Eur. Polym. J.* **19**, 721 (1983).

33. J. Roovers and P. M. Toporowski, *Macromolecules* **16**, 843 (1983).

34. K. Koniaris and M. Muthukumar, *Phys. Rev. Lett.* **66**, 2211 (1991).

35. A. Y. Grosberg, *Phys. Rev. Lett.* **85**, 3858 (2000).

36. N. T. Moore and A. Y. Grosberg, *Phys. Rev. E* **72**, 0161803 (2005).

37. G. B. McKenna and D.J. Plazek, *Polymer Commun.* **27**, 304 (1986)

38. H. C. Lee, H. Lee, W. Lee, T. Chang and J. Roovers, *Macromolecules* **33**, 8119 (2000).

39. D. Cho, S. Park, K. Kwon, T. Chang and J. Roovers, *Macromolecules* **34**, 7570 (2001).

40. W. Lee, H. Lee, H. C. Lee, D. Cho, T. Chang, A. A. Gorbunov and J. Roovers, *Macromolecules* **35**, 529 (2002).

41. M. Kapnistos, M. Lang, D. Vlassopoulos, W. Pyckhout-Hintzen, D. Richter, D. Cho, T. Chang and M. Rubinstein, *Nature Mater.* **7**, 997 (2008).

42. J. Roovers, *J. Polym. Sci.: Polym. Phys. Ed.* **23**, 1117 (1985).

43. S. T. Milner and J. D. Newhall J.D., *Phys. Rev. Lett.* **105**, 208302 (2010).

44. D. Cho, K. Masuoka, K. Koguchi, T. Asari, D. Kawaguchi, A. Takano and Y. Matsushita, *Polym. J.* **37**, 506 (2005).

45. D. Kawaguchi, K. Masuoka, A. Takano, K. Tanaka, T. Nagamura, N. Torikai, R. M. Dalgliesh, S. Langridge and Y. Matsushita, *Macromolecules* **39**, 5180 (2006).

46. S. Geyler and T. Pakula, *Makromol. Chem., Rapid Commun.* **9**, 617 (1988).

47. M. Müller, J. P. Wittmer and M. E. Cates, *Phys. Rev. E.* **53**, 5063 (1996).
48. S. Brown and G. Szamel, *J. Chem. Phys.* **109**, 6184 (1998).
49. J. Suzuki, A. Takano, T. Deguchi and Y. Matsushita, *J. Chem. Phys.* **131**, 144902 (2009).
50. G. Tsolou, N. Stratikis, C. Baig, P.S. Stephanou and V. G. Mavrantzas, *Macromolecules* **43**, 10692 (2010).
51. J. D. Halverson, W. B. Lee, G. S. Grest, A.Y. Grosberg and K. Kremer, *J. Chem. Phys.* **134**, 204905 (2011).
52. T. Pakula and S. Geyler, *Macromolecules* **21**, 1665 (1988).
53. S. Brown, T. Lenczycki and G. Szamel, *Phys. Rev. E.* **63**, 052801 (2001).
54. K. Hur, C. Jeong, R. G. Winkler, N. Lacevic, R. H. Gee and D. Y. Yoon, *Macromolecules* **44**, 2311 (2011).
55. A. R. Bras, R. Pasquino, T. Koukoulas, G. Tsolou, O. Holderer, A. Radulescu, J. Allgaier, V. G. Mavrantzas, W. Pyckhout-Hintzen, A. Wischnewski, D. Vlassopoulos and D. Richter, *Soft Matter* **7**, 11169 (2011).
56. T. Chang, personal communication (2011).
57. S. Nam, J. Leisen, V. Breedveld and H. W. Beckham, *Polymer* **49**, 5467 (2008).
58. H. Watanabe, T. Inoue, Y. Matsumiya, *Macromolecules* **39**, 5419 (2006).
59. P. G. Santangelo, C. M. Roland, T. Chang, D. Cho and J. Roovers, *Macromolecules* **34**, 9002 (2001).
60. J. D. Halverson, G.S. Grest, A.Y. Grosberg and K. Kremer, *Phys. Rev. Lett.* **108**, 038301 (2012).
61. R. Vasquez and S. Shanbhag, *Macromol. Theory Simul.* **20**, 205 (2011).
62. B. V. S. Iyer, A. K. Lele and S. Shanbhag, *Macromolecules* **40**, 5995 (2007).
63. G. Subramanian and S. Shanbhag, *Phys. Rev E.* **80**, 041806 (2009).
64. C. A. Helfer, G. Xu, W. L. Mattice and C. Pugh, *Macromolecules* **36**, 10071 (2003).
65. G. Subramanian and S. Shanbhag, *Phys. Rev E.*, **77**, 011801 (2008).
66. E. von Meerwall, R. Ozisik, W. L. Mattice and P. M. Pfister, *J. Chem. Phys.* **118**, 3867 (2003).
67. S. Nam, J. Leisen, V. Breedveld and H. W. Beckham, *Macromolecules* **42**, 3121 (2009).
68. X. Chen, M. S. Rahman, H. Lee, J. Mays, T. Chang and R. Larson, *Macromolecules* **44**, 7799 (2011).
69. F. Snijkers, E. van Ruymbeke, P. Kim, H. Lee, A. Nikopoulou, T. Chang, N. Hadjichristidis, J. Pathak and D Vlassopoulos, *Macromolecules* **44**, 8631 (2011).
70. M. Lang and M. Rubinstein, private communication (2011).
71. R. Pasquino, T. Vassilakopoulos, Y. Cheol, S. Rogers, D. Vlassopoulos, G. Sakellariou, N. Hadjichristidis, T. Chang and M. Rubinstein, unpublished data (2012).
72. A. Takano, Y. Ohta, K. Masuoka, K. Matsubara,T. Nakano, A. Hieno, M. Itakura, K. Takahashi, S. Kinugasa, D. Kawaguchi, Y. Takahashi and Y. Matsushita, *Macromolecules* **45**, 369 (2012).
73. A. Takano, K. Matsubara, Y. Ohta, Y. Matsushita, Y. Takahashi and H. Watanabe, Viscoelastic properties of highly-purified ring-shaped polymers, *PRCR-5, 5th Pacific Rim Conference on Rheology*, Sapporo, Japan (2010).
74. T. McLeish, *Nature Mater.* **7**, 933 (2008).
75. K. J. Meaburn and T. Misteli, *Nature* **445**, 379 (2007).
76. S. Jun and B. Mulder, *PNAS* **103**, 12388 (2006).

77. T. Cremer, M. Cremer, S. Dietzel, S. Müller, I. Solovei and S. Fakan, *Curr. Opin. Cell Biol.* **18**, 307 (2006).
78. M. H. Chisholm, J. C. Gallucci and H. Yin, *PNAS* **103**, 15315 (2006).
79. D. Marenduzzo, E. Orlandini, A. Stasiak, D. W. Sumners, L. Tubiana and C. Micheletti, *PNAS* **106**, 22269 (2009).
80. B. Chen, K. Jerger, J. M. Frechet and F. C. Szoka, Jr., *J. Controlled Release* **140**, 203 (2009).
81. H. Iatrou, N. Hadjichristidis, G. Meier, H. Frielinghaus and M. Monkenbusch, *Macromolecules* **35**, 5426 (2002).
82. E. Baba, S. Honda, T. Yamamoto and Y. Tezuka, *Polym. Chem.*, in press (2012).
83. M. M. Santore, C. C. Han and G. B. McKenna, *Macromolecules* **25**, 3416 (1992).
84. S. Singla and H. W. Beckham, *Macromolecules* **41**, 9784 (2008).
85. J. Cooke, K. Viras, G.-E.. Yu, T. Sun, T. Yonemitsu, A. J. Ryan, C. Price and C. Booth, *Macromolecules* **31**, 3030 (1998).
86. E. J. Shin, W. Jeong, H. A. Brown, B. J. Koo, J. L. Hedrick and R. M. Waymouth, *Macromolecules* **44**, 2773 (2011).
87. M. E. Córdova, A. T. Lorenzo, A. Müller, J. N. Hoskins and S. M. Grayson, *Macromolecules* **44**, 1742 (2011).
88. K. Schäler, E. Ostas, K. Schröder, T. Thurm-Albrecht, W. H. Binder and K. Saalwächter, *Macromolecules* **44**, 2743 (2011).
89. Y. Tezuka, T. Ohtsuka, K. Adachi, R. Komiya, N. Ohno and N. Okui, *Macromol. Rapid Commun.* **29**, 1237 (2008).
90. G. Witz, K. Rechendorff, J. Adamcik and G. Dietler, *Phys. Rev. Lett.* **106**, 248301 (2011).
91. R. M. Robertson, S. Laib and D. E. Smith, *PNAS* **103**, 7310 (2006).
92. R. M. Robertson and D. E. Smith, *PNAS* **104**, 4824 (2007).
93. S. Habuchi, N. Satoh, T. Yamamoto, Y. Tezuka and M. Vacha, *Angew. Chem. Int. Ed.* **49**, 1418 (2010).
94. B. V. S. Iyer, S. Shanbhag, V. A. Juvenar and A. K. Lele, *J. Polym. Sci: Part B: Polym. Phys.* **46**, 2370 (2008).
95. W. Carl, *J. Chem. Soc. Faraday Trans.* **91**, 2525 (1995).
96. J. M. Wiest, S. R. Burdette, T. W. Liu and R. B. Bird, *J. Non-Newtonian Fluid Mech.* **24**, 279 (1987).
97. J.G. Hernandez Cifre, R. Pamies, M.C. Lopez Martınez and J. Garcıa de la Torre, *Polymer* **46**, 267 (2005).
98. Y. Ohta, M. Nakamura, Y. Matsushita, A. Takano, *Polymer* **53**, 44 (2012).

CHAPTER 15

CRYSTALLIZATION OF
CYCLIC AND BRANCHED POLYMERS

Hiroki Takeshita and Tomoo Shiomi

Department of Materials Science and Technology
Nagaoka University of Technology
1603-1 Kamitomioka, Nagaoka, Niigta 940-2188, Japan
E-mail: takeshita@mst.nagaokaut.ac.jp, shiomi@vos.nagaokaut.ac.jp

It has been observed that cyclic polymers have an even number of crystalline stems or chain folding times in the crystalline lamella, as expected from their specific molecular shape. The molecular arrangement in the lamellar layers including the folding part is constrained for cyclic polymers. The contrastive results on the growth rate of spherulites have been observed: the growth of spherulites for cyclic poly(tetrahydrofuran) and polyethylene is slower than that for the respective linear ones, while that for cyclic poly(ε-caprolactone) is faster than that for the linear one. The former and latter results have been attributed to the entropic deficit in the lamellar formation and the faster diffusion in the melt, respectively. Clearer theoretical explanations for these conflicting results are desired. Crystallization of branched polymer chains is also presented. The grafting effects can be seen in the nucleation and growth and the melting temperature.

1. Introduction

The remarkable feature of cyclic polymers is a conformational restriction, which affects crystallization as well as other molecular and material properties. Since cyclic polymers have no chain ends, which means lack of flexibility of chain ends, numerous conformations are excluded and therefore the molecular arrangement into the crystalline lamellae is constrained. The crystallization kinetics is also affected by the topological effect on the chain conformation. Both rates of nucleation and growth in the crystallization of polymers have classically and usually been expressed by the combination of two free energies of the transport and of the primary nucleation or secondary nucleation (growth) as seen typically in the Turnbull-Fischer equation.[1] The cyclic molecules are collapsed and their radius of gyration is smaller than that for linear polymers[2]. These give fast diffusion of molecules in the melt, which leads to fast crystallization. The restricted molecular arrangement into the crystalline lamellae as well as in the melt results in the entropic deficit. This has an entropic effect on the free energy of the primary and secondary nucleations. Furthermore, the conformational constraint due to no chain ends would be related to the crystallization process in which the molecular chain must change conformationally. This factor in the crystallization process may be included in the transport free energy, such as adsorption and reeling-in on the growth front, other than the molecular diffusion in the melt.

Contrastive results of the spherulite growth rate and melting behavior have been reported for cyclic polymers. The discrepancy in these results has not been explained clearly yet. We will focus on these results in this chapter, and also describe lamellar structure and chain folding manners characteristic of cyclic polymers.

Crystallization of branched chains is also attractive even if the topological effect is not expected so much, compared with cyclic polymers. One chain end of branched polymers is bound, which gives high chain densities and low diffusion in the melt. The grafting effect on the crystallization will also be presented.

2. Crystallization of Cyclic Polymers

2.1. *Chain folding and conformation in crystalline lamellae*

The most striking feature of crystalline lamellar structure in cyclic polymers is that the number of folds per molecule must be an even number. This may usually correspond to an even number of crystalline stems, as shown in Fig. 1.

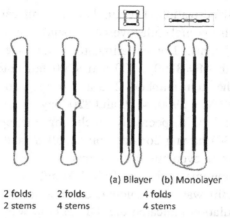

(a) Bilayer   (b) Monolayer

| 2 folds | 2 folds | 4 folds |
| 2 stems | 4 stems | 4 stems |

Fig. 1. Schematic drawing of possible conformations of cyclic polymers in crystal lamellae. Above are the views perpendicular to the stems, in which dark and light lines indicate the folds near and far surfaces, respectively. The figures of 4 folds/4 stems were reproduced from Ref. 4 by permission of American Chemical Society.

Yu *et al.*[3] found that the lamellar spacing d_l of a cyclic polymer was almost equivalent to half that of a linear polymer with the same molecular weight for poly(oxyethylene)s (PEO) with M = 1000-3000. Cooke *et al.*[4] (in the same group) evaluated the folding number from the discrete change of d_l with crystallization temperature observed by small-angle X-ray scattering (SAXS) measurements for PEOs with M = 4000, 6000, 10,000, and confirmed that the cyclic PEO took only even numbers while the linear one could take both odd and even numbers.

Takeshita *et al.*[5] also estimated the number of crystalline stems for poly(tetrahydrofuran) (PTHF) in another method described below. Approximating that three-dimensional degree of crystallinity (DC) is equivalent to one-dimensional DC along the molecular axis, the crystalline thickness d_c can be estimated from

$$d_c = d_l \times DC \tag{1}$$

Alternatively, the calculated crystalline thickness d_{cc} can be obtained using

$$d_{cc} = (\text{monomer length}) \times (DP) \times (DC) / (\text{number of stems}) \tag{2}$$

because the numerator in Eq. (2) expresses the total length of the crystalline stems in a molecule regardless of chain folding, where DP is the degree of polymerization. The number of stems can be estimated by comparing d_c with the molecular-weight dependence of d_{cc} calculated in each number of stems. The experimental crystalline thickness d_c for linear PTHFs with M = 3100, 5700 and 9100 fell onto the lines of d_{cc} calculated with the stem numbers 2, 3 and 4, respectively, while d_c for cyclic PTHFs with M = 2900, 4500 and 8200 lay on the lines of the stem numbers 2, 2 and 4, respectively, including the cyclic PTHF with M = 2900 (Cyc030) having considerably small d_c and DC. Namely the cyclic PTHF takes an even number in stems.

Amorphous monomer units per fold, DPam/Fold, in the folding part of a polymer chain was also calculated for a series of cyclic PTHFs above[5]. The calculation indicated that DPam/Fold was almost a constant irrespective of the stem number. This suggests that for the above cyclic PTHFs with comparatively low molecular weights, one polymer chain is not put in any adjacent crystalline lamellae but exists within only one crystalline lamella, and that the chain length of the folding part is the same irrespective of chain folding times. This means that the same chain length is required for the folding, and therefore the short d_c and low DC in Cyc030 can be attributed to a long folding part relative to a whole chain length.

As shown in Fig. 1, two conformations are possible in four-fold structure:[4] (a) bilayer and (b) monolayer. The bilayer conformation consists of four adjacent (short) folds, and the monolayer has three adjacent and one nonadjacent (long) folds. Cooke et al.[4] concluded that the bilayer structure was most likely under consideration of the Gibbs free energy for the secondary nucleation; the Gibbs energy of forming an adjacent fold is smaller than that of forming a nonadjacent fold. The same conclusion on the four-fold conformation can be derived from the above result about the folding part.[5] If the monolayer conformations are contained in four-fold structure, DPam/Fold should be larger for the four-fold conformation than for the two-fold conformation because the chain of the folding part should be longer for a nonadjacent fold than for an adjacent fold.

2.2. Melting behavior

Yu *et al.*[3] and Cooke *et al.*[4] showed for cyclic and linear PEOs described in section 2.1 that the plots of the melting temperature T_m vs. the reciprocal of the lamellar thickness d followed the equation

$$T_m = T_m^{\,\circ}(1 - 2\gamma/\Delta_{fus}H^{\circ}d) \qquad (3)$$

where $T_m^{\,\circ}$ is the equilibrium melting temperature, γ is the excess surface Gibbs free energy, $\Delta_{fus}H^{\circ}$ is the thermodynamic enthalpy of fusion. The plots fell onto the same straight line among those for the higher-molecular-weight cyclic and linear PEOs (M = 4000–10,000) and the lower-molecular-weight linear PEOs (M = 1000–3000), while the plots for the lower-molecular-weight cyclic PEOs lay on another straight line. These two lines were extrapolated to the same melting temperature $T_m^{\,\circ}$~70 °C. For the lower-molecular-weight PEOs, compared at the same lamellar thickness, T_m was higher for the cyclic than for the linear ones. High T_m means the low entropy of fusion $\Delta_{fus}S$, according to

$$T_m = \Delta_{fus}H/\Delta_{fus}S. \qquad (4)$$

Yu *et al.*[3] and Cooke *et al.*[4] explained the higher T_m, i.e., smaller $\Delta_{fus}S$, in the cyclic PEO as follows: the linear and cyclic polymers are similarly conformationally restricted in the crystalline state but the cyclic polymer is more conformationally restricted in the melt state, which gives lower $\Delta_{fus}S$ for the cyclic polymer.

For poly(ε-caprolactone) (PCL), no significant difference between the cyclic and linear polymers was detected in $T_m^{\,\circ}$ obtained by the extrapolation of the Hoffman-Weeks plots; the $T_m^{\,\circ}$ for the cyclic polymer (M = 110,000) was higher than that of the linear one (M = 120,000) by only 2 °C[6] and by 1 °C for M = 7500.[7]

In contrast to the above results for PEO and PCL, Takeshita *et al.*[5] found that $T_m^{\,\circ}$s obtained from the Hoffman-Weeks plots for cyclic PTHFs with M = 3000 and 6000 were lower than those for the linear analogues by 6 and 14 °C, respectively. Tezuka *et al.*[8] also obtained a lower T_m by 5 °C for the cyclic PTHF with M = 5100.

These contrastive results of the melting temperature are related to growth rates of spherulites as discussed in the next section.

2.3. *Crystallization kinetics*

Conflicting results have been reported for growth rates of spherulites in cyclic polymers, as shown in Fig. 2. Tezuka *et al.*[8] and Kitahara *et al.*[9] reported that the cyclic polymer has a low growth rate for PTHF and polyethylene, compared with their linear ones, respectively. Takeshita *et al.*[5] also found the similar result for PTHF. In contrast, Córdova *et al.*[7] reported that the spherulite growth of the cyclic PCL was faster than that of the corresponding linear one.

The growth rate of crystallization is expressed by the free energy terms of the transport and secondary nucleation according to the Turnbull-Fischer equation,[1]

$$G = G_0 \exp(-\Delta E/RT_c)\exp(-\Delta \Phi/RT_c) \qquad (5)$$

where G_0 is a constant, ΔE the energy of transport, and $\Delta \Phi$ the free energy for nucleation which is written as

$$\Delta \Phi = A\sigma - V\Delta f \qquad (6)$$

with the surface free energy $A\sigma$ and the free energy difference $V\Delta f$

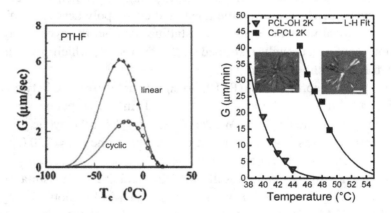

Fig. 2. Temperature dependence of spherulite growth rates for PTHF (left) and PCL (right). The cyclic PTHF has a low growth rate compared to the linear one, while the cyclic PCL (C-PCL 2K) has a higher growth rate than the linear one (PCL-OH 2K). M_n's are 5.1k (cyclic PTHF), 4.9k (linear PTHF), 2.3k (cyclic PCL) and 2.0k (linear PCL). Modified from Refs. 7 and 8 by permission of American Chemical Society and John Wiley & Sons, Inc., respectively.

between the melt and crystal in bulk:

$$\Delta f = \Delta_{fus}H^\circ - T\Delta_{fus}S^\circ \qquad (7)$$

Therefore, the growth rate is determined by the interplay or competition between the transport and entropic contributions.

Although very small cyclic alkanes (carbon numbers: 6–16) diffuse slower than the corresponding linear ones because of higher densities of cyclic ones[10,11] and such a density effect becomes negligible with increasing molecular weight, diffusion of cyclic polymers is faster than that of linear polymers even at much higher molecular weight than the critical entangle molecular weight.[12] In low-molecular-weight polymers with M = 400–1500, the diffusion coefficient was found to be larger for cyclic PEOs.[13]

The equilibrium melting temperature T_m° is affected by the entropic contribution when the enthalpy of fusion ΔH is the same, according to Eq. (4).

In the case of PCL, T_m° of the cyclic is not different from that of the linear. Therefore, Córdova et al. attributed the faster growth rate of the cyclic PCL to the easier diffusion of the cyclic polymer in the melt,[7] though usually the secondary nucleation rather than the diffusion may be the dominant factor in high crystallization temperatures. On the other hand, T_m° of the cyclic PTHF is considerably reduced, compared with that of the linear PTHF. This means that the entropy of fusion is larger for the cyclic PTHF as described in section 2.2. Tezuka et al.[8] ascribed the lower growth rate of the cyclic PTHF to entropic factors based on the lower melting temperature, and proposed some explanations: the conformational entropy, adsorption mechanism on the crystal growth front, or the chain folding surface free energy. Takeshita et al.[5] also explained the reduction of spherulite growth rate for the cyclic PTHF as follows: arrangement of the cyclic polymer chain into the lamella including folding parts is conformationally restricted as seen in an example of bilayer four-fold structure (Fig. 1), therefore the entropy difference between the crystalline and molten states is large even if the conformation in the melt is restricted. In addition, since the cyclic chain must behave as if it was a pair of chains tied with each other, conformational change may be difficult in the process of crystallization, which is related to the transport energy ΔE (except viscous diffusion in the melt) such as adsorption and reel-in on the growth front.

Considering the behavior of the melting temperature and the diffusion, the growth rate in the crystallization of the cyclic polymer appears to be governed by entropic factors for PTHF, while by diffusions for PCL. The experimental facts say that the crystal structure of the cyclic polymer is the same as that of the linear one in both PCL[6] and PTHF,[5] and that as seen in Fig. 2, the growth of spherulites for PCL is slower than that for PTHF even if it is taken into account that the growth rate depends on the degree of supercooling. It is not clear in the present stage what causes the differences in the growth rate and melting temperature between PTHF and PCL.

It was observed[5] that considerably many spherulites generated for cyclic PTHFs, compared with that for the linear one at the same crystallization temperature, and it was attributed to high chain density of cyclic polymers in the same way as that of graft chains. The overall crystallization rate obtained by DSC was lower for cyclic PTHF than for linear one.[14] For PCL, Córdva *et al.*[7] found that the cyclic PCL had a higher overall crystallization rate in the molecular weights $M_n = 2.3k$ and 7.0k, and Shin *et al.*[6] also obtained the same result for PCLs with $M_w = 110–140k$ but for the $M_w = 75k$ PCL the overall crystallization rate of the cyclic was almost the same as that of the linear.

3. Branched Polymers

The molecular characteristics of branched polymers is that one chain end of polymers is jointed with one another in star-like polymers and with a main chain in graft polymers. Therefore, flexibility and diffusion of branched chains are limited, while chain densities are high. In graft *co*polymers, in addition, crystallization of graft chains which are miscible with main chains is affected by the main chain in the same way as that of miscible polymer blends, and in the immiscible case the microphase separation may play an important role in the crystallization like block copolymers.[15–18] In this section, crystallization of branched chains in star polymers and graft (co)polymers is presented, where graft chains miscible with main chains are adopted to avoid the influence of the microphase separation.

3.1. *Lamellar structure*

The experimental and calculated crystalline thicknesses (d_c and d_{cc}, respectively) were compared with each other for three-arm star PTHFs with M = 3100, 5700 and 9200 of the arm[5] in the same method as described in section 2.1. The experimental d_c deviates from the lines calculated with integral stem numbers for the star PTHFs having the lower molecular weight of M = 3100 and 5700. This suggests that the junctions of the arm chains disturb the formation of the uniform crystalline stem length.

Ungar and Zeng[19] investigated crystalline lamellar structure of Y-shaped alkane ($C_{120}H_{241}CH(C_{61}H_{123})C_{119}H_{239}$) which has two long and one shorter arms. In the high temperature range, one or two long arms in the molecule crystallized and the crystalline lamellar thickness was determined by the length of the long arms. In subsequent cooling, the short and long arms remaining in the amorphous layer crystallized to form a double layer crystalline structure with different layer thicknesses. The crystalline lamella formed in cooling contained both extended short arms and folded long arms due to the overcrowding effect at the crystalline-amorphous interface.

3.2. *Kinetics of crystallization*

It was found that many spherulites generated in star PTHFs[5] and in the graft polymers[20] (GRPEG) prepared by the polymerization of ethylene glycol macromonomers, namely many nucleations occurring, compared with those in the respective linear ones. This may be caused by a high graft density. On the other hand, the growth rate of spherulites was much depressed in both star PTHF and GRPEG. The reduced growth rate may be attributed to low mobility or diminished diffusion rates of the graft chains.

The nucleation density and the spherulite growth rate were investigated in the blends of star PTHF with linear PTHF.[5] The nucleation density in the large composition range (>50%) of the star polymer was as high as that in the neat star polymer, while that was low in the linear polymer-rich composition. On the other hand, the growth rate of the spherulite varied linearly with the blend composition. These show that the nucleation is governed by the major component contained in the blend, and that the spherulite grows cooperatively once the nucleation occurs.

The similar behavior of the growth rate can be seen in the blend of poly(methyl methacrylate-*graft*-ethylene glycol) (PMMA-*graft*-PEG) with linear PEG.[20] In the case that an amorphous component, which is miscible with a crystalline one, is contained in the system, the crystal growth rate includes a dilution effect. According to the Turnbull-Fischer equation extended to dilute systems,[21] the growth rate G can be given by insertion of the composition term into Eq. (5) as follows:

$$G = \phi_2 G_0 \exp(-\Delta E/RT_c)\exp(-\Delta\Phi/RT_c) \qquad (8)$$

where ϕ_2 is the volume fraction of the crystallizable component and $\Delta\Phi$ the free energy for the secondary nucleation[22] which is expressed as

$$\Delta\Phi = k_i b_0 \sigma_e \sigma_s T_m^{\ o}/\Delta_{fus}H^{\ o}\Delta T - 2\sigma_e RT_c \ln \phi_2/(b_0\Delta T) \qquad (9)$$

in which k_i is the constant depending on Regime, b_0 is the distance between two adjacent fold planes, σ_s and σ_e are the lateral and end surface energies, respectively, $\Delta_{fus}H^{\ o}$ is the thermodynamic enthalpy of

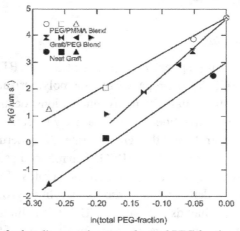

Fig. 3. Dependence of spherulite growth rate on the total PEG fraction at the constant degree of supercooling ($1/\Delta T = 3.5 \times 10^{-2}K^{-1}$) for (linear PEG)/(linear PMMA) blends, (PMMA-*graft*-PEG)/(linear PEG) blends and neat PMMA-*graft*-PEG. The linear dependence indicates that the graft PEGs are incorporated cooperatively with linear ones once nucleation occurs. Reproduced from Ref. 20 by permission of Nature Publishing Group.

fusion per unit volume and ΔT is the supercooling degree. At a constant ΔT, Eq. (8) with (9) can be expressed as a function of $\ln \phi_2$ as follows:

$$\ln G = A + B \ln \phi_2 \qquad (10)$$

when T_c is approximated to be the same at a constant ΔT.

As shown in Fig. 3,[20] the plots of $\ln G$ vs. $\ln \phi_2$ is a linear relation for the (linear PMMA)/(linear PEG) homopolymer blend as expected from Eq. (10). The (PMMA-*graft*-PEG)/(linear PEG) blend also has a linear relation and the extrapolated value to $\phi_2 = 1$ ($\ln \phi_2 = 0$) agrees with that of the linear PEG homopolymer, which suggests that the spherulite grows cooperatively between the graft and linear chains, in the same way as PTHF described above. However, although $\ln G$ for the neat PMMA-*graft*-PEG also varies linearly with the graft composition, the value extrapolated to $\ln \phi_2 = 0$ deviates from that of the neat linear PEG. This difference may be caused by the grafting effect.

3.3. *Melting temperature*

The equilibrium melting temperature obtained from the Hoffman-Weeks relation for the graft PEG in the above graft copolymers PMMA-*graft*-PEG was considerably low, compared with that for the linear PEG in the PEG/PMMA blend having the corresponding composition.[20] The melting temperature in the graft chain is reduced more than the thermodynamic melting depression in usual dilute systems. This also shows the grafting effect.

4. Concluding Remarks

Crystallization of cyclic polymers has not been understood clearly, as seen in the difference between PTHF and PCL for the growth rate and melting temperature. It should be elucidated whether this discrepancy is caused by the respective characteristics of PTHF and PCL or by any other factor. Moreover, it is desired that the topological effect of cyclic polymers could be put in the kinetic theory of crystallization more clearly.

References

1. D. Turnbull, J. C. Fisher, *J. Chem. Phys.*, **17**, 7 (1949).
2. J. Roovers, Section 3, 4, Ch. 11 in this book.
3. G.-E. Yu, T. Sun, Z.-G. Yan, C. Price, C. Booth, J. Cook, A. J. Ryan, K. Viras, *Polymer*, **38**, 35 (1997).
4. J. Cooke, K. Viras, F.-E. Yu, T. Sun, T. Yonemitsu, A. J. Ryan, C. Price, C. Booth, *Macromolecules*, **31**, 3030 (1998).
5. H. Takeshita, M. Poovarodom, T. Kiya, F. Arai, K. Takenaka, M. Miya, T. Shiomi, *Polymer*, **53**, 5375 (2012).
6. E. J. Shin, W. Jeong, H. A. Brown, B. J. Koo, J. L. Hedrick, R. M. Waymouth, *Macromolecules*, **44**, 2773 (2011).
7. M. E. Córdova, A. T. Lorenzo, A. J. Müller, J. N. Hoskins, S. M. Grayson, *Macromolecules*, **44**, 1742 (2011).
8. Y. Tezuka, T. Ohtsuka, K. Adachi, R. Komiya, N. Ohno, N. Okui, *Macromol. Rapid Commun.*, **29**, 1237 (2008).
9. T. Kitahara, S. Yamazaki, K. Kimura, *Kobunshi Ronbunshu (Jpn. J. Polym. Sci. Tech.)*, **68**, 694 (2011).
10. E. D. von Meerwall, R. Ozisiki, W. L. Mattice, P. M. Pfister, *J. Chem. Phys.*, **118**, 3867 (2003).
11. R. Ozisiki, E. D. von Meerwall, W. L. Mattice, *Polymer*, **43**, 629 (2002).
12. D. Kawaguchi, K. Masuoka, A. Takano, K. Tanaka, T. Nagamura, N. Torikai, R. M. Dalgliesh, S. Langridge, Y. Matsushita, *Macromolecules*, **39**, 5180 (2006).
13. S. Nam, J. Leisen, V. Breedveld, H. W. Beckham, *Polymer*, **49**, 5467 (2008).
14. H. Takeshita, M. Poovarodom, T. Kiya, T. Shiomi, unpublished data.
15. I. W. Hamley, Chapter 5 in *The Physics of Block Copolymers* (Oxford University Press, Oxford, 1998).
16. T. Shiomi, H. Takeshita, H. Kawaguchi, M. Nagai, K. Takenaka, M. Miya, *Macromolecules*, **35**, 8056 (2002).
17. T. Shiomi, H. Tsukada, H. Takeshita, K. Takenaka, Y. Tezuka, *Polymer*, **42**, 4997 (2001).
18. H. Takeshita, N. Ishii, C. Araki, K. Takenaka, M. Miya, T. Shiomi, *J. Polym. Sci., Part B, Polym. Phys.*, **42**, 4199 (2004).
19. G. Ungar, X.-B. Zeng, *Chem. Rev.*, **101**, 4157 (2001).
20. H. Takeshita, G. Sasagawa, K. Takenaka, M. Miya, T. Shiomi, *Polym. J.*, **42**, 482 (2010).
21. J. Boon, J. M. Azcue, *J. Polym. Sci., A-2*, **6**, 885 (1968).
22. J. I. Lauritzen, Jr., J. D. Hoffman, *J. Res. Nat. Bur. Stand.*, **64A**, 73 (1960).

CHAPTER 16

SELF-ASSEMBLY AND FUNCTIONS OF CYCLIC POLYMERS

Eisuke Baba and Takuya Yamamoto

Department of Organic and Polymeric Materials, Tokyo Institute of Technology
2-12-1, S8-41 O-okayama, Meguro-ku, Tokyo 152-8552, Japan
E-mail: yamamoto.t.ay@m.titech.ac.jp

This chapter focuses on recent studies on the self-assembly of cyclic block copolymers and their functions. The properties of cyclic polymers have been known to differ from corresponding linear counterparts caused by *topology effects*. On the basis of *topology effects*, the morphology and properties of self-assemblies can be controlled. Applications including drug delivery system are the subjects of intensive researches by taking advantage of the unique properties of cyclic polymers.

1. Introduction

Cyclic polymers, including monocyclic and multicyclic structures as well as linear–cyclic topological block copolymers, with programmed chemical structures, have become accessible thanks to the recent development of novel synthetic protocols.[1–10] By making use of these ring-containing polymers, unusual properties and functions of polymeric materials have now been revealed based on the cyclic topologies, i.e., *topology effects*, unattainable either by traditional linear or branched[11–13] counterparts.[14–17] Furthermore, the properties of a self-assembled cyclic amphiphilic block copolymer attract great attention for the amplification of the *topology effect*.[18] Recent studies presented in this chapter cover diverse fields including self-assembly of monocyclic and tadpole-shaped copolymers as well as their properties, functions, and applications.

2. Control of the Morphology of Self-Assemblies by the Topology of Polymeric Components

2.1. *Cyclic poly(ethylene oxide)-b-poly(propylene oxide) and poly(ethylene oxide)-b-poly(butylene oxide)*

A *topology effect* by a cyclic polymer was reported to be amplified in the assembled states rather than in the individually separated forms.[18] Therefore, micelles, the simplest self-assembled structure, formed from monocyclic block copolymers are of remarkable interest. A series of micelles were prepared from cyclic AB-type amphiphilic block copolymers of poly(ethylene oxide)-*b*-poly(propylene oxide) and poly(ethylene oxide)-*b*-poly(butylene oxide) along with their linear AB- and ABA-type counterparts.[19–23] The cyclic block copolymers employed in these studies were prepared by the *bimolecular* cyclization process[1,3–5,8,17] using hydroxy-terminated linear precursors and dichloromethane as a linking reagent in the presence of KOH to form an acetal linkage.

The critical micelle concentration (cmc) of the cyclic AB-type block copolymers was similar to or marginally smaller than that with the linear ABA-type counterparts.[20] In contrast, the corresponding micelle from the linear AB-type counterpart showed a significantly smaller cmc. Moreover, the association number (N_w), thermodynamic radius (r_t), and hydrodynamic radius (r_h) of the micelles from the cyclic AB-type and linear ABA-type copolymers were comparable but distinctively smaller than those from the linear AB-type counterparts (Table 1). These results were consistent with the geometric estimation of the size of the micelles; the parameters for the cyclic AB- and linear ABA-type block copolymers were estimated to be approximately a half of those for the linear AB-type counterparts. Moreover, the cyclic AB-type block copolymer was expected to form a stiff hairpin-like conformation, resulting in a slightly larger hydrophobic core compared to the linear ABA-type counterpart having free chain ends.[23] The enthalpy of micellization ($\Delta_{mic}H°$) was smaller for the cyclic AB-type block copolymer than the corresponding linear ABA-type counterpart, presumably because of the reduced exposure of the hydrophobic segment to water in the molecularly dispersed state.[20] Given that the cmc's were similar, the smaller $\Delta_{mic}H°$

agreed with the smaller entropy of micellization ($\Delta_{mic}S°$) for the cyclic copolymer.

Table 1. M_w, N_w, r_t, r_h, and $\Delta_{mic}H°$ at the indicated temperature for micelles formed from cyclic AB-, linear ABA-, and linear AB-type poly(ethylene oxide)-*b*-poly(propylene oxide) and poly(ethylene oxide)-*b*-poly(butylene oxide).[23]

	T/°C	M_w/10^5 g mol^{-1}	N_w	r_t/nm	r_h/nm	$\Delta_{mic}H°$/kJ mol^{-1}
cyclic P$_{34}$E$_{104}$	55	1.6	22	5.7	7.5	126 ± 40
linear E$_{52}$P$_{34}$E$_{52}$	55	1.0	15	4.9		190 ± 20
linear E$_{102}$P$_{37}$	45	4.6	67	10.2	13.7	210 ± 20
cyclic B$_8$E$_{42}$	50	0.4	16	2.9	4.4	55 ± 10
linear E$_{21}$B$_8$E$_{21}$	50	0.15	6	2.2	4.0	95 ± 20
linear E$_{41}$B$_8$	50	1.1	44	4.9	7.1	80 ± 10
cyclic B$_{27}$E$_{144}$	40	5.6	62	10.4	12.8	
linear E$_{72}$B$_{27}$E$_{72}$	40	4.5	49	10.0	11.3	

E: poly(ethylene oxide).
P: poly(propylene oxide).
B: poly(butylene oxide).

2.2. *Cyclic polyisoprene-b-polystyrene*

Borsali and coworkers reported that self-assembly of linear and cyclic polyisoprene-*b*-polystyrene in *n*-heptane and *n*-decane, which are good solvents for polyisoprene.[24-26] The self-assembled structures were characterized by DLS and cryogenic TEM to determine the morphology and size depending on temperature and the concentration of the copolymer. In the results, the cyclic copolymer showed the concentration-dependence of the morphology of the assembled structure. At the concentration lower than 0.1 mg/mL, planar sunflower-shaped particles formed. In the concentration ranges of 0.1 mg/mL < *c* < 2 mg/mL and 2 mg/mL < *c* < 5 mg/mL, the copolymer gave a giant worm-like micelle and vesicle, respectively. Furthermore, the process of the formation of the vesicular structure on the basis of the polymer concentration was successfully observed by cryogenic TEM (Fig. 1).

Especially, at the concentration of 5 mg/mL, the incomplete vesicular structure may provide new insight for the transition process (Figs. 1c – e). On the other hand, the linear counterpart formed spherical particles (Fig. 2a) for the concentration lower than 35 mg/mL, above which cylindrical micelles were also observed (Fig. 2b). The concentration- and temperature-dependence of the r_h of the self-assembled structures was determined by DLS. As the temperature was lowered from 25 to 10 °C, the r_h of the assemblies from the cyclic copolymer became larger. A similar effect was observed by increasing the polymer concentration for a range lower than 5 mg/mL. In sharp contrast, the morphology of the assemblies from the linear counterpart was independent of temperature and the concentration in the experimental ranges. In the subsequent report, however, the authors showed that the concentration-dependence

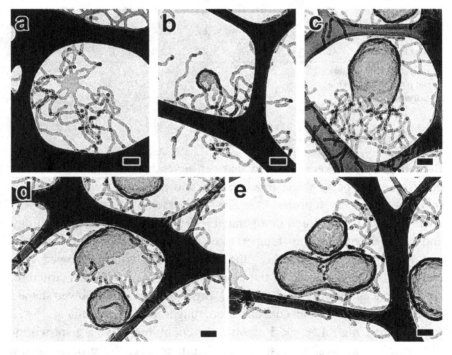

Fig. 1. Process of the formation of vesicles from cyclic polyisoprene-*b*-polystyrene copolymer in *n*-heptane observed by cryogenic TEM. (a), (b) 2 mg/mL; (c) – (e) 5 mg/mL. Scale bar: 100 nm. Reprinted with permission from Ref. 24. Copyright 2004 Royal Society of Chemistry.

of the morphology of an assembly from linear polyisoprene-*b*-polystyrene while a cyclic counterpart was insensitive to the concentration and temperature.[25] Small-angle X-ray scattering studies were also performed on the assemblies from the linear and cyclic copolymers.[26] In relation to these, Hadjichristidis and coworkers reported a comprehensive study on the synthesis and self-assembly of cyclic polybutadiene-*b*-polystyrene.[27] Moreover, a theoretical investigation on the cyclic and linear topologies of block copolymers was made by Jo and coworkers.[28]

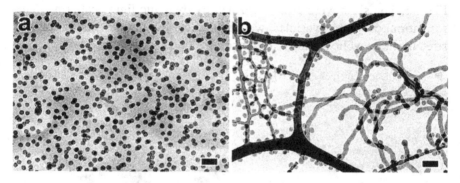

Fig. 2. Formation of spherical particles and cylindrical micelles from linear polyisoprene-*b*-polystyrene in *n*-decane observed by cryogenic TEM. (a) 0.1 and (b) 35 mg/mL. Scale bar: 100 nm. Reprinted with permission from Ref. 24. Copyright 2004 Royal Society of Chemistry.

2.3. *Tadpole-shaped linear poly(ethylene oxide)-cyclic polystyrene*

Li and coworkers prepared a tadpole-shaped amphiphilic block copolymer, linear poly(ethylene oxide)-cyclic polystyrene, where a poly(ethylene oxide) tail was attached to a cyclic polystyrene segment, which was prepared by click chemistry in the last step in the synthesis.[29] Click chemistry allowed for the synthesis of a variety of ring-containing topologies such as not only tadpole[29,30] but also 8,[31–33] θ,[34] α,[33,35] P,[33] and Q.[33]

For the synthesis of a tadpole-shaped copolymer, linear AB-type poly(ethylene oxide)-*b*-polystyrene was first prepared. Via click chemistry, the polystyrene segment was cyclized by connecting the azido terminus with the ethynyl group located between the hydrophilic and hydrophobic segments. The molecular weight of a tadpole and corresponding linear prepolymer was determined to be 9470 and 9160, respectively, including a poly(ethylene oxide) segment of $M_n = 2000$. When self-assembled in water, the linear copolymer gave a vesicular structure of 50 to 100 nm in diameter with a wall thickness of 15 nm coexisted with micelles of 30 nm in diameter (Fig. 3a). On the other hand, the tadpole copolymer formed significantly larger vesicles of 300 nm in diameter along with smaller ones (60 – 100 nm in diameter) as shown in Fig. 3b. The sizes were also confirmed by dynamic light scattering. The formation of larger vesicles was explained by the restriction of the stretch of the cyclic segment as the core.

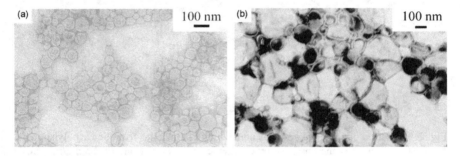

Fig. 3. TEM pictures of vesicles formed from (a) linear and (b) tadpole poly(ethylene oxide)-polystyrene block copolymers. Reprinted with permission from Ref. 29. Copyright 2009 American Chemical Society.

3. Control of the Thermal Properties by the Topology of Polymers

3.1. Cyclic poly(NIPAM)

Temperature-responsive properties of poly(N-isopropylacrylamide), poly(NIPAM),[36] have extensively been studied for their potential applications in controlled release devices and cell-sheets.[37,38] Recently, two groups independently reported *topology effects* by cyclic poly(NIPAM).[39–42] Both groups prepared linear poly(NIPAM) with ethynyl and azido end groups by reversible addition–fragmentation chain transfer (RAFT) polymerization or atom transfer radical polymerization

(ATRP). Subsequently, the asymmetric *unimolecular* cyclization process by click chemistry was applied.[39,41]

Winnik and coworkers reported that the LCST of the obtained cyclic poly(NIPAM) was several degrees higher than that of the corresponding linear counterpart (Fig. 4).[39,40] Additionally, distinctively slower phase transition took place upon heating for the cyclic poly(NIPAM). Liu and coworkers, on the other hand, observed cyclic poly(NIPAM) showing a few degrees lower cloud point (CP) against the linear counterpart at a polymer concentration of 2.0 g/L (Fig. 5).[41,42] At a lower polymer concentration (0.2 g/L), the CP was a few degrees higher for the cyclic poly(NIPAM) as in the Winnik and coworkers' study, indicating the concentration-dependence of the *topology effect*. Furthermore, they reported that the smaller enthalpy change (ΔH) during the phase transition for the cyclic poly(NIPAM). These results were accounted for by the suppression of the hydrogen bond interaction between the cyclic poly(NIPAM) with water molecules due to the topological constraints in comparison with the flexible linear counterpart with free chain ends.

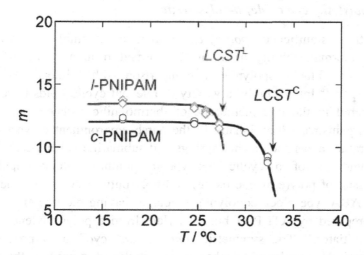

Fig. 4. Plots of the hydration number per NIPAM unit (m) versus temperature for linear and cyclic poly(NIPAM)s. Reprinted with permission from Ref. 40. Copyright 2009 American Chemical Society.

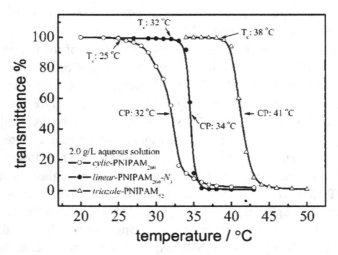

Fig. 5. Plots of the LCST of linear and cyclic PNIPAMs. Adapted with permission from Ref. 42. Copyright 2008 American Chemical Society.

3.2. *Cyclic poly(ethylene oxide)-b-poly(butyl acrylate) and poly(ethylene oxide)-b-polystyrene*

Recently, a significant *topology effect* upon self-assembly was revealed in the thermal stability of a micelle formed from a cyclic AB-type amphiphilic block copolymer in comparison with a linear ABA-type counterpart.[18] Interestingly, a variety of relevant cyclic lipids have been discovered in the cell membrane of thermophilic archaea, single-cell microorganisms, which inhabit in the harsh environment of very high-temperature areas such as hot springs and submarine volcanoes.[43-46] For the synthesis of a cyclic AB-type amphiphilic block copolymer comprised of poly(ethylene oxide) and poly(butyl acrylate) segments, a linear ABA-type block copolymer precursor having olefinic end groups was prepared by ATRP of butyl acrylate from a poly(ethylene oxide) macroinitiator.[47] The symmetric *unimolecular* cyclization process by metathesis of the ethenyl telechelics was employed to produce the cyclic product (Scheme 1).

The obtained cyclic and relevant linear amphiphilic block copolymers were self-assembled to form corresponding micelles (Scheme 1), where

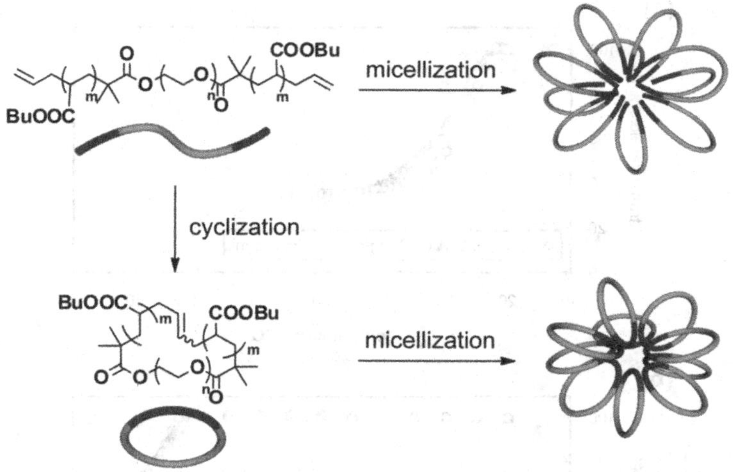

Scheme 1. Chemical structures of linear and cyclic amphiphilic block copolymers and schematic representation for the formation of micelles.

their cmc's and the number-average hydrodynamic diameter were measured to be closely comparable to each other.[18] The size and morphology of the micelles were directly observed by atomic force microscopy (AFM) and transmission electron microscopy (TEM) to show a spherical shape with consistent diameters determined by DLS.

Interestingly, the thermal stability of the micelle was drastically enhanced by the linear-to-cyclic topological conversion of the copolymer. Thus, the micellar solutions were heated in stepwise fashion from 20 °C, and the ones from the linear block copolymer became turbid at around 25 °C (Fig. 6a), which was defined as a cloud point (T_c), while those from the cyclic counterpart was stable till above 70 °C (Fig. 6b).[18]

On the basis of the large deviation, the tuning of the T_c was achieved by coassembly of the linear and cyclic block copolymers.[18] Thus, the T_c was able to be systematically altered by the relative contents of the linear and cyclic copolymer components in the micelle, as shown in Fig. 6c. This *topology effect* would be utilized to design temperature-responsive devices to encapsulate and release appropriate guests from the micelle. In particular, the topology-based polymer materials design should be

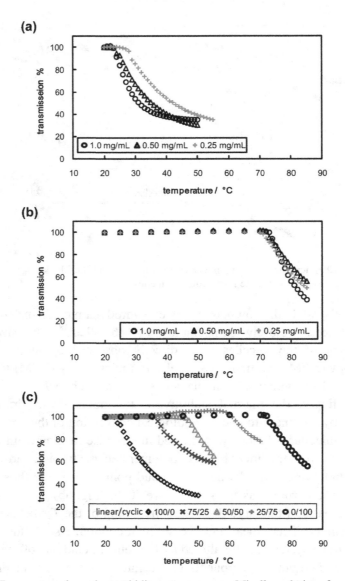

Fig. 6. Temperature-dependent turbidity measurements. Micellar solutions from (a) linear and (b) cyclic copolymers (1.0, 0.50, 0.25 mg/mL). (c) Homoassemblies and coassemblies (linear/cyclic = 100%/0%, 75%/25%, 50%/50%, 25%/75%, and 0%/100%). The total concentration of copolymers in a solution was 0.50 mg/mL.

suitable for human body-related applications including drug delivery system, food, and cosmetics, in which the modification of the chemical structure and molecular weight cause significant concerns on toxicity and biocompatibility.

In the meantime, amphiphilic block copolymers consisting of polystyrene components are of versatility and importance, and self-assembly of linear poly(ethylene oxide)-*b*-polystyrene has been extensively studied to show the formation of micelles, vesicles, and further complex aggregates.[48,49] Relevant cyclic copolymers may provide vast opportunities to reveal novel *topology effects* upon self-assembly. Thus, cyclic poly(ethylene oxide)-*b*-polystyrene was recently reported to be synthesized by a similar procedure to cyclic poly(ethylene oxide)-*b*-poly(butyl acrylate).[50] Thus, ATRP of styrene from a poly(ethylene oxide) macroinitiator, and the Br-termini were allylated by allyltrimethylsilane in the presence of $TiCl_4$ without decomposition of the main chain. Subsequent cyclization was performed by the second-generation Grubbs' catalyst to produce cyclic amphiphilic poly(ethylene oxide)-*b*-polystyrene.

4. Potential Application as Drug Delivery System

4.1. *Topology effect on blood circulation time*

A pharmacological application of a cyclic polymer was recently demonstrated by elucidating the permeation mechanism of the nanoporous channels in a kidney.[51] Thus, it was anticipated that a cyclic polymer remains longer in a body than a linear counterpart, since the former should be reluctant to transform its conformation into a folded shape to traverse through the nanoporous channel.[15] To examine this *topology effect*, linear and cyclic random copolymers of ε-caprolactone and α-chloro-ε-caprolactone were synthesized.[52,53] After the azidation of the α-chloro groups, a phenolic functionality was introduced for radiolabeling by ^{125}I, and subsequently poly(ethylene glycol) with a series of DP_n were grafted to tune the molecular weight of the linear and cyclic polymers by click chemistry. In comparison with the linear polymer, lower secretion into urine and higher accumulation in various organs in turn were indeed confirmed for the cyclic polymer having an appropriate molecular weight (Fig. 7). This finding may provide new design methodologies for drug carriers and imaging agents via the topology-based control of a blood circulation time.

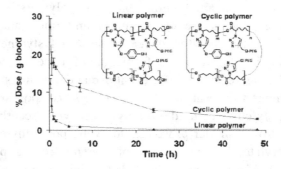

Fig.7. Time-dependent concentration of linear and cyclic polymers in the blood. Reprinted with permission from Ref. 51. Copyright 2009 American Chemical Society.

4.2. Self-assembled tadpole-shaped cyclic poly(NIPAM)-linear poly (ε-caprolactone)

Another application intended for drug delivery system was reported by Liu and coworkers using a self-assembled amphiphilic tadpole-shaped block copolymer.[54] First, a linear poly(NIPAM) was prepared by ATRP from an initiator containing ethynyl and hydroxy functionalities. By click chemistry, the prepolymer was intramolecularly reacted to give cyclic poly(NIPAM) with a hydroxy group, from which ε-caprolactone was then polymerized to form a tadpole-shaped cyclic poly(NIPAM)-linear poly(ε-caprolactone) (Scheme 2). The cyclic poly(NIPAM) had a molecular weight of M_n(GPC) = 4700, and that after the copolymerization with ε-caprolactone increased to 12400. The M_n(GPC) value of a linear counterpart was 12800 with an uncyclized PNIPAM segment of M_n(GPC) = 5500.

When self-assembled, the tadpole-shaped copolymer gave a micelle with a r_h of approximately 70 nm at 20 °C. Upon heating a micellar solution to 40 °C, the radius decreased to ca. 56 nm due to the dehydration from the poly(NIPAM) segment. For the linear counterpart, on the other hand, the hydrodynamic radius was 62 and 54 nm at 20 and 40 °C, respectively. The present result implies that the lower stabilization capability of the cyclic topology of the poly(NIPAM) segment in comparison with the linear counterpart. The cloud point (T_c) of the micelles from the tadpole-shaped and linear copolymers was determined to be 25 and 30 °C, respectively, by transmittance at 600 nm, where Yamamoto and coworkers reported that the T_c of the micelle from a cyclic AB-type block copolymer was significantly higher (>40 °C) than that from a linear ABA-type counterpart.[18]

The micelles were loaded with doxorubicin (Dox), and their release behavior was studied. The loading content and encapsulation efficiency were determined to be 26.3% and 6.2%, respectively, for the micelle from the tadpole-shaped copolymer, which were slightly higher than those of the linear counterpart (26.1% and 5.1%, respectively). The release from the micelle formed from the tadpole-shaped copolymer at 20 h was determined to be 50% at 20 °C and 78% at 37 °C. After 50 h, the release was slightly increased to 55% at 20 °C and 84% at 37 °C. Since, the release from the micelle formed from the linear copolymer was a few percent lower (ca. 53% at 20 °C and 80% at 37 °C after 50 h), this *topology effect* was likely due to that the cyclic segment of the tadpole-shaped copolymer does not entangle and thus facilitates the release. Furthermore, a cytotoxicity test confirmed that the micelle from the tadpole-shaped copolymer is essentially nontoxic, and the Dox-loaded micelles effectively killed HeLa cells.

4.3. *Unimer micelle by a cyclic amphiphile*

Grayson and coworkers recently reported that the synthesis of a cyclic amphiphilic homopolymer.[55] First, they synthesized α-alkynyl-ω-bromo poly(acetoxystyrene) via ATRP from propargyl bromo-isobutyrate (Scheme 3). The Br-terminus was then converted to azido functionality by using NaN$_3$, and the resulting α-alkynyl-ω-azido poly(acetoxystyrene) was cyclized via click chemistry by making use of CuBr and PMDETA in DMF under "pseudo high dilution" conditions, in which the linear

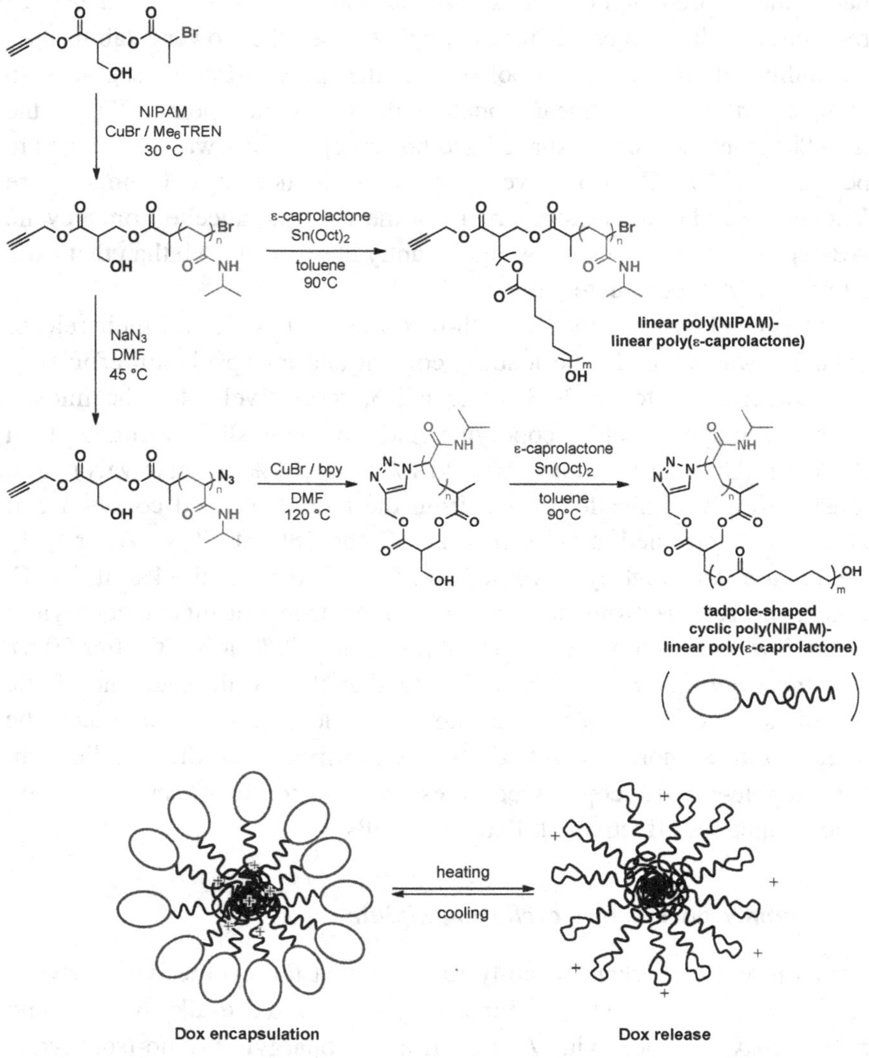

Scheme 2. Synthesis and self-assembly of linear poly(NIPAM)-linear poly(ε-caprolactone) and tadpole-shaped cyclic poly(NIPAM)-linear poly(ε-caprolactone) and schematic representation of the encapsulation and release of Dox from a micelle formed from the tadpole-shaped copolymer.

prepolymer was added dropwise to a CuBr solution. This synthetic sequence for the construction of a cyclic structure has been applied to a variety of ATRP-based polymers.[39,41,56-64] Subsequently, the acetyloxy groups were quantitatively hydrolyzed by a treatment of an acid, converting the esters into hydroxyphenyl groups with the cyclic structure intact. The hydroxyphenyl groups were subsequently functionalized to ethynyl groups by esterification to give an alkyne-appended cyclic polymer with M_n = 6750 and PDI = 1.11 in high yield. A linear

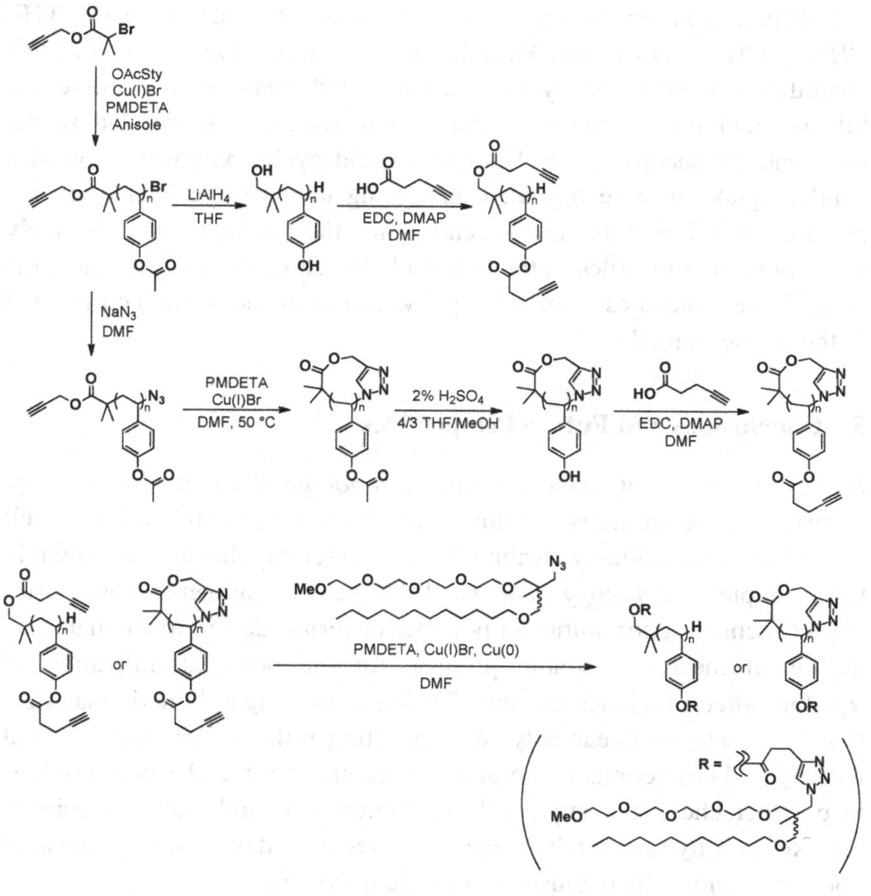

Scheme 3. Synthesis of cyclic and linear amphiphilic homopolymers.

counterpart was similarly prepared by the simultaneous reduction of the Br-terminus and acetyloxy protection groups. An azido-functionalized amphiphile, possessing methoxy-terminated tetra(ethylene glycol) and dodecyl groups, was separately synthesized. Click chemistry between the alkyne-appended cyclic or linear prepolymers and amphiphilic azide resulted in high grafting yield (92 – 96%). The produced cyclic and linear amphiphilic homopolymers had $M_n = 13000$, PDI = 1.10 and $M_n = 14240$, PDI = 1.08, respectively.

Due to the amphiphilicity, the cyclic and linear polymers showed solubility in a variety of organic solvents such as toluene, n-hexane, THF, CH_2Cl_2, $CHCl_3$, DMF, and MeOH, but not in water. On the basis of this solubility behavior, the cyclic and linear polymers were dissolved in toluene, and rose bengal, a water-soluble dye, was suspended in the solutions for encapsulation. Both linear and cyclic polymers showed a similar uptake of rose bengal. Considering that the encapsulation took place even at low polymer concentrations, the amphiphiles were likely forming unimer micelles as the encapsulation should not occur below the cmc. Three molecules of rose bengal were calculated to be incorporated in the unimer micelle.

5. Conclusions and Future Perspectives

A variety of recent achievements in *topological polymer chemistry*, focusing on the synthesis and novel properties of cyclic polymers as well as the functions uniquely attributed to self-assembly, have been shown in this chapter. *Topology effects* by cyclic polymers now offer unprecedented opportunities in polymer materials design unattainable by traditional means. A viable promise for the practical utilization of *topology effects* includes the modification of existing polymeric materials mostly relying on linear polymers, inflicting little concern on chemical toxicity and environmental pollution. Furthermore, broader opportunities in polymer chemistry, supramolecular chemistry, and materials science are expected by taking advantage of the recently developed synthesis of not only monocyclic but also multicyclic polymers.

Acknowledgments

This work was supported by Integrated Doctoral Education Program at TokyoTech (E.B.), Global COE Program (Education and Research Center for Material Innovation), MEXT, Japan (T.Y.), Challenging Research President's Honorary Award, Tokyo Institute of Technology (T.Y.), The Kurata Memorial Hitachi Science and Technology Foundation (T.Y.), Yazaki Memorial Foundation for Science and Technology (T.Y.), and KAKENHI (23685022 and 23106709, T.Y.).

References

1. J. A. Semlyen, *Cyclic Polymers* (Kluwer Academic, Dordrecht, 2000).
2. A. Deffieux and R. Borsali, in *Macromolecular Engineering: Precise Synthesis, Materials Properties, Applications*, Eds. K. Matyjaszewski, Y. Gnanou and L. Leibler (Wiley-VCH, Weinheim, 2007), p. 875.
3. K. Endo, *Adv. Polym. Sci.*, **217**, 121 (2008).
4. S. M. Grayson, *Nat. Chem.*, **1**, 178 (2009).
5. B. A. Laurent and S. M. Grayson, *Chem. Soc. Rev.*, **38**, 2202 (2009).
6. H. R. Kricheldorf, *J. Polym. Sci., Part A: Polym. Chem.*, **48**, 251 (2010).
7. T. Yamamoto and Y. Tezuka, in *Complex Macromolecular Architectures: Synthesis, Characterization, and Self-Assembly*, Eds. N. Hadjichristidis, A. Hirao, Y. Tezuka and F. Du Prez (Wiley, Singapore, 2011), p. 3.
8. T. Yamamoto and Y. Tezuka, *Eur. Polym. J.*, **47**, 535 (2011).
9. T. Yamamoto and Y. Tezuka, in *Synthesis of Polymers*, Eds. D. Schlüter, C. Hawker and J. Sakamoto (Wiley-VCH, Weinheim, 2012), p. 531.
10. Z. Jia and M. J. Monteiro, *J. Polym. Sci., Part A: Polym. Chem.*, **49**, 2085 (2012).
11. N. Hadjichristidis, H. Iatrou, M. Pitsikalis and J. Mays, *Prog. Polym. Sci.*, **31**, 1068 (2006).
12. N. V. Tsarevsky and K. Matyjaszewski, *Chem. Rev.*, **107**, 2270 (2007).
13. A. Hirao, T. Watanabe, K. Ishizu, M. Ree, S. Jin, K. S. Jin, A. Deffieux, M. Schappacher and S. Carlotti, *Macromolecules*, **42**, 682 (2009).
14. T. McLeish, *Nat. Mater.*, **7**, 933 (2008).
15. M. E. Fox, F. C. Szoka and J. M. J. Fréchet, *Acc. Chem. Res.*, **42**, 1141 (2009).
16. S. Perrier, *Nat. Chem.*, **3**, 194 (2011).
17. T. Yamamoto and Y. Tezuka, *Polym. Chem.*, **2**, 1930 (2011).
18. S. Honda, T. Yamamoto and Y. Tezuka, *J. Am. Chem. Soc.*, **132**, 10251 (2010).
19. G.-E. Yu, Z.-K. Zhou, D. Attwood, C. Price, C. Booth, P. C. Griffiths and P. Stilbs, *J. Chem. Soc. Faraday Trans.*, **92**, 5021 (1996).
20. G.-E. Yu, Z. Yang, D. Attwood, C. Price and C. Booth, *Macromolecules*, **29**, 8479 (1996).

21. G.-E. Yu, Z. Yang, M. Ameri, D. Attwood, J. H. Collett, C. Price and C. Booth, *J. Phys. Chem., Part B*, **101**, 4394 (1997).
22. H. Altinok, G.-E. Yu, S. K. Nixon, P. A. Gorry, D. Attwood and C. Booth, *Langmuir*, **13**, 5837 (1997).
23. G.-E. Yu, C. A. Garrett, S.-M. Mai, H. Altinok, D. Attwood, C. Price and C. Booth, *Langmuir*, **14**, 2278 (1998).
24. J.-L. Putaux, E. Minatti, C. Lefebvre, R. Borsali, M. Schappacher and A. Deffieux, *Faraday Discuss.*, **128**, 163 (2005).
25. E. Di Cola, C. Lefebvre, A. Deffieux, T. Narayanan and R. Borsali, *Soft Matter*, **5**, 1081 (2009).
26. E. Minatti, R. Borsali, M. Schappacher, A. Deffieux, V. Soldi, T. Narayanan and J.-L. Putaux, *Macromol. Rapid Commun.*, **23**, 978 (2002).
27. H. Iatrou, N. Hadjichristidis, G. Meier, H. Frielinghaus and M. Monkenbusch, *Macromolecules*, **35**, 5426 (2002).
28. K. H. Kim, J. Huh and W. H. Jo, *J. Chem. Phys.*, **118**, 8468 (2003).
29. Y.-Q. Dong, Y.-Y. Tong, B.-T. Dong, F.-S. Du and Z.-C. Li, *Macromolecules*, **42**, 2940 (2009).
30. G.-Y. Shi, X.-Z. Tang and C.-Y. Pan, *J. Polym. Sci., Part A: Polym. Chem.*, **46**, 2390 (2008).
31. G.-Y. Shi and C.-Y. Pan, *Macromol. Rapid Commun*, **29**, 1672 (2008).
32. G.-Y. Shi, L.-P. Yang and C.-Y. Pan, *J. Polym. Sci., Part A: Polym. Chem.*, **46**, 6496 (2008).
33. B. Schmidt, N. Fechler, J. Falkenhagen and J.-F. Lutz, *Nat. Chem.*, **3**, 234 (2011).
34. G.-Y. Shi and C.-Y. Pan, *J. Polym. Sci., Part A: Polym. Chem.*, **47**, 2620 (2009).
35. M. Zamfir, P. Theato and J.-F. Lutz, *Polym. Chem.*, **3**, 1796 (2012).
36. T. Tanaka, *Phys. Rev. Lett.*, **40**, 820 (1978).
37. T. Okano, *Adv. Polym. Sci.*, **110**, 179 (1993).
38. M. Nakayama and T. Okano, *J. Drug Delivery Sci. Technol.*, **16**, 35 (2006).
39. X.-P. Qiu, F. Tanaka and F. M. Winnik, *Macromolecules*, **40**, 7069 (2007).
40. Y. Satokawa, T. Shikata, F. Tanaka, X.-P. Qiu and F. M. Winnik, *Macromolecules*, **42**, 1400 (2009).
41. J. Xu, J. Ye and S. Liu, *Macromolecules*, **40**, 9103 (2007).
42. J. Ye, J. Xu, J. Hu, X. Wang, G. Zhang, S. Liu and C. Wu, *Macromolecules*, **41**, 4416 (2008).
43. M. De Rosa and A. Gambacorta, *Prog. Lipid Res.*, **27**, 153 (1988).
44. M. Kates, in *The Biochemistry of Archaea (Archaebacteria)*, Eds. M. Kates, D. J. Kushner and A. T. Matheson (Elsevier, Amsterdam, 1993), p. 261.
45. A. Yamagishi, *Biol. Sci. Space*, **14**, 332 (2000).
46. K. Arakawa, T. Eguchi and K. Kakinuma, *Bull. Chem. Soc. Jpn.*, **74**, 347 (2001).
47. K. Adachi, S. Honda, S. Hayashi and Y. Tezuka, *Macromolecules*, **41**, 7898 (2008).
48. K. Yu, L. Zhang and A. Eisenberg, *Langmuir*, **12**, 5980 (1996).
49. N. S. Cameron, M. K. Corbierre and A. Eisenberg, *Can. J. Chem.*, **77**, 1311 (1999).

50. E. Baba, S. Honda, T. Yamamoto and Y. Tezuka, *Polym. Chem.*, **3**, 1903 (2012).
51. N. Nasongkla, B. Chen, N. Macaraeg, M. E. Fox, J. M. J. Fréchet and F. C. Szoka, *J. Am. Chem. Soc.*, **131**, 3842 (2009).
52. H. Li, A. Debuigne, R. Jérome and P. Lecomte, *Angew. Chem., Int. Ed.*, **45**, 2264 (2006).
53. R. Mehvar, M. A. Robinson and J. M. Reynolds, *J. Pharm. Sci.*, **84**, 815 (1995).
54. X. Wan, T. Liu and S. Liu, *Biomacromolecules*, **12**, 1146 (2011).
55. B. A. Laurent and S. M. Grayson, *Polym. Chem.*, **3**, 1846 (2012).
56. A. S. Goldmann, D. Quemener, P.-E. Millard, T. P. Davis, M. H. Stenzel, C. Barner-Kowollik and A. H. E. Muller, *Polymer*, **49**, 2274 (2008).
57. D. M. Eugene and S. M. Grayson, *Macromolecules*, **41**, 5082 (2008).
58. B. A. Laurent and S. M. Grayson, *J. Am. Chem. Soc.*, **128**, 4238 (2006).
59. Z. Ge, Y. Zhou, J. Xu, H. Liu, D. Chen and S. Liu, *J. Am. Chem. Soc.*, **131**, 1628 (2009).
60. J. N. Hoskins and S. M. Grayson, *Macromolecules*, **42**, 6406 (2009).
61. S. Binauld, C. J. Hawker, E. Fleury and E. Drockenmuller, *Angew. Chem., Int. Ed.*, **48**, 6654 (2009).
62. D. E. Lonsdale, C. A. Bell and M. J. Monteiro, *Macromolecules*, **43**, 3331 (2010).
63. E. D. Pressly, R. J. Amir and C. J. Hawker, *J. Polym. Sci., Part A: Polym. Chem.*, **49**, 814 (2011).
64. B. A. Laurent and S. M. Grayson, *J. Am. Chem. Soc.*, **133**, 13421 (2011).

INDEX